Toxic Organic Chemicals in Porous Media

Edited by
Z. Gerstl, Y. Chen, U. Mingelgrin,
and B. Yaron

With 88 Figures

Springer-Verlag Berlin Heidelberg New York
London Paris Tokyo Hong Kong

Dr. ZEV GERSTL
Institute of Soils and Water
The Volcani Center, ARO
P.O. Box 6
Bet Dagan, 50250, Israel

Prof. Dr. Y. CHEN
Dept. of Soils and Water Sciences
Faculty of Agriculture
P.O. Box 12
Rehovot, 76100, Israel

Prof. Dr. U. MINGELGRIN
Dr. BRUNO YARON
Institute of Soils and Water
The Volcani Center, ARO
P.O. Box 6
Bet Dagan, 50250, Israel

ISBN 3-540-50799-X Springer-Verlag Berlin Heidelberg New York
ISBN 0-387-50799-X Springer-Verlag New York Berlin Heidelberg

Library of Congress Cataloging-in-Publication Data. Toxic organic chemicals in porous media / edited by Z. Gerstl ... [et al.]. p. cm. – (Ecological studies; v. 73) Includes index. Papers presented at the second International Workshop on Behavior of Pollutants in Porous Media, sponsored by IUPAC and IAHS and held at Bet Dagan, Israel during June, 1987. ISBN 0-387-50799-X (U.S.) 1. Organic compounds – Environmental aspects – Congresses. 2. Zone of aeration – Congresses. I. Gerstl, Z. (Zev), 1947– . II. International Workshop on Behavior of Pollutants in Porous Media (2bd: 1987: Bet Dagan, Israel) III. International Union of Pure and Applied Chemistry. IV. International Association of Hydrological Sciences. V. Series. TD196.073T69 1989 628.5'5 – dc20 89-6396

This work is subject to copyright. All rights are reserved, whether the whole or part of the material is concerned, specifically the rights of translation, reprinting, reuse of illustrations, recitation, broadcasting, reproduction on microfilms or in other ways, and storage in data banks. Duplication of this publication or parts thereof is only permitted under the provisions of the German Copyright Law of September 9, 1965, in its version of June 24, 1985, and a copyright fee must always be paid. Violations fall under the prosecution act of the German Copyright Law.

© Springer-Verlag Berlin Heidelberg 1989
Printed in the United States of America

The use of registered names, trademarks, etc. in this publication does not imply, even in the absence of a specific statement, that such names are exempt from the relevant protective laws and regulations and therefore free for general use.

Typesetting: International Typesetters Inc., Makati, Philippines
2131/3145-543210 – Printed on acid-free paper

Preface

In March, 1983 a workshop on Pollutants in Porous Media was hosted by the Institute of Soils and Water of the Agricultural Research Organization in Bet Dagan, Israel. At this workshop, the unsaturated zone between the soil surface and groundwater was the focal point of discussions for scientists from various disciplines such as soil chemists, physicists, biologists and environmental engineers. Since then, the problem of soil and water pollution has only worsened as more and more cases of pollution caused by human activities including agriculture and industry have been revealed. A great deal of work has been carried out by environmental scientists since 1983 in elucidating the behavior of the many classes of pollutants and the complex physical, chemical, and biological transformations which they undergo as they move through the soil to the vadose zone and, in many cases, the groundwater. In light of this, it was felt that another meeting of specialists from the many disciplines which deal with this subject was necessary and so a Second International Workshop on the Behavior of Pollutants in Porous Media, sponsored by IUPAC (the International Union of Pure and Applied Chemistry) and IAHS (the International Association of Hydrological Sciences), was organized and held in the Institute of Soils and Water of the Agricultural Research Organization in Bet Dagan, Israel during June, 1987.

The present volume is a selection of the talks presented at this second workshop and deals only with toxic organic chemicals in porous media. Inorganic chemicals are treated in a companion volume[1]. The book is divided into five parts. Part one provides us with two general, and different, outlooks as to the scope and extent of the problem. The following four sections deal with different aspects of this problem, but due to the complex and multidisciplinary nature of the pollution of soils and water by toxic organic chemicals, these chapters cannot be all inclusive. Rather, they reflect the ideas and opinions of the participants in the workshop. It is because of this interchange of views amongst the selected specialists from diverse disciplines which took place at the workshop, and which is reflected in their contributions to this book, that we are confident that

---

[1] Bar-Yosef B, Barrow NJ, Goldshmid Y (1989) Inorganic Contaminants in the Vadose Zone. Springer, Berlin Heidelberg New York (Ecological Studies Vol. 74)

all those involved in the study, management or regulation of pollution by toxic organics will find this book useful.

We would like to acknowledge the support given to the workshop by various institutions and specifically by the US-Israel Binational Agricultural Research and Development Fund (BARD), the Segram Center for Soil and Water Research of the Hebrew University, the Water Commissioner of the Ministry of Agriculture, and the National Council for Research and Development. Finally we would like to thank the participants for preparing their written contributions and Springer-Verlag for the publication of this book.

Bet Dagan
Spring, 1989

Z. GERSTL
Y. CHEN
U. MINGELGRIN
B. YARON

Contents

Part IV: Petroleum Hydrocarbons

Introductory Comments 209

10   On the Behavior of Petroleum Hydrocarbons in the Unsaturated Zone: Abiotic Aspects
B. YARON (With 11 Figures) 211
10.1 Introduction 211
10.2 Properties of the Interacting Materials 211
   10.2.1 Crude Oils and Petroleum Products 211
10.3 Partitioning Between Gaseous, Aqueous, and Solid Phases 213
   10.3.1 Migration into the Gaseous Phase 214
   10.3.2 Dissolution in the Water Phase 218
   10.3.3 Retention by the Solid Phase 219
10.4 Surface Degradation 223
10.5 Multiphase Transport 225
10.6 Conclusions 228
References 229

11   Transport of Organic Pollutants in a Multiphase System
H. RUBIN and E. MECHREZ (With 6 Figures) 231
11.1 Introduction 231
11.2 Classification of LH According to Contamination Criteria   231
11.3 Introduction of the Pollutant into the Subsurface Water .. 234
11.4 Migration of LH in the Unsaturated Zone 238
11.5 Migration of LH in the Saturated Zone 240
11.6 Modelling Procedures 240
   11.6.1 Gross Control Volumes 240
   11.6.2 Differential Control Volumes Referring to the Whole Flow Field 241
   11.6.3 Differential Control Volumes Referring to Various Parts of the Flow Field 242
11.7 Determination of the Flow Field Characteristics 244
11.8 Practical Aspects 247
References 248

12   Biochemical Aspects of Hydrocarbon Biodegradation in Sediments and Soils
T. HÖPNER, H. HARDER, K. KIESEWETTER, U. DALYAN, I. KUTSCHE-SCHMIETENKNOP, and B. TEIGELKAMP (With 16 Figures) 251
12.1 Introduction 251
12.2 Biochemical Considerations 252
12.3 Application of Results of Biochemical Considerations ... 254

12.4 Laboratory Experiments with Fresh Marine Sediment 256
12.5 Laboratory Experiments with a Standardized and
     Inoculated Sediment Mixture 258
12.6 Anaerobic Degradation 263
12.7 Hydrocarbon Biodegradation in Soils 266
12.8 Conclusions 268
References 270

Part V: Restoration of the Unsaturated Zone and Groundwater

Introductory Comments 273

13 Manipulation of the Vadose Zone to Enhance Toxic Organic Chemical Removal
R.S. BOWMAN (With 5 Figures) 275
13.1 Introduction 275
13.2 Physical Methods 275
     13.2.1 Leaching 275
     13.2.2 Enhanced Volatilization 276
     13.2.3 Electrokinetics 280
13.3 Chemical Methods 282
13.4 Biological Methods 283
     13.4.1 Stimulation of Native Organisms 283
     13.4.2 Addition of Altered Organisms 284
13.5 Conclusions and Future Research Needs 286
References 286

14 The Relative Importance of pH, Charge, and Water Solubility on the Movement of Organic Solutes in Soils and Ground Water
R.L. MALCOLM (With 6 Figures) 288
14.1 Introduction 288
14.2 Experimental 288
     14.2.1 Overview of the Injection Site 288
     14.2.2 Properties of the Ogallala Aquifer near Stanton, Texas 289
     14.2.3 Design and Construction of Injection and Observation Wells 289
     14.2.4 Experimental Sequence 290
     14.2.5 Solute or Tracer Addition and Observation Well Sampling 293
     14.2.6 On Site Analyses in the Mobile Laboratory ... 294
     14.2.7 Laboratory Analyses 294

14.3	Hypotheses to be Tested	294
14.4	Results and Discussion	295
14.5	Conclusions	300
	References	301

15 Photochemical Inactivation of Organic Pollutants from Water
A. ACHER and S. SALTZMAN (With 3 Figures) 302

15.1	Introduction	302
	15.1.1 Dye-Sensitized Photooxidation Reactions	302
15.2	Photochemical Inactivation of Industrial Organic Pollutants	303
	15.2.1 Photolysis of Uracil Derivatives	305
	15.2.2 Photolysis of s-Triazine Derivatives	308
	15.2.3 Photolysis of Anilide Derivatives	311
15.3	Photochemical Inactivation of Biological Pollutants	312
	15.3.1 Laboratory Experiments	312
	15.3.2 Pilot Plant Experiments	314
15.4	Summary and Conclusions	317
	References	318

16 Restoration of Aquifers Polluted with Hydrocarbons
H. SONTHEIMER, G. NAGEL, and P. WERNER
(With 14 Figures) 320

16.1	Introduction	320
16.2	The Water Works "Durlacher Wald"	321
16.3	Treatment Results	328
16.4	Microbiological Aspects of the Treatment Used	331
16.5	Summary and Conclusions	333
	References	334

Subject Index 335

List of Contributors

You will find the addresses at the beginning of the respective contribution.

Acher, A. 302
Aharonson, N. 193
Bowman, R.S. 275
Chen, Y. 37
Chiou, C.T. 163
Dalyan, U. 251
Gerstl, Z. 151
Harder, H. 251
Höpner, T. 251
Katan, J. 193
Kiesewetter, K. 251
Kutsche-Schmietenknop, I. 251
Lee, L.S. 176
Lehr, J.H. 3
Malcolm, R.L. 288
Mechrez, E. 231
Mingelgrin, U. 91

Muszkat, L. 16
Nagel, G. 320
Nkedi-Kizza, P. 176
Perry, A.S. 16
Perry, R.Y. 16
Prost, R. 91
Rao, P.S.C. 176
Rubin, H. 231
Saltzman, S. 302
Senesi, N. 37
Sontheimer, S. 320
Teigelkamp, B. 251
Werner, P. 320
Wolfe, N.L. 136
Yalkowsky, S.H. 176
Yaron, B. 211

Part I. Overview of the Problem

Introductory Comments

How serious is the problem of groundwater pollution? How much effort and funding should authorities allocate to the prevention and solution of groundwater pollution? There are no easy answers to these questions. Many different factors have to be considered and, given similar situations, different agencies will decide upon differing solutions.

Even amongst the scientific community there exists a difference of opinion as to how severe and how pressing the pollution of our groundwaters really is. The first approach feels that we can easily cope with the problem while the second approach is much more cautious and concerned with what we do not yet know. Two such approaches are presented in this section as an introduction to the more detailed papers which appear in later sections.

The more optimistic approach of Lehr (Chapter 1) does not dispute the existence of groundwater pollution. The author claims that many cases of pollution are localized and do not necessarily entail shortages in water supplies since in many cases either alternatives are available or because the necessary steps required to clean up the polluted aquifers are known and can be readily implemented. More effort, the author claims, should be spent in education as to the importance of our groundwater resources and in prevention of their pollution. Money spent in prevention of groundwater pollution goes a lot farther than larger sums spent on restoration.

The more guarded approach is concerned with the fact that more and more cases of groundwater pollution are coming to light, not only because of more sensitive analytical instrumentation but because the process of groundwater contamination is a slow one and many toxicants are only now reaching the groundwater after their transport through the unsaturated zone. Many compounds which we previously thought of as "safe", meaning that for one reason or another they would not reach the aquifer, have suddenly turned up in well water in surprisingly high concentrations. Perry and coauthors (Chapter 2) present their case for the vulnerability of aquifers and call for more research on the processes which govern the transport of pollutants, specifically toxic organic chemicals, through the vadoze zone into the underlying groundwater. Concern over the effects of even minute amounts of contaminants on humans is expressed and requires lengthy toxicological studies.

We will not take sides in the controversy. Rather, we feel that anyone concerned with the problem of groundwater quality will have to decide which approach is the correct one for his specific case based upon the seriousness of the problem and budgetary constraints.

1 Dimensions of Ground Water Pollution: A Global Perspective

J.H. LEHR[1]

1.1 Water Management Misfortunes

Historically, the world has depended on surface water supplies since man's earliest history. He has dammed up his natural water courses and formed reservoirs for immediate use or guided transport. Even before then he took his water directly from pristine streams. We have always utilized that which is visible with little consideration or even awareness of alternative sources out of sight. While well-drilling dates back 4,000 years to the Chinese, well-digging has been common in rural areas for small water sources throughout our history. For the past century over 90% of the water used throughout the world has come from surface sources. Expanding pressure by agricultural irrigation, water power, industrialization, our growing cities, and of course domestic needs have changed for the better our ultimate view of potential water supplies. We are at last looking below the land as well as upon it for valuable water resources.

Strangely, working against common-sense ground-water alternatives has always been the fact that they represent simple solutions. Little opportunity for man's inclination toward grandiosity existed in this area. The more convoluted an engineering scheme man could develop to solve a problem, the more enthusiastic constituencies can be persuaded to support the plan. The public, the press, the government, and the engineering community revel in their ability to match complex answers to simple questions. Thus, aqueducts, towering dams, canals, and pipelines criss-cross the globe moving water from Point A to Point B frequently disregarding schemes to develop self-sufficient water supplies at Point B.

A parallel can be drawn here to man's desire to bend nature to his wishes, to bring her to her knees, to prove man's intellectual supcriority over nature's once omnipotent forces. There are those who view going with the flow of the natural order of things as taking the easy way out, giving in to wimpish philosophies.

Reinforcing these absurd attitudes has always been the political motivations which also promote complex solutions that afford opportunities to spend moneys: the expenditure of moneys produces power for the spender and supporter from the recipients of the largesse. This type of politics has also lead to economic incentives to waste and a lack of incentives to conserve and thus, a development of local, regional and world water shortages that are more apparent than real but

---

[1]Executive Director, National Water Well Association, 6375 Riverside Drive, Dublin, Ohio 43017, USA.

are in fact, with few exceptions, indicators of gross mismanagement rather than natural shortfalls in available water.

There will always be vested interests in constituencies that will support inappropriate projects and oppose sound water development, protection and management. Concurrently political control of water utilization will always offer dangerous amounts of power to legislative entities that can artificially control and disperse water in ways that invariably lead to inefficiency and waste even in the best of all worlds. In spite of these obstacles over the past 30 years, ground-water utilization has grown to 25% of total water use and is expected to grow at twice the rate of surface water in the coming decades because of its clearly recognized economic advantages as well as the obvious reduction in available dam and surface reservoir sites throughout the world today. Thus, at long last and in spite of man's tendency toward grandiose schemes to defeat natural forces together with political disincentives to conserve and economically develop water, the world is at last on the right track toward balancing its water budget and minimizing dislocations caused by water shortfalls, ultimately eliminating the paranoid fear that we are running out of water.

1.2 The Emerging Perception of Quality Decline

Water quality is rapidly emerging as a primary problem throughout the world. Although the quantities of this resource that have been polluted are an infinitesimal percentage of the whole of available water, the multiplicity of activities that have created the problem, coupled with man's ignorance of his role in the problem, make water quality the primary water supply issue today. Recognition of the causes of surface water pollution and efforts to eliminate those causes preceded by decades the evaluation of potential problems with ground water as a result of man's activities on the surface of the earth. Ground water has historically been considered to be a pristine, pure resource of all but magical powers. Misconceptions about its occurrence and movement carry over to the belief that its quality is safe from surface activities. It was once true. In areas uninhabited by man, there are few natural forces that reduce the quality of available ground water supplies. Ground water emanating from rainfall normally carries few contaminants into the ground and the minerals picked up in transit through subterranean rock are generally innocuous; they contain no threat to human health. But as man inhabited the land he impacted ground water, first by indiscriminate disposal of his own domestic waste, then by that produced by animals and eventually by his treatment of the crops he grew on the surface. Today on a well populated earth, the major causes of ground water pollution encompass all of man's diverse activities.

Water supplied from ground-water sources now comprises over 25 percent of our world's water usage and exceeds 50 percent of our drinking water. As the potential for economic capture of additional surface supplies declines, the potential for further reliance on ground water increases. In light of these developments, greater attention has been placed on ground-water quality and the activities that impact it. The resulting desire to monitor the quality of our ground water, coupled with the availability of new precise instrumentation to measure chemical components in parts per billion and even parts per trillion, has uncovered the existence of a diverse suite of objectionable contaminants in our water supplies. Their sources are now recognized as daily human activities running the gamut between safety (highway salting), religion (human burial), and commerce (chemical handling), to all the more obvious waste disposal practices. In all, as many as three dozen separate source categories can be delineated with varying undesirable impacts on human health dependent on physiographic area and demographic concentration. Among the most severe problems are surface impoundments, landfills, open dumps, underground storage tanks and agricultural chemicals.

1.3 Man's Impact on Ground Water

Holding ponds and lagoons are shallow excavations that range from a few square feet to many acres in area. They are commonly used to hold municipal sewage, a variety of industrial chemicals and oil field brines. If adjacent ground water is to be protected, holding ponds and lagoons should be prohibited unless it can be shown that fluids will not leak from them. In practice, this necessitates an impermeable lining that has been proven effective in retaining the fluids to be contained.

Ground water pollution results from landfills constructed by placing wastes in excavations, then the material is covered with soil daily – the term "sanitary" is used to indicate that garbage and other materials are not left exposed to produce odors, smoke, vermin and insects. Even though a landfill is covered, leachate almost certainly forms by the infiltration of rain. Pollution of ground water by landfills can be prevented by proper site location to prevent their use in places where leachate can readily enter ground water reservoirs, by regulation of their construction and operation, and by prohibiting the disposal therein of highly toxic wastes.

Stockpiles and wastepiles are an often ignored source of ground water pollution. Perhaps the prime example of ground water pollution caused by stockpiles is related to the storage of salt used for highway snow and ice control. In most instances, it is possible to control or reduce ground water pollution from these sources by providing facilities that eliminate or reduce infiltration of rainfall into them and control surface runoff. Impermeable covers are the best examples.

Individually, effluent from septic tanks, cesspools and privies is of little significance, but these devices are important collectively because they are so abundant in areas not served by sewage treatment systems. Ground water pollution by septic tanks can be controlled by regulating their construction and installation. Requirements setting satisfactory subsurface soil and rock conditions and adequate land use regulation, including limitations on the number of septic tanks per unit area, are needed.

Literally thousands of miles of buried pipelines criss-cross our lands. Leaks, of course, do occur, but it is exceedingly difficult to detect and locate them. A growing problem of substantial consequence is leakage from storage tanks and their associated piping. Ground water pollution from this source could be minimized by periodic inspections of tanks and pipelines, including pressure testing, and careful monitoring of fluid levels and inflow-outflow comparisons. Metal tanks will be replaced with noncorrodable material, such as fiberglass, and safeguards will include double liners and leak collection capability.

Hundreds of thousands of abandoned exploratory wells or test holes dot the country side. Many of these were drilled to determine the presence of underground mineral resources. The open holes permit water to migrate freely from one aquifer to another. A fresh water aquifer could thus be joined hydraulically with a polluted aquifer or a deeper saline aquifer. The only realistic and practical solution to ground water pollution caused by abandoned wells and exploratory holes is timely and adequate well plugging procedures.

Toxic materials are transported throughout most countries by truck, rail and aircraft. Accidental spills of toxic materials due to transportation accidents are becoming increasingly common. No amount of laws and regulations will ever cause accidents to cease, but inspection of transportation equipment and travel ways, adequate insurance provisions and the placing of damage responsibilities on the hauler can reduce the number of "accidents."

Animal feedlots cover relatively small areas, but provide a high volume of wastes. Feedlots should be designed, operated and maintained in such a manner that the liquid runoff is not allowed to infiltrate or flow into streams.

The use of both fertilizers and pesticides is increasing each year. A reduction in fertilizer application rates and adoption of sediment runoff control measures will reduce ground water pollution. In many instances, the volume of pesticide application can be lowered without reducing the effects of the pesticide.

In certain situations, pumping of ground water can induce significant water quality problems. The principal causes include interaquifer leakage, induced infiltration and landward migration of seawater in coastal areas. Reduction of ground water pumping, water conservation and artificial recharge of ground water reservoirs are partial solutions to man-induced water quality problems.

1.4 A Current Assessment of Ground Water Pollution

Having touched on some of the most significant sources of ground water pollution, let me state that these problems of ground-water pollution are

1) somewhat less catastrophic than many believe,
2) better understood than some are willing to admit,

3) manageable with a few exceptions, and
4) have a special ally that with time allows this temporal concern to be overcome
   – that ally is Mother Nature.

Rapidly advancing hydrogeologic knowledge is fostering scientific efforts to eliminate or contain contaminants now reaching each nation's ground water. Technologies are improving both in terms of their potential for success and economic feasibility. However, the long-term goal of maintaining a still largely uncontaminated water resource must rely on the future prevention of pollution. A thorough re-evaluation of all contaminant-producing activities must be conducted. Environmentally sound waste disposal, chemical handling, and application procedures must replace those of the past if we are to protect our ground water as a legacy for future generations.

We have finally reached our tolerance level for the indiscriminant, ignorant and wrong-headed manner by which we have disposed of industrial waste and applied, stored and transported chemicals these past 40 years. The scaremongering stories of oozing gunk generating disease, chromosomal damage and potential death are vastly exaggerated. Yet, like it or not, the media's excesses have drawn our attention to the mess we have created.

Were it not for their rubbing our noses in it, we might have turned our attention to more mundane affairs of the world. But we were forced to look, and in so looking, there are few left among us who cannot be rallied to oppose the continuation of our past stupidity. There are few among us not willing to join the ranks of those who would work to reverse the past through calculated though costly efforts.

1.5 Changing Patterns of Waste Disposal

The very wastes which have created our serious ground-water pollution problems these past three or four decades are at last being significantly reduced by some of the industries who were previously among the worst, though largely unwitting parts of the problem. Though few people appear to be satisfied with the environmental regulatory matrix intended to serve as a safety net, the bottom line for industry is quickly being perceived as an increasingly high cost for waste disposal. Just as necessity is well known as the mother of invention, economics is the driving force for industrial innovation. Today it is becoming apparent to many industries that it really does make "cents" to reduce the waste stream from every and any plant and factory. Disposal costs and the cost of raw materials are finally forcing plant managers to be innovative. Many have been very successful. A few cases in point:

1) U.S. Steel estimates that it has reduced the generation of hazardous waste by nearly 50% in the past few years while concurrently reducing its reliance on landfills by as much as 80%. Heavy sludge residues that were once discarded by that company are now mixed with tars and converted into fuel. They have also

substituted nonleaded greases and oils for leaded lubricants in their rolling mills, thus improving the nature of their disposal problem.
2) Dupont has begun removing organic chemicals from its waste streams with a new membrane separation technology not previously thought economically possible.
3) 3-M Company is saving nearly a million dollars a year in the application of abrasive resins to its sandpaper by creating a continuous feed process that no longer requires the emptying and cleaning of used-up containers of these materials. They reduced their hazardous waste stream by 400 tons a year in the process.
4) CIBA-GEIGY used a similar technique in the manufacturing of dyes that netted them an increased product yield of over 25% from their basic raw materials. That same company recently abandoned the use of zinc in its process for making a plastics additive and substituted hydrogen which eliminated the handling of a strictly-regulated hazardous material.

Thus, today's waste stream is being dramatically reduced by industry which is seeing economic benefits in maximizing the use of their raw material and in turn minimizing their costs of waste disposal. By the century's end, new contaminants entering the ground will likely be less than 10% of what we still contribute to ground water today. That is to say it is entirely possible that we can one day reduce our waste stream to a manageable volume. But will we have the intelligence to manage it? That question remains unanswered. Not because the answer is not apparent but because the public, the government, and even many in our scientific fraternity resist the answers.

1.6 Legislative Controls

I honestly believe that there is less need today for national regulation of our water than at any time in history. Fifteen years ago when a company was likely to locate where it could best get away with dumping its wastes in the nation's waters, a federal mandate upon each and every State was a strategic necessity. Today an environmentally conscious citizenry which has resisted the industrial double talk that blames all the economic ills on the environmental movement no longer needs such a strong big brother in Washington. "We" really can do more ourselves – we being the States acting independently with merely the moral and financial support of the feds.
What we need in national government today is logistical and organizational support for each and every State professional water scientist and regulator so that they can make the solution fit the problem at the most pragmatic level.
These are not easy words for me to write. From 1969 to 1974, I fought zealously for a Safe Drinking Water Act in the United States which would recognize that half the water we drink is from underground sources and that these

sources were given virtually no protection under the law as was afforded our surface waters.

Following the passage of the Safe Drinking Water Act there was a virtual avalanche of legislation attempting to control ground water. These included the Resource Conservation and Recovery Act, the Toxic Substances Control Act, the Surface Mining and Reclamation Act, the Clean Water Act and our Comprehensive Emergency Response Compensation and Liability Act, commonly called Superfund. But enough is enough.

It is an interesting fact that the legislative process probably functions at its best during the period when a country and its government debates the need for legislation. In that exercise a process of education occurs allowing people to become aware of a problem so that it can be addressed with a sincere desire to achieve a solution without need for a legal blackjack held against the public head.

I am not so naive as to think that the good of humanity never requires a push in the right direction, but at our present state of environmental awareness, none of us either wants to drown in our wastes or be poisoned by our water.

What does it all mean? Simply that we may be approaching the point where the least rigid regulation is the best regulation. As nation after nation and state after state becomes concerned about their potential ground-water pollution problems, good things are happening without the prod of federal legislation.

A number of major issues have however recently been resolved in the United States in the passage of amendments to our "Superfund" waste cleanup laws. The "how clean is clean?" issue was one of the most important. When our initial laws were passed it was broadly stated as the need to mitigate the "imminent and substantial endangerment to human health, welfare, and the environment in a cost-effective manner."

We are now required "to select, to the maximum extent practicable, remedial actions that utilize permanent solutions and alternative treatment technologies or resource recovery technologies." Further, offsite transport and disposal without treatment is to be the least preferred option where practicable treatment technologies are available.

A section of these amendments offer the leverage that has long been needed to pry loose major obstacles to siting new hazardous waste disposal facilities. Effective three years after enactment of this new legislation in the USA, our President shall not provide any remedial actions unless the State in which the release of contamination occurs first enters into a contract or cooperative agreement with the President assuring the availability of hazardous waste treatment or disposal facilities.

This is perhaps the first of what may be a long series of legislative efforts to overcome the "not-in-my-backyard" attitude that prevails in most communities around the world.

It is very difficult to demonstrate that commitment to the environment is associated with some particular slice of the partisan political spectrum. The environment as an issue is a many-splendored thing, and trying to co-opt it, to make it part of some formal party, is almost as foolish as pretending it doesn't exist. Some of the strangest bedfellows in all politics are found together on environmental issues – people with patches on their sleeves and shoulders who shoot ducks joined with people who grow pot and live on bean sprouts. And, with few exceptions, the environmental battles played out in national political arenas have been non-partisan in nature, consistent with this broad and eclectic public support.

We are advancing environmental protection at pretty close to the maximum rate it can be advanced given current resources and the fact that environmental progress is locked into the pace of developing science, technology and institutional development. But the politicians who make our national environmental policy and never have to turn their pronouncements into action are not so restricted. They can demand an instant end to longstanding problems. They can pick the scientific evidence that supports their position and ignore the rest, which we cannot do. Above all, they can promise more than government can reasonably deliver.

We will not be able to avoid emotions and tensions during this cleanup. People are extremely sensitive about ground water, and I believe I know why. Ground water has enormous symbolic value to people. The air, the lakes, streams and the surface of the land are all touched by the hand of man, often in unpleasant ways. But ground water is pure, the last bit of untrammeled paradise available to people in their daily lives. I think this is why we will shrug off involuntary risks thousands of times worse than those we experience through drinking ground water, and yet demand absolute purity in that one area. We have said it so often that we have come to believe it: "We don't worry about our water because it comes from a well." It is something to think about as the cleanup rolls on and we must make increasingly tougher choices.

1.7 The Cleanup

The majority of ground-water pollution incidents reported in scientific literature were discovered some time after subsurface contamination began and, in most cases, contamination of water-supply wells was the first indication of the ground-water pollution problem. Most individuals and an increasing number of communities whose well water is found to be contaminated are abandoning the use of the affected well and turning to an alternate water supply. If more and more wells are abandoned, the stress on other means of water supply will increase to an intolerable level. We must develop some alternate means of dealing with the problem of encroaching and/or apparent ground-water contamination.

Fortunately, some technology has already been developed to deal with contaminated ground water. Whereas several years ago the prevailing opinion was that once ground water was fouled it was nearly impossible to restore, today there are means of containing and rehabilitating contaminated ground water, though they are not inexpensive.

In areas where contamination is pervasive, and where the population relies on ground water as its only potable water supply, control of the contamination or decontamination of the ground water may in fact be the safest and most cost-effective remedy.

Designing a program for containment or control of ground-water contamination is further complicated by the fact that not all contaminants behave in the same manner in the subsurface. Many contaminants move at different rates in the ground-water system, and may occupy different levels in aquifers according to their solubility in water, their density, and other physical properties peculiar to the contaminants. Heavy metals may migrate in ground water at one rate, water-soluble organics at another rate, and separate-phase liquid organics at yet another rate. Some contaminants, including most petroleum products, are immiscible and float on top, or otherwise move independently of the ground water.

Earth materials may attenuate certain contaminants to a sufficient degree as to reduce their mobility and concentration. Various materials, most notably certain types of clays, act as very efficient adsorbents, and may aid in retarding the migration of selected contaminants. In addition, natural microbial action may help to break down toxic organic contaminants to simpler, nontoxic forms. Thus, some chemical constituents of waste products deposited at the surface are absorbed, attenuated, or transformed in the subsurface, while others pass through with little or no attenuation or change.

The containment and/or control of contaminated ground water can generally be accomplished using one or a combination of several available techniques. The alternatives available for remedial action can be broken down into three broad categories:

1) Physical containment measures, including
   - slurry trench cutoff walls,
   - grout curtains,
   - sheet piling, and
   - hydrodynamic control.
2) Aquifer rehabilitation, including
   - withdrawal, treatment, reinjection (or recharge), and
   - in-situ treatment, such as
     - chemical neutralization, and
     - biological neutralization.
3) Withdrawal, treatment and use.

1.8 Public Perception

Not long ago, few people knew what ground water was and fewer, still, cared to learn. Yet, we have been actively developing our ground-water assets and, in so doing, uncovering its liabilities for many, many decades. Those of us who began our careers with the U.S. Geological Survey always knew of the threats to ground water posed by many of man's activities. We erred, however, in being less vocal about the issue than our public responsibility dictated, but few citizens were interested in optimizing the use of the resource and even less were concerned over its well-being.

A snapshot taken today would indicate the dismal dimensions of the ground-water pollution problem caused by dozens of activities thoughtlessly carred out by citizens and industries in past decades. A snapshot of this scenario taken a few years ago would be little different even though less of the iceberg had been uncovered. Thus, a comparison of the pictures leads one to despair and frustration over the apparent lack of progress; but as smoke and mirrors skew the visual impression of magical illusions, a lack of proper and extensive historical perspective offers a badly distorted and inaccurate perception of where we actually are in space and time.

Those of us with more than three decades of experience on the learning curve realize that in public awareness we have come from ground zero 30 years ago to high on the list of public concerns. In problem assessment, we have moved well beyond iceberg tips and ticking time bombs to a reasonable view of the ugly monster from all angles. In waste-disposal techniques and other human activities that impact ground water – be these industrial waste ponds, agricultural chemical use, petroleum and chemical storage tanks, highway salting, or things as mundane as septic tanks, we are achieving visible progress and impressive results on all fronts. Perhaps most exciting of all, technologies employed in ground water restoration are increasingly effective. While we will not eliminate the problem in this decade or the next, we can, with continuing energy, effort, and financing, win the day in the lifetime of the young among us.

1.9 The Professional's Responsibility to the Public

Comparison of articles in the popular press and articles and editorials in technical journals suggest a wide difference of opinion between the public and our community of ground-water scientists regarding the problem of ground-water pollution. The public's intense concern about ground-water pollution stems partly from the fact that, to most people, ground water is mysterious and partly from widespread acceptance of media scare stories. Hydrogeologists, on the other hand, may tend to be less concerned than the general public because of their awareness of the slow rate of movement of ground water and the beneficial reactions that may modify or immobilize pollutants as they move through the ground.

The difference in perception of the pollution problem by professionals and the public has, up to now, probably been beneficial. The public's concern has resulted in more effective regulations, increased efforts by industry to reduce waste generation, and larger appropriations for investigations and pollution abatement. All of these are benefits which professionals can and do appreciate.

Unfortunately, the public does not seem to have gained as much from us as we have gained from them. Instead of acquiring some of the professionals' optimism, the public is becoming increasingly concerned. As a result, demands that wastes be removed from sites that might best be left alone, proliferate, and public opposition to the establishment of new waste-disposal sites has brought this activity nearly to a halt. Public concern is also causing regulatory agencies to adopt technically unsound policies related to waste disposal and pollution-monitoring requirements.

Before an already chaotic situation gets worse, it is essential that hydrogeologists do a better job of communicating their knowledge and their optimism to the public. In the process, the public must be convinced that ground-water professionals really understand ground-water pollution and are capable of dealing effectively with it. This will require convincing the public that we know the extent and the nature of ground-water pollution, know how to reduce pollution where it is a serious threat to public health, and know how to select waste-disposal sites that will not cause future problems. This is no small order.

Many of today's ground-water pollution problems result from a surprisingly widespread ignorance of the underground environment among people responsible in the past for waste disposal. Only in the last decade or so have most of these people learned that the surface of the Earth is not impervious and that all liquids placed in depressions on the land surface do not disappear through evaporation. Recognizing now that wastes do seep into the ground and pollute ground water and recognizing that as long as wastes are generated, places must be found to dispose of them, hydrogeologists are intensely involved in identifying areas in which waste disposal will have the least adverse impact.

So, in spite of the historical perspective which can only lead the experienced ground-water scientists to be more optimistic than the public at large, we are yet to learn the lesson of geologic history which dictates that we must eventually go with the flow and work with Mother Nature, not against her.

1.10 Reasonable Risk

There are activities, such as the disposal of domestic/municipal wastes and the development of mineral resources which involve the consumption of large tracts of land. Society's goal in such instances is to minimize the environmental impacts, recognizing that complete prevention of impacts is neigther economically nor technically feasible. Other activities, such as the production and distribution of industrial chemicals and fuels, result in the need to dispose of highly toxic wastes. Society's goal in these instances is also to minimize the environmental impacts, but the efforts to do so are less constrained by economic and technical feasibility arguments. The latter situation stems from a greater fear of potential health hazards associated with industrial chemicals and solvents. In this context, U.S. society has become nearly risk-averse; it is tempted to spend more to prevent such problems than it will cost to remedy them. This parallels the attitudes that have developed historically for other potentially hazardous activities, such as the construction of bridges and dams, where large margins of safety are integral to design.

Ground-water contamination – compared to other environmental problems – poses fewer risks than the public perceives, concluded a new "comparative assessment" issued by the Environmental Protection Agency. In four risk assessment areas examined by 75 agency experts, ground water "consistently ranked medium or low," the report found, and this finding may shape the future of EPA's resource priorities.

Charged by EPA Administrator Lee Thomas, the group found that areas of relatively high risk include: indoor radon; indoor air pollution; ozone depletion; non-point sources; and worker exposures. Areas of medium to low risk but high EPA effort include: Resource Conservation and Recovery Act-permitted facilities; Superfund; underground storage tanks; and municipal non-hazardous waste facilities.

Although the public is more concerned with chemical waste disposal and its impact on ground water, the report found health data "do not appear to match public concern in these areas." The group found that underground storage tanks containing petroleum pose a relatively low health risk because "people generally avoid drinking water known to be contaminated with motor fuel . . . [although] the public ranks this problem as a moderate risk." Pesticide risk was ranked medium due to the large populations that are potentially exposed from runoff.

Industry officials generally applauded Thomas for undertaking the study, while environmental groups and Sen. Robert Stafford (R.-Vt.) voiced concern that the report could be misused to minimize attention in certain areas. The report, "Unfinished Business: A Comparative Assessment of Environmental Problems," was released by U.S. EPA's Office of Policy Analysis.

One of the problems is that knowledge desensationalizes ground water contamination stories. This is not beneficial for the news industry or for the bureaucrats who try to hype problems for their own benefit.

1.11 The Future

Rapidly advancing hydrogeologic knowledge is fostering scientific efforts to eliminate or contain contaminants already reaching the world's ground water. Technologies are improving both in terms of their potential for success and economic feasibility. However, the long-term goal of maintaining a still largely uncontaminated water resource must rely on future prevention of pollution. A thorough re-evaluation of all contaminant-producing activities must be carried out. More environmentally sound waste disposal, chemical handling and application procedures must replace those of the past if we are to protect our ground water as a legacy for future generations.

Over the last 10 years, considerable activity at national levels involving regulation and legislation have begun to develop a safety net against future ground-water contamination. Concurrently, these activities and their educational byproducts have catalyzed considerable movement within government and private industry. Significant programs are underway to correct some of the existing ground-water pollution that resulted largely from past ignorance and commonly persists due to lack of visibility.

Now that recognition of the existing problems and development of alternate technologies are coming collectively into focus, the world is ready to establish a long-term strategy to contain and reduce existing pollution by the end of this century through detection, correction and prevention methodologies. It warrants the highest priority action by the appropriate national agencies. However, it should be recognized that while ground-water's glacially slow movement produces long-term problems, it also affords adequate time to achieve the most reasonable approach to managing these residual problems of our past. Within that time-frame, the legislative programs being forged must be refined and maintained to support the elimination of ground-water pollution through adequate funding and technical assistance. National initiatives in these areas will make it possible to detect sources yet undiscovered, correct those now recognized, and prevent the future from repeating the past.

We have likely polluted less than 1 percent of the earth's ground water, which has impacted less than 5 percent of our population. In the next 10 years, these numbers will increase but with greater awareness in the following decade, they can decline.

We can reduce new hazardous, toxic, and troublesome emissions into our ground water in the year 2000 by well over 90% of its present level. We will still be left with decades of work to contain, control, and remove the mess we have been making throughout most of our lifetimes, but it will be seen clearly as a manageable job no longer worth screaming about on TV or in tabloid. We have not been sitting on our hands, and the number of us with competence to deal with these ground-water quality problems is multiplying in a healthy manner. More enthusiasm can be generated toward tackling the tremendous tasks facing us by recognizing the good things we have going for us and the strong possibility that we are equal to the challenge ahead — rather than by joining the ranks of the legion of doomsayers who know little of what they speak but revel in the pessimism of ignorance.

2 Pollution Hazards from Toxic Organic Chemicals

A.S. PERRY[1], L. MUSZKAT[2] and R.Y. PERRY[1]

2.1 Introduction

Until recently it was widely held that groundwater is relatively uncontaminated with toxic chemicals as compared to surface water, and in cases where contamination was detected it was assumed that the contaminants posed no immediate hazard to people's health.

To be sure, the process of groundwater contamination is not direct. The pollutant has to percolate through a top layer of soil, then penetrate through a protective layer (the unsaturated zone), eventually reaching groundwater either in its original form or as a breakdown product. This eventuality is believed to be slow and time consuming during which the pollutant may be subjected to various processes such as adsorption − desorption on soil constituents, volatilization, chemical breakdown, aerobic and anaerobic microbial degradation (including oxidation, reduction, hydrolysis and conjugation), complex formation and ion exchange.

Each of these processes may influence the fate of the pollutant and its subsequent transfer to groundwater. Even after reaching groundwater many physicochemical processes may operate to neutralize or diminish the toxic effects of pollutants. The rate and amount of pollutants penetration through the unsaturated zone is controlled by several factors:

1) the extent of sorption to soil particles (Karickhoff 1981);
2) the thickness and nature of the unsaturated zone: A thick unsaturated zone with high levels of clay and organic matter reduces the potential for groundwater contamination. However, it has been shown that occasionally the rate of movement of contaminants could be much higher than expected due to rapid channeling through tunnels, voids, gravel, and through sandy or other porous layers (Rothschild et al. 1982);
3) the chemical properties of pollutants, particularly their water solubilities;
4) rainfall and water management of the field.

---

[1]George S. Wise Faculty of Life Sciences, Institute for Nature Conservation Research, Tel Aviv University, Ramat Aviv 69978, Israel
[2]Department of Chemistry of Pesticides and Natural Products, ARO, Volcani Center, Bet Dagan, Israel

Pollutants of moderate to high water solubilities are of higher penetration potential. However, many groundwater contaminants including volatile halogenated organics have low water solubilities. Such chemicals migrate in the subsurface as a non-aqueous liquid phase and its mobility is largely governed by density and viscosity of the components (Mackay and Cherry 1985). The low density organic liquids float on the water table (Nielson 1982).

The biotic and abiotic processes which occur in the groundwater zone are not yet well understood, being the subject of increasing research. In the meantime, since a detailed experimental evaluation of the effect of each pollutant is almost impossible, mathematical modeling of the pollution hazards from organic toxicants is attractive, offering a relatively rapid and inexpensive way to assess the potential of such hazards. However, despite the intensive effort to develop new models, it appears that more attention should be given to their verification through laboratory and field studies. When field data are limited, there is no unique model solution and the conclusions from models may be imprecise (Cohen 1986).

The events characterizing the behavior of herbicides in soils (Weber and Weed 1974) and in irrigated soils (Yaron et al. 1984) have recently been reviewed. A comprehensive up-to-date evaluation of pesticides in groundwater edited by Garner et al. (1986) deals with all aspects of groundwater contamination including physicochemical parameters, monitoring techniques, field monitoring, modeling and model validation, risk assessment (toxicological significance) and regulatory aspects.

The present report is by no means a review of the extensive literature available. It is intended to bring to focus the highlights of groundwater contamination by organic chemicals with emphasis on problematic compounds of high risk to human and environmental health.

2.2 The Facts

In contrast with the assumption that pesticides and other organic chemicals do not contaminate groundwater, recent reports from different parts of the world repeatedly show that under favorable conditions of soil type, temperature, pH and organic content, groundwater can be contaminated with toxicants as easily as surface water (Page 1981).

2.2.1 Pesticides in Groundwater

2.2.1.1 Insecticides and Nematicides

Aldicarb (Temik), [2-Methyl-2-(methylthio)proprionaldehyde-*O*-(methyl-carbamoyl)-oxime] is a carbamate compound used as a systemic insecticide –

acaricide — nematicide. It is extremely toxic to mammals with an oral LD_{50} of 0.8–1.0 mg/kg, dermal LD_{50} of 3 mg/kg and is rapidly absorbed through the skin. It is used as a granular formulation for incorporation in the soil where the plant absorbs the toxicant through the root system. Its high water solubility (0.4% at 10° C — 0.9% at 30° C) results in high mobility through the upper layers of irrigated soil. Recent tests conducted in Wisconsin showed that traces of aldicarb residues may appear in potable water from shallow wells located near agricultural areas. The appearance of such residues in drinking water was brought about by a combination of factors including extensive aldicarb usage for control of potato insect pests, highly permeable and acidic soils, high water recharge rate, low soil organic matter content, low soil and water temperatures, low soil microbial activity and a high water table (Wyman et al. 1985). On the other hand, no detectable residues of aldicarb, its sulfoxide and sulfone and carbofuran (2,3-dihydro-2,2-dimethyl-7-benzofuranyl methylcarbamate) above the 1 ppb detection limit were observed in 14 samples of well water adjacent to potato fields treated with aldicarb (Cochrane et al. 1982).

Aldicarb concentrations in groundwater beneath three main study fields and two subsidiary fields were monitored during 1980–1981 (Rothschild et al. 1982). A total of 67 well points, 25 private wells and 7 irrigation wells were sampled. The highest concentration of aldicarb among private wells was 17 ppb. Higher concentrations (as high as 190–210 ppb and an average of 80 ppb) were found in several of the shallow and very shallow wells. The authors concluded that:

1) Most aldicarb under treated fields was concentrated near the water table;
2) No aldicarb was detected in any of the deep monitoring wells, roughly 60 ft below the water table;
3) Aldicarb seems to be concentrated in a 5 ft thick layer near the water table;
4) Marked seasonal fluctuations in aldicarb concentration occurred in several wells;
5) There may be a time lag of 10–20 months after application before peak concentrations occur in the aquifer;
6) Aldicarb is a good indicator of potential groundwater contamination by other organic chemicals due to its high water solubility and poor adsorbability.

One of the major factors favoring leaching is the water recharge rate from rainfall and irrigation minus evapotranspiration. A change in irrigation schedules can dramatically affect the downward movement of water soluble pesticides. A rating system for predicting the appearance of aldicarb residues in potable water has been proposed by Back et al. (1984). Under favorable soil temperature ($> 15°$ C) and pH (> 8.0) conditions aldicarb undergoes hydrolysis to toxic metabolites, i.e. aldicarb sulfoxide (LD_{50} 0.9 mg/kg) and slower to aldicarb sulfone (LD_{50} 25 mg/kg) and to several other nontoxic oximes and nitriles. Soil type has a great influence on rates of aldicarb persistence, with sandy soil showing the highest persistence (Coppedge et al. 1967). It has been shown that, occasionally, the rate of movement of contaminants through the soil could be much higher than expected due to rapid channeling through tunnels, voids, gravel, and sandy or other porous layers (Rothschild et al. 1982). Long term studies in sterile water (Hansen and Spiegel 1983) established chemical hydrolysis rates for aldicarb and its metabolites with a half life of 1900 days for aldicarb, 360 days for sulfoxide and 125 days for the sulfone at 15° C and pH 7.5. The rate of hydrolysis is positively correlated with both pH and temperature. These rates can be used to establish upper bounds for the half lives of aldicarb sulfoxide and sulfone in groundwater, although, under natural conditions, the concentrations may be further reduced by dilution, dispersion in the moving groundwater and aerobic and anaerobic microbial degradation. Other researchers who simulated aldicarb behavior in the Long Island unsaturated zone assumed chemical decomposition rates for aldicarb of five years or more (Intera 1980, quoted by Hansen and Spiegel 1983). It must be emphasized that these rates are valid only for specified and highly controlled conditions under laboratory testing. In the field, interactions of moisture, temperature, pH, soil characteristics and soil biota will largely determine the actual residues of aldicarb and its metabolites in groundwater.

Traces of aldicarb residues were also found in potable groundwater in Florida (Jones and Back 1984) following application of granular formulations of Temik for nematode and mite control in citrus groves. Rapid hydrolysis of the sulfoxide and sulfone by soil biota occurred several weeks after application. Computer calculations indicated that less than 1% of the aldicarb applied will leach more than 1 m beneath the soil surface.

Because of its high water solubility aldicarb is weakly adsorbed by soil particles and its oxidation products are even less so. Thereafter, their mobility in soil may be rather high. Laboratory experiments with different types of soils (Smelt et al. 1978) showed the half life of aldicarb sulfone to be 18 days in clay loam and 154 days in peat-sand soil. Conversion in deeper layers was considerably slower, and in a sandy layer at 90–110 cm no clear loss occurred during the 294 days of incubation at 15° C.

Factors conducive to the appearance of aldicarb residues in potable well water include high aldicarb usage, highly permeable and acidic soils, high water recharge rate, low soil and water temperatures, low soil organic matter, low soil microbial activity and shallow potable groundwater. The glacial sands of Long Island, N.Y., and the Central Sands area of Wisconsin are two areas where such conditions occur and where traces of aldicarb residues have been found in potable groundwater (Zaki et al. 1982; Rothschild et al. 1982).

The insecticide *methomyl* (Lannate), [S-Methyl-N (methyl-carbamoyl)oxy thio-acetimidate] is a carbamate compound used as a systemic insecticide-nematicide with an LD_{50} of 17 mg/kg and a water solubility of 10%. It is used on a large scale in greenhouses for the control of caterpillars, greenhouse white flies, leaf miners, thrips and aphids. Under laboratory conditions 48% and 31% of the applied dose in two different soils remained after 42 days. Under field conditions only 2% of the methomyl applied remained after one month (Harvey and Pease 1973). Transformation rates of methomyl under greenhouse conditions were higher (Leistra et al. 1984), due perhaps to the higher content of organic matter and greater microbial activity in the soil. Leaching of methomyl from greenhouse soil is probably faster than from ordinary soil due to the greater amount of water used for irrigation. The comparatively higher rates of water percolation increases the risk of residues reaching groundwater and water courses, although this was not established in the case of methomyl (Leistra et al. 1984).

DBCP, EDB and 1,2-D Liquid DBCP (1,2-Dibromo-3-chloropropane) was used extensively as a soil nematicide in California from 1950 to 1977 when it was banned because of its identified potential as a carcinogen and testicular toxin in test animals and man. Water samples collected in 1979 from several wells near a pesticide dump site were found to contain 4–68 ppb of DBCP. Ninety four additional samples of well water from agricultural areas contained from 0.1 to more than 20 ppb of DBCP (Peoples et al. 1980). About 43% of California relies on groundwater for their water supplies. In a recent study, 53 pesticides were detected in 512 wells in half the State counties. Among the most pervasive toxic pesticide chemicals found in drinking water have been DBCP, ethylene dibromide (EDB) and 1,2-dichloropropane (1,2-D). Contamination by DBCP has resulted in the closure of about 1,000 drinking water wells (Berteau and Spath 1986). In another study (Cohen 1986) 2,500 California wells were found to be contaminated with DBCP, and this number continues to increase seven years after its use was banned. Recent findings of 1,2-D contamination in more than 60 wells, as well as other chemicals in California groundwater poses a potential danger of widespread contamination and movement from rural areas to urban water supplies. The precise mechanism whereby DBCP contaminates groundwater has not been established, but considering its widespread nature of contamination further use of this compound should be banned the world over. The recommended safety level (acceptable daily intake) of 1 ppb by U.S. EPA (1975) has recently been readjusted to 50 ppt.

Organochlorine Insecticides. Organochlorine insecticides are, by and large, lipophilic compounds of limited water solubility. It is, therefore, predictable that such compounds will remain in the upper layers of the soil with little downward movement. Lichtenstein et al. (1971) studied the persistence and vertical distribution of DDT, lindane and aldrin residues, 10 and 15 years after a single soil application. The results showed that all three insecticides are metabolized in loam soils and disappear at relatively slow rates. Fifteen years after application 10.6% DDT, 5.8% as dieldrin of the aldrin applied and 0.2% lindane were recovered from the soils. Although the insecticides had been worked into the soil to a depth of 5 in., the 6–9 in. soil layer contained 10 years later about 30% of the total DDT residues and 18% of the aldrin residues.

The direct movement of these compounds through the unsaturated zone is unlikely, but there is a possibility that water, under certain conditions, may transport soil particles containing pesticides. Evidence that organochlorine insecticides reach groundwater is given by the works of Sandhu et al. (1978) and Page (1981) (cf. Table 1). Supporting evidence comes from the investigations of Achari et al. (1975) who analyzed pesticide residues in 27 wells, 21 to 100 ft deep, in South Carolina and found lindane, aldrin and DDT (averages of 1.19, 7.11 and

Pollution Hazards from Toxic Organic Chemicals 21

Table 1. Organochlorine pesticide residues in potable water supplies of Chesterfield and Hampton counties, S.C. (Sandhu et al. 1978) and from New Jersey groundwater (Page 1981)

Compound	Chesterfield (ng/L)		Hampton (ng/L)		New Jersey Groundwater
	Mean	Maximum	Mean	Maximum	Maximum (ng/L)
BHC-alpha	–	–	–	–	800
BHC-beta	–	–	–	–	8700
Lindane	23	193	143	319	900
Heptachlor	15	159	9	44	1000
Heptachlor epoxide	8	90	18	87	600
Aldrin	17	191	9	56	1200
Dieldrin	65	153	204	771	900
Endrin	–	–	–	–	200
Chlordane	–	–	–	–	400
p,p'-DDT	261	3307	264	812	900
p,p'-DDE	17	200	8	30	–
p,p'-DDD (TDE)	284	4333	129	779	400
o,p-DDT	–	–	–	–	500
o,p-DDE	–	–	–	–	1000
Methoxychlor	33	312	23	100	–
Mirex	2	230	83	437	–
Total residue/sample	725	9168	890	3435	–

37.7 ng/L, respectively) in groundwater. Federal limits for these insecticides in groundwater are 5.0, 1.0 and 50.0 ppb, respectively.

The Ogallala aquifer in Texas was recharged at 350 gal/min for a period of 10 days and p,p'-DDT was injected into the recharge stream at a concentration of 74 ppb (Scalf et al. 1969). Following a 3 hour stabilization period the water of the recharge well was pumped at a rate of 500 gal/min for 12 days and the chemical was monitored. The DDT concentration in the pumped water initially was over 16 times that of the recharge concentration but decreased to below the recharge level after 1 hour and continued to decrease thereafter. The aquifer material was shown to have a considerable capacity for adsorption of DDT. A major portion of the DDT remained in the aquifer and was not recovered during pumping. It was concluded that a well used for domestic purposes should not be recharged with water containing DDT because of the possible release of DDT-contaminated aquifer material into water subsequently pumped from the well.

2.2.1.2 Herbicides

Herbicides are, to a large extent, more water soluble than most insecticides. They vary widely in terms of their mammalian toxicity with relatively low and high

LD_{50}'s, e.g. endothal 38 mg/kg, paraquat 150 mg/kg, 2,4-D 375 mg/kg and as high as 15,000 mg/kg for prodiamine and 20,000 mg/kg for naproanilide. Generally speaking, most herbicides fall in the range of 1,000–10,000 mg/kg oral LD_{50}. There is ample evidence that some herbicides are highly persistent in soil while others are quite mobile.

One of the most widely investigated herbicides is atrazine [2-chloro-4-(ethylamino)-6-(isopropylamino)-S-triazine]. Hall and Hartwig (1978) studied the mobility, dissipation rate and residual activity of atrazine in two different soils at various depths. Maximum penetration was 76 cm with smaller concentrations detected in lysimeter leachates up to 122 cm. The authors concluded that application of atrazine to fine textured conventionally tilled soils at rates of 1.0–4.5 kg/ha would not seriously affect groundwater supplies as a result of vertical displacement of the herbicide by internal drainage. Harris (1967) and Rodgers (1968) have shown atrazine to be relatively immobile, whereas Helling (1970) has classified atrazine as being intermediate in leaching potential compared to other herbicides. Burnside et al. (1963) showed that atrazine leached to 30–45 cm depth 4 months after application and 45–60 cm depth or more after 16 months. These authors indicated that atrazine would continue movement through the soil profile below the 60 cm depth. Spalding et al. (1979) analyzed water samples from 10 tail water recovery (reuse) pits and 18 irrigation wells for NO_3-N and atrazine concentrations. Atrazine in the reuse pits ranged from 1.02 to 23.1 µg/L and in groundwater from 0.005 to 6.96 µg/L. A significant correlation (r = +0.48) was found between NO_3-N and atrazine concentrations in groundwater. Junk et al. (1980) reported atrazine in amounts of 0.01–3.28 µg/L in 64 samples of water with one sample showing 88 µg/L. Highest concentrations were found in shallow well water downgradient from irrigated fields. Spalding et al. (1979) detected small amounts of atrazine in all groundwater samples analyzed. Concentrations ranged from 0.06 to 3.12 µg/L. Significant correlation was found between atrazine and NO_3-N levels. It was estimated that 1% of the applied atrazine migrated vertically through medium textured soils and this movement was not halted by unsaturated layers of 5.6–10.7 m thickness. Atrazine degradation was slow.

Atrazine is one of the most widely used herbicide in the U.S. In many instances, atrazine is applied with nitrogen fertilizers giving rise to chemical reaction conditions that might favor N-nitrosamine formation. Due to its low basicity (pKa 1.68), the chemical environment of the stomach may offer a suitable site for N-nitrosamine formation (Walters et al. 1973) from atrazine ingested through drinking water. Wolfe et al. (1976) studied the N-nitrosation of atrazine under laboratory conditions and isolated and characterized the 2-chloro-4-(N-nitroso-N-ethylamino)-6-isopropylamino-S-triazine (NNA). Maximum rate of formation occurred at pH 1.8. NNA was stable in water at pH values greater than 4.0 and in river water at pH 7.1. At pH values lower than 4.0 NNA slowly decomposed to atrazine. In contrast to its hydrolytic stability in solution NNA was rapidly decomposed by light yielding atrazine and 4-amino-2-chloro-6-isopropylamino-S-triazine (desethyl atrazine). The latter was also found in New Orleans drinking water (U.S. EPA 1974) and can also form via microbial or free radical oxidation of triazine (Kaufman and Kearney 1970; Plimmer et al. 1971). The half life for photolysis of NNA in water under sunlight conditions is less than 10 minutes. However, the most likely location for environmental formation of NNA is in soil where no light is present. Triazines have been found to be widespread in groundwater drinking supplies in the Midwest USA, presumably as a result of leaching from the soil. Since NNA is more water soluble (227 ppm at 25° C) than atrazine (70 ppm at 25° C), it is likely that it would also be leached more readily into groundwater (Wolfe et al. 1976).

Following application of bromacil (5-bromo-3-sec-butyl-6-methyluracil) along railroad tracks in the Netherlands, highest concentrations of the herbicide were found in the 10–20 cm soil layer. However, within the first two years following several applications, bromacil was detected down to a depth of 100 cm. From field and laboratory experiments it was concluded that even deeper penetration of bromacil was probable and due to its low conversion rate this herbicide could very likely leach to groundwater (Zandvoort et al. 1980).

Bromacil and Norflurazon [4-chloro-5-(methylamino)-2-(α,α,α-trifluoro-m-tolyl)-pyridazin-3-one] were subject to considerable leaching in the porous (90% or more sand) soils of Florida having low organic matter content. Leaching was further enhanced by rainfall or irrigation leading to groundwater contamination (Singh et al. 1985). Over a period of 8 weeks the amount of bromacil dropped from 7 ppm to 1 ppm at a depth of 7.5 cm. Samples obtained after 4 weeks at depths of 22.5–30 cm had about 5 ppm and after 8 weeks 6.5 ppm bromacil indicating a rapid downward movement. A similar trend was noted with norflurazon at 2 weeks after application but its downward movement, subsequently, was slower than bromacil due perhaps to the lower water solubility of norflurazon (28 ppm vs 815 ppm for bromacil).

The distribution of bromacil and napropamide [2-(α-naphthoxy)-N,N-diethyl-propionamide] in soils after application via drip irrigation was studied by Gerstl and Yaron (1983). Behavior of the herbicides was affected mainly by soil type and application rate. High application rates gave a wider and shallower distribution of water and herbicides while lower rates gave a narrower and deeper distribution. Continued irrigation almost completely leached napropamide out of the emitter zone in the lighter soils but in finer textured soils the herbicide remained concentrated around the emitter zone. Bromacil was leached out of the emitter zone to a greater extent than napropamide. Under unsaturated flow conditions bromacil was found to have considerable mobility, greater than atrazine, prometon or diuron. Experiments in an orchard soil (Leistra et al. 1975) indicated that under specified field conditions a concentration of 0.044 kg of bromacil/ha would be leached from the top 1 m layer per year. This corresponds to 2.8 and 1.8% of the annual dosages of 1.6 and 2.4 kg/ha, respectively. These and similar calculations made under different conditions can provide first approximation to the upper limit of possible contamination of groundwater by leaching.

2.2.2 Industrial Chemicals in Groundwater

Groundwater accounts for over 90% of the fresh water in the U.S. including all streams and reservoirs (Ward et al. 1984) and more than 40% of the U.S. population use groundwater for drinking, often with no other treatment than disinfection (McCarty et al. 1981). The extent of groundwater contamination is presently unknown, but there is no doubt that this has the potential of becoming the country's most serious problem in the years to come.

Contamination of groundwater is thought to be largely irreversible due to the absence of microbiological processes responsible for biodegradation of organic compounds. However, recent studies (Howard and Banerjee 1984) indicate that the microorganism flora is quite abundant in certain aquifers and that some organic chemicals are biodegraded in these waters, but this is not a general rule.

The U.S. Environmental Protection Agency estimates that 2.4×10^8 tons of industrial wastes end up in land disposal sites each year (U.S. EPA 1977). This clearly represents a threat to groundwater resources unless improvement in the handling and disposal of these toxic wastes is fully implemented.

Examination of toxic substances contamination of groundwater in New Jersey (Page 1981) provides a demonstration of the extent of the problem in a heavily industrialized area, as well as in urban and rural areas. The data includes samples of groundwater collected from more than 1,000 different wells and samples of surface water collected from 600 different sites for comparison. The data were analyzed for 56 toxic substances of which 27 were light chlorinated hydrocarbons, 20 heavy chlorinated hydrocarbons and 9 heavy metals. Selected examples of these chemicals are shown in Tables 2-3. The maximum concen-

Table 2. Comparison of toxicant concentrations in groundwater and surface water from New Jersey, 1977-1979 (Adapted from Page 1981)

Chemical	No. sampled		% positive		Highest conc. (ppb)	
	Ground water	Surface water	Ground water	Surface water	Ground water	Surface water
Methylene chloride	1047	605	23.5	45.4	1900.0	743.3
Carbon tetrachloride	1073	608	64.3	67.7	263.9	20.6
1,1,2-Trichloro-ethylene	669	462	58.0	56.5	635.1	32.6
1,1,2-Trichloroethane	1069	603	6.7	8.8	31.1	18.7
Dichlorobromoethane	543	431	34.2	43.1	43.0	10.0
1,2-Dibromoethane	421	175	8.1	6.3	48.8	0.2
1,1,2,2-Tetrachloro-ethylene	421	179	41.3	86.0	90.6	4.5
Bromoform	1072	235	56.3	83.8	34.3	3.7
p-Dichlorobenzene	685	463	2.8	5.6	995.1	30.5
o-Dichlorobenzene	685	463	2.9	3.2	6800.0	8.2
Dichlorobenzene (total)	1090	615	3.4	7.1	8031.9	241.5
gem-dichloroethylene	378	305	44.4	64.6	17,288.3	2071.5

Table 3. Comparison of toxicant concentrations in groundwater and surface water from New Jersey, 1977–1979 (Adapted from Page 1981)

Chemical	No. sampled		% positive		Highest conc. (ppb)	
	Ground water	Surface water	Ground water	Surface water	Ground water	Surface water
Fluoroform	949	431	3.0	7.6	3.5	2178.2
Methyl chloride	1058	605	0.3	3.9	6.0	222.4
Vinyl chloride	1060	606	0.4	3.5	9.5	566.0
1,2-Dichloroethane	1066	606	9.7	11.7	36.5	304.9
1,1,1-Trichloroethane	1071	606	77.9	78.9	607.8	1016.8
Dibromochloromethane	1070	606	13.8	17.8	2.4	8.2
1,1,2,2-Tetrachloro-ethane	1072	608	6.0	11.0	2.7	3.0
Dibromomethane	377	282	11.9	28.0	44.9	358.6
Diiodomethane	1071	608	6.0	1.8	2.0	3.2
trans-Dichloroethylene	378	273	50.5	63.0	818.6	1307.5
Aroclor 1242	662	612	6.5	7.5	3.4	117.3
Aroclor 1248	668	612	6.3	13.5	5.4	109.1
Aroclor 1254	1040	612	2.9	14.4	0.4	127.0

trations reveal considerable variation between groundwater and surface water. For 32 of the toxic compounds (64% of the total) the highest concentrations were found in groundwater. This may represent a significant threat to the health of people consuming water from groundwater sources (Page 1981). In the survey, extremely high correlation between groundwater and surface water was found with respect to pesticides, and a high correlation between the two water systems was indicated with regard to the light chlorinated hydrocarbons. It was concluded that groundwater in New Jersey is as contaminated with toxic compounds as is surface water.

Wilson et al. (1981) investigated the transport and fate of 13 organic pollutants in a sandy soil with low organic matter content (0.078% organic C) in glass columns up to a depth of 140 cm. Chloroform, 1,2-dibromo-3-chloro-propane, dichlorobromomethane, 1,2-dichloroethane, tetrachloroethane, 1,1,2-trichloroethane and trichloroethene were not degraded and percolated rapidly through the soil. Between 19 and 65% of the material applied to the surface percolated to a depth of 140 cm. Chlorobenzene, 1,4-dichlorobenzene and 1,2,4-trichlorobenzene also percolated through the soil but at a slower rate. Between 26 and 49% reached 140 cm. 60 to 80% of nitrobenzene, 13% of toluene, and 86% of bis(2-chloroethyl)ether also reached 140 cm. It was concluded that groundwaters underlying soils with low organic matter are highly vulnerable to contamination by the above chemicals.

Supposedly, PCBs and numerous other organics do not dissolve in water. Actually, PCB (Aroclor 1254) can dissolve in water to the extent of 0.044 mg/L at 25° C. Moreover, benzene can concentrate in water to 1780 mg/L, chloroform to 9300 mg/L and phenol to as much as 82,000 mg/L, all at 25° C (Josephson 1980).

These and other substances could find their way to groundwater because of improper disposal or illegal dumping.

Organic contaminants were inadvertantly discovered in a public supply well and in additional 27 private wells in South Brunswick, N.J. The principal contaminants were 1,1,1-trichloroethane and tetrachloroethylene. The former exceeded 1000 ppb in some wells and were consistently in the range of 150 to 550 ppb. The latter compound ranged from 100 to 300 ppb (Roux and Althoff 1980). It is worth noting that the concentration of tetrachloroethylene (100–300 ppb, as shown in Table 5) far exceeds the health criteria of 0–0.800 ppb established by the EPA (Argo 1985).

2.2.2.1 Detergents

Anionic detergents such as alkyl benzene sulfonates (ABS), linear alkyl benzene sulfonates (LAS) and sodium dodecyl sulfonates (NaLS) are water soluble compounds and have the potential to be transported in groundwater. The ABS detergents appear to be transported in the aquifer at the same rate as major cations and anions and boron. There is little or no biological degradation of ABS in the aquifer. On the other hand, the LAS and NaLS detergents degrade fairly rapidly in the first 300 m downgradient of the sand beds. Generally, the concentration of detergents in the contaminated plume of the area studied (Thurman et al. 1986) reaches 2 mg/L. The question arises whether or not detergents transport hydrophobic organic compounds in groundwater that otherwise would be strongly sorbed by aquifer sediments. The hazards involved from such interactions are not known at present. The behavior of anionic surfactants in a soil – sewage effluent system was also studied by Acher and Yaron (1977). Detergents may also increase the persistence and toxicity of various insecticides in soil (Lichtenstein 1966).Addition of the detergents ABS and LAS to insecticide-treated soils increased the persistence of parathion (13 times) and diazinon (5 times) two months after application. A synergistic effect on toxicity was indicated.

2.2.3 Wastewater

It is estimated that surface impoundments of municipal, industrial and agricultural wastewater in the U.S. receive about 1.7×10^{12} gallons of liquid waste products per year (Johnson 1978). Land application is an innovative technology which offers a practical and economical method for the treatment of domestic wastewater. Soil surface, soil matrix and plants are able to effectively remove many wastewater constituents. Rapid infiltration has the greatest potential to influence groundwater quality due to the high application rates possible. In this system most of the applied wastewater percolates through the soil, eventually reaching groundwater. Removal of suspended solids, biochemical oxygen demand and fecal coliforms is excellent. However, a study at Rice University demonstrated that trace organic compounds in primary and secondary effluents treated in rapid infiltration systems are detected in associated groundwater (Tomson et al. 1981). Field studies in a rapid infiltration system in Louisiana have indicated that characteristic trace organics are mostly absent from groundwater underlying the infiltration site. On the other hand, in a laboratory study using topsoil packed columns, Hutchins et al. (1983) have shown that several trace organics in unchlorinated secondary effluents are capable of migrating through the soil profile, although at reduced concentrations (Table 4). These results indicate that groundwater contamination at rapid infiltration sites can occur, albeit with a significant reduction in the concentrations of specific trace organics.

In water scarce areas of the world reclaimed water offers the best source for barrier injection. An ambitious project of this kind was undertaken in 1976 in Southern California which has since produced nine trillion gallons of reclaimed water (Argo 1985). The comparison of the reclaimed water from effluent injection with the health criteria of the EPA for priority pollutants (Table 5) shows that most of the criteria were met at the 10^6 lifetime cancer risk level.

Improper disposal of industrial wastes containing the carcinogen 3,3'-dichlorobenzidine (DCB) has resulted in contamination of soil, groundwater and surface water near a manufacturing site in Michigan (Boyd et al. 1984). DCB has also been shown to be mutagenic in the Ames Salmonella assay and is a demonstrated animal carcinogen, also causing bladder cancer in humans (Jones 1980). DCB becomes rapidly and strongly bound in clay loam soil forming, perhaps, covalent linkages with soil humic components. However, when present in soils that are low in organic matter, DCB residues may be less bound to humus and may percolate more freely through the unsaturated zone. Humus-bound DCB residues may be potential sources of future contamination. The list of case

Table 4. Wastewater trace organics in groundwater at four land application sites (from Hutchins et al. 1983)

Trace organic chemical	Range of concentrations at four sites (ppb)
Tetrachloroethylene	0.07–0.63
Toluene	0.02
Xylenes	0.05–1.14
m-Dichlorobenzene	0.05–0.56
p-Dichlorobenzene	0.07–0.50
Naphthalene	0.03–0.22
2,6-Di-t-butyl-p-benzoquinone	0.03–1.51
Dimethyl phthalate	0.01–0.19
2-(Methylthio)benzothiazole	0.03–0.15
Benzophenone	0.05–2.13
p-(1,1,3,3-Tetramethylbutyl)phenol	0.06–1.57
N-Butylbenzene sulfonamide	0.01–0.48
Dibutyl phthalate	0.73–2.38
Bis(2-Ethylhexyl)phthalate	0.13–1.40

Table 5. Health criteria for priority pollutants (EPA) vs. concentrations in groundwater

Compound	Health criteria			Reference
	Maximum (ppb)	10^6 lifetime risk[a] (ppb)	Conc. in ground water (ppb)	
Carbon tetrachloride	0	0.400	1900[b]	Page 1981
			0.060[c]	Argo 1985
Dichlorobenzene	400	–	8031[b]	Page 1981
			0.060[c]	Argo 1985
			0.050–0.560	Hutchins et al. 1983
1,1,1-Trichloroethane	18,400	–	31	Page 1981
			150–550	Roux and
			0.050[c]	Althoff 1980
			0.070[c]	Argo 1985
Monochlorobenzene	488	–	0.040[c]	Argo 1985
Ethylbenzene	1,400	–	0.050[c]	Argo 1985
Phenol	300	–	0.050[c]	Argo 1985
Pentachlorophenol	30	–	1.700[c]	Argo 1985
bis(2-Ethylhexyl)phthalate	15,000	–	0.730–2.380	Argo 1985
Dibutyl phthalate	> 15,000	–		Hutchins et al.
			0.400–5.400[b]	1983
Polychlorinated biphenyls(PCBs)	0	0.079	635[b]	Page 1981
Trichloroethylene	0	2.700	90.6[b]	Page 1981
Tetrachloroethylene	0	0.800	100–300[b]	Page 1981
				Roux and
			0.080[c]	Althoff 1980
			0.130[c]	Argo 1985
Toluene	14,000	–	0.20	Argo 1985
				Hutchins et al. 1983

a – one in a million lifetime cancer risk level
b – exceeds health criteria
c – reclaimed water from effluent injection

studies given above is by no means complete. Only a few typical examples are shown to illustrate the magnitude of the problem and emphasize the risk involved in groundwater contamination.

2.2.4 Mathematical Modelling of Groundwater Contamination

Prediction of the extent of subsurface contamination relies on data from experimental and mathematical models. Experimental tests include both laboratory and field studies. In the laboratory, the column chromatography model is widely used to examine persistence and leaching potentials of pollutants (Jarczyk 1978).

The reliability of laboratory models needs to be tested in the field at full scale. Field studies frequently require soil core sampling and installation of wells and are usually very costly (Cohen et al. 1986).

Much activity is currently invested in the development and large scale application of mathematical models for predicting the fate and transport of organic chemicals in the subsurface. Such models are attractive because they offer a relatively rapid and inexpensive way to assess potential environmental hazards. Furthermore, in many cases no monitoring data exist, and mathematical models provide the only means of predicting complex environmental events. Pesticide transport simulation models may be either deterministic or stochastic in their approach. In deterministic models pesticide leaching is the outcome of the operation of processes such as volatilization, adsorption, degradation, convection, and diffusion, along with physical and chemical characteristics of the soil medium. The mutual interaction of these factors results in a complex system and often it is necessary to make assumptions which are at best a first approximation of what occurs under field conditions. Nevertheless, most pesticide transport models are of this type. They are of general use and their analytical solutions provide preliminary estimates of the distribution of contaminants in the subsoil and in groundwaters. Stochastic models emphasize the randomness associated with water flow and other environmental processes governing transport. Although on a small homogeneous scale (e.g. in the laboratory) pesticide movement occurs according to single valued physical laws; in the case of a heterogeneous system (e.g. in the field), the random component of these laws would override their deterministics behavior. However, the application of a purely stochastic model for predicting the distribution of pesticides which relies on probability functions seem to be very limited (Helling and Gish 1986). Many models provide descriptions of solute transport, both in the unsaturated and saturated zones (Enfield et al. 1982; Leistra 1980; Aharonson, in press). The recent Handbook of Mathematical Models of Groundwater Transport (Javandel et al. 1984) reviews the principles governing the formulation of some analytical, semi analytical, and numerical models, and discusses the relative merits and drawbacks of each. The results of a mathematical model need certainly to be critically examined, in particular when their application is considered, such as prior to any regulatory decision. When field data are limited, there is no unique model solution, and the information obtained from models could be very imprecise.

In general, in spite of the tremendous activity in the field of modelling of pollutant transport, our knowledge as well as our prediction capability about the extent of organic chemicals occurrence in groundwater are still very rudimentary. Obviously, the predictive power of a given model is completely determined by our detailed knowledge of the fundamental processes that govern the fate and transport of contaminants. Therefore, there is an urgent need for intensive research to provide full understanding of the critical processes to allow their correct incorporation in models that can be tested on both laboratory and field scales.

2.3 Conclusions

One should be cognizant of the fact that the toxicity of a chemical compound is not the only criterion of its hazard to health. Mention should be made of several pesticides with low, intermediate and high oral LD_{50} values which have been classified as carcinogens or of potential carcinogenic risk to man by the International Agency for Research on Cancer sponsored by the World Health Organization. Among these may be cited: Aldrin and dieldrin (38 mg/kg, 40 mg/kg, respectively); Heptachlor (40 mg/kg); Toxaphene (49 mg/kg); Mirex (306–600 mg/kg); Dibromochloropropane (170–300 mg/kg); Captan (9000 mg/kg); Perthane (8000 mg/kg); Picloram (8200 mg/kg); Monuron (2300 mg/kg); Trifluralin (3700 mg/kg); Nitrofen (2630 mg/kg), to mention only a few. Several of the above compounds have been found in groundwater.

If a chemical is known to be non-carcinogenic the safe level for humans is considered to be the acceptable daily intake level (ADI). These values can be used to estimate "safe" levels of chemicals in groundwater (Stara 1985). For compounds that have the potential to cause cancer and for which adequate quantitative data are available, the groundwater concentrations associated with lifetime carcinogenic risks are estimated by the improved multi-stage evaluation model of the U.S. Environmental Protection Agency (Stara 1985). Of course, known carcinogens should be banned altogether.

This brief review emphasizes the vulnerability of groundwater contamination by chemicals and the toxic hazards associated with these contaminants.

2.4 Recommendations

Recognition of the vulnerability of aquifers to chemical contamination has been intensified over the past decade as more and more positive identification of contaminants in groundwater aided by more sensitive analytical instrumentation comes to our attention. Research efforts should now focus on the transport and fate of these and other contaminants through the unsaturated zone and down to groundwater. A thorough understanding of these processes is essential to the development of realistic groundwater protection and management plans.

While the long-term health effects of low level exposure to toxic chemicals in water supplies is at present unknown, public policy should give at least equal emphasis to the control of toxic substances in groundwater as is given to the control of such chemicals in surface water (Page 1981). Threshold Limit Values (TLV) for chemicals in groundwater should be established. Contaminated groundwater would be difficult if not impossible to clean up, and restoration of the aquifer to its original composition with assurance of reasonable water quality cannot meet with success. Hence, the strategy should emphasize the prevention of groundwater contamination. Contaminants which find their way to groundwater have virtually no natural antidotes (Josephson 1980). The old saying "an ounce of prevention is worth a pound of cure" can go much further in the case of groundwater contamination, for an ounce of prevention is worth a lot more than a pound of cure where there is no cure.

References

Achari RG, Sandhu SS, Warren WJ (1975) Chlorinated hydrocarbon residues in groundwater. Bull Environ Contam Toxicol 13:94–96
Acher AJ, Yaron B (1977) Behavior of anionic surfactants in a soil-sewage effluent system. J Environ Qual 6:418–420
Argo DG (1985) Water reuse: where are we headed? Environ Sci Technol 19(3)
Back RC, Remine RR, Hansen JL (1984) A rating system for predicting the appearance of Temik[R] Aldicarb residues in potable water. Environ Toxicol Chem 3:589–597
Berteau PE, Spath DP (1986) Toxicological and epidemiological effects of pesticide contamination in California groundwater. ACS symposium on Evaluation of Pesticides in Groundwater. Abstr No 75, 189th Natl Meeting, Am Chem Soc, Div Pestic Chem, Miami Beach, FL
Boyd SA, Kao CW, Suflita JM (1984) Fate of 3,3'-Dichlorobenzidine in soil: Persistence and binding. Environ Toxicol Chem 3:201–208
Burnside OC, Fenster CR, Wicks GA (1963) Dissipation and leaching of monuron, simazine, and atrazine and Nebraska soils. Weeds 11:209–213
Cochrane WP, Lanouette M, Trudeau S (1982) Determination of aldicarb, aldicarb sulfoxide, aldicarb sulfone and carbofuran residues in water using high performance liquid chromatography. J Chromat 243:307–314
Cohen DB (1986) Groundwater contamination by toxic substances: A California assessment. ACS symposium on Evaluation of pesticides in Groundwater. Abstr. No. 91, 189th Natl Meeting, Am Chem Soc, Pestic Chem, Miami Beach, FL
Cohen SZ, Eiden C, Lorber MN (1986) Evaluation of Pesticides in Groundwater. Am Chem Soc Ser 315 (Garner WY et al., eds), Washington DC, pp 170–196
Coppedge JR, Lindquist DA, Bull DL, Dorough HW (1967) Fate of 2-methyl-2-(methylthio)-proprionaldehyde 0-(methylcarbamoyl)oxime (Temik) in cotton plants and soil. J Agric Food Chem 15:902–910
Enfield CG, Carsel RF, Cohen SZ, Phan R, Walter DM (1982) Approximating pollutant transport to groundwater. Ground Water 20:711–722
Garner WY, Honeycutt RC, Nigg H (eds) (1986) Evaluation of Pesticides in Groundwater. ACS Symp Ser No 315, Washington DC
Gerstl Z, Yaron B (1983) Behavior of bromacil and napropamide in soils. II. Distribution after application from a point of source. Am J Soil Sci 47:478–483
Hall JK, Hartwig NL (1978) Atrazine mobility in two soils under conventional tillage. J Environ Qual 7:63–68
Hansen JL, Spiegel MH (1983) Hydrolysis studies of aldicarb, aldicarb sulfoxide and aldicarb sulfone. Environ Toxicol Chem 2:147–153
Harris CI (1967) Movement of herbicides in soil. Weed Sci 15:214–216
Harvey J Jr, Pease HL (1973) Decomposition of methomyl in soil. J Agric Food Chem 21:784–786
Helling CS (1970) Movement of S-triazine herbicides in soils. Residue Rev 32:175–210
Helling CS, Gish JT (1986) Evaluation of Pesticides in Groundwater. Am Chem Soc Ser 315 (Garner et al., eds), Washington DC, pp 14–38
Howard PH, Banerjee, S (1984) Interpreting results from biodegradability tests of chemicals in water and soil. Environ Toxicol Chem 3:551–562

Hutchins SR, Tomson MB, Ward CH (1983) Trace organic contamination of groundwater from a rapid infiltration site: a laboratory-field coordinated study. Environ Toxicol Chem 2:195-216

Intera Environmental Consultants Inc (1980) Mathematical simulation of aldicarb behavior on Long Island: Unsaturated flow and groundwater transport. Prepared for the hazard evaluation division, US Environmental Protection Agency

Jarczyk RJ (1978) Fourth Intl Congr Pesticide Chemistry (IUPAC), Zürich

Javandel I, Doughly C, Tsang CF (1984) Groundwater transport: Handbook of Mathematical Models. Water Resources Monograph Series No. 10. American Geophysical Union, Washington DC

Johnson CC (1978) Resources for the future. In "Safe Drinking Water: Current and Future Problems", Chapt. 1 (Russel CS Ed). Washington DC

Jones TC (1980) Benzidine, its congeners and their derivative dyes and pigments. TSCA Chemical Assessment Series. EPA 440/9-76-018. US Environmental Protection Agency, Washington DC

Jones RL, Back RC (1984) Monitoring of aldicarb residues in Florida soil and water. Environ Toxicol Chem 3:9-20

Josephson J (1980) Safeguards for groundwater. Environ Sci Technol 14:38-44

Junk GA, Spalding RF, Richard JJ (1980) Areal, vertical and temporal differences in groundwater chemistry. II. Organic constituents. J Environ Qual 9:479-483

Karickhoff SW (1981) Semi empirical estimation of sorption of hydrophobic pollutants on natural sediments and soils. Chemosphere 10:833-845

Kaufman DD, Kearney PC (1970) Microbial degradation of S-triazine herbicides. Residue Rev 32:235-265

Leistra M, Smelt JH, Zandvoort R (1975) Persistence and mobility of bromacil in orchard soils. Weed Res 15:243-247

Leistra M, Bromilow RH, Boesten JJTI (1980) Measured and Simulated Behavior of Oxamyl in Fallow Soils. Pestic Sci 11:379-388

Leistra M, Dekker A, Burg AMM, Van der (1984) Computed and measured leaching of the insecticide methomyl from greenhouse soils into water courses. Water Air Soil Pollut 23:155-167

Lichtenstein EP (1966) Increase of persistence and toxicity of parathion and diazinon in soils with detergents. J Econ Entomol 59:985-993

Lichtenstein EP, Fuhremann TW, Schultz KR (1971) Persistence and vertical distribution of DDT, lindane and aldrin residues, 10 and 15 years after a single soil application. J Agric Food Chem 19:718-721

Mackay DV, Cherry JA (1985) Transport of organic contaminants in groundwater. Environ Sci Technol 19:384-392

McCarty PL, Reinhard M, Rittmann BE (1981) Trace organics in groundwater. Environ Sci Technol 15:40-51

Neilson DM (1982) In: Proceedings of the second national symposium on aquifer restoration, Ohio, pp 58-68

Page GW (1981) Comparison of groundwater and surface water for patterns and levels of contamination by toxic substances. Environ Sci Technol 15:1475-1481

Peoples SA, Maddy IT, Cusick W, Jackson T, Cooper C, Frederickson AS (1980) A study of samples of well water collected from selected areas in California to determine the presence of DBCP and certain other pesticide residues. Bull Environ Contam Toxicol 24:611-618

Plimmer J, Kearney PC, Klingebiel UI (1971) S-Triazine herbicide dealkylation by free radical generating systems. J Agric Food Chem 19:572-573

Rodgers EG (1968) Leaching of seven S-triazines. Weed Sci 16:117-120

Rothschild ER, Manser RJ, Anderson MP (1982) Investigation of aldicarb in groundwater in selected areas of the Central Sand plain of Wisconsin. Ground Water 20:437-445

Roux PH, Althoff WF (1980) Investigation of organic contamination of groundwater in South Brunswick Township, New Jersey. Ground Water 18:464-471

Sandhu SS, Warren WJ, Nelson P (1978) Pesticidal residue in rural potable water. J Am Water Works Assoc 70:41-45

Scalf MR, Dunlap WJ, McMillion LG, Keeley JW (1969) Movement of DDT and nitrates during groundwater recharge. Water Res Res 5:1041-1052

Singh M, Castle WS, Achnireddy NR (1985) Movement of Bromacil and Norflurazon in a sandy soil in Florida. Bull Environ Contam Toxicol 35:279–284

Smelt JM, Leistra M, Houx NWH, Dekker A (1978) Conversion rates of aldicarb and its oxidation products in soils. I Aldicarb sulfone. II Aldicarb sulfoxide. III Aldicarb. Pestic Sci 9:279–300

Spalding RF, Exner ME, Sullivan JJ, Lyon PA (1979) Chemical seepage from a tail water recovery pit to adjacent groundwater. J Environ Qual 8:374–383

Stara JF (1985) US-EPA risk assessment models for groundwater. ACS symposium on Evaluation of Pesticides in Groundwater. Abstr. No. 73, 189th Nat'l Meeting, Am Chem Soc, Div Pestic Chem, Miami Beach, FL

Thurman EM, Barber LB Jr, LeBlanc D (1986) Movement and fate of detergents in groundwater. A field study. J Contam Hydrol 1:143–161

Tomson MB, Dauchy J, Hutchins SR, Curran C, Cook CF, Ward CH (1981) Groundwater contamination by trace level organics from a rapid infiltration site. Water Res 15:1109–1116

US Environmental Protection Agency. Draft Analytical Report, New Orleans Area Water Supply Study. Region VI (Nov 1974)

US Environmental Protection Agency. National Interim Primary Drinking Water Regulations. Fed Reg IV 59566–88, Dec 1975. Safe Drinking Water Act (Publ 93–523)

Walters CL, Archer MC, Tannenbaum SR (1973) Nitrosation in the Environment: Can It Occur? Science 179:96–97

Ward CH, Durham NN, Center LW (1984) Guest editorial: Groundwater – A National Issue. Ground Water 22:138–140

Weber JB, Weed SB (1974) Effects of soil on the biological activity of pesticides. In "Pesticides in Soil and Water" (Guenzi WD, ed). Soil Sci Soc Am Madison, Wisc pp 223–256

Wilson JT, Enfield CG, Dunlap WJ, Cosby RL, Foster DA, Basmin LB (1981) Transport and fate of selected organic pollutants in a sandy soil. J Environ Qual 10:501–506

Wolfe NL, Zepp RG, Gordon JA, Fincher RC (1976) N-nitrosamine formation from atrazine. Bull Environ Contam Toxicol 15:342–347

Wyman JA, Jensen JO, Curwen D, Jones RL, Marquardt TE (1985) Effects of application procedures and irrigation on degradation and movement of aldicarb residues in soil. Environ Toxicol Chem 5:641–651

Yaron B, Gerstl Z, Spencer WF (1984) Behaviour of herbicides in irrigated soils. Adv Soil Sci 2:1–143

Zaki MH, Moran D, Harris D (1982) Pesticides in groundwater. The aldicarb story in Suffolk county. NY Am J Publ Health 72:1391–1395

Zandvoort R, Born GW Van den, Braber JM (1980) Leaching of the herbicide Bromacil after application on railroads in the Netherlands. Water Air Soil Pollut 13:363–372

Part II. Physicochemical and Biological Interactions with Porous Media

Introductory Comments

Organic pollutants are found in every phase of our environment and move readily from one phase to another. Their ultimate fate depends upon the reactions they undergo in each phase and their interaction with other components in each phase. These components can be classified basically as either organic or inorganic materials and their interactions with toxic organic compounds differ due to the difference in their intrinsic properties. In this section several specific cases emphasizing the problems posed by toxic organic chemicals in the soil and groundwater environment are presented.

Humic substances are the most ubiquitous and widespread natural non-living organic materials in terrestrial and aquatic environments and their effects on the fate of toxic organic chemicals in the environment are of fundamental importance. On the other hand, the assumption that the organic fraction dominates the interactions of many toxic organics has obscured to a certain measure the role that mineral surfaces play in determining the fate of these pollutants. Two papers in this section treat each fraction clearly defining the importance and complexity of the interaction of toxic organics with both humic substances and mineral surfaces.

Senesi and Chen (Chapter 3) provide us with an in depth survey of all aspects of humic substances and toxic organics. They begin with brief outlines on the nature of humic substances and the different types of organic pollutants to give us the proper perspective on the scope of the problem confronting us. This is followed by a detailed literature review which focuses on the role of organic matter in accumulation phenomena, mobilization and transport processes, phytotoxicity and bioavailability, residue persistence and monitoring of toxic organic chemicals in soils and water. The multiple modes of interaction which humic substances exhibit toward organic pollutants are discussed at length. The authors bring evidence for the occurrence of hydrophobic adsorption and partitioning, physical and chemical binding, enzyme-mediated binding, solubilization effects, hydrolysis and dealkylation catalysis as well as photosensitization action. Some media effects which may affect the interaction are examined.

The importance of the mineral fraction is presented by Mingelgrin and Prost (Chapter 4). Mineral surfaces are especially important at sufficiently large depths where the organic matter content is rather low and even in cases of high organic matter content the presence of clay- organic matter complexes can significantly alter the interaction of organic toxicants and organic matter. The nature of the compound and their adsorption and/or tendency to undergo heterogeneously catalyzed transformations at surfaces is described. A unified description of these interactions defining the role of the exchangeable cations and their hydration status in the interaction of polar and polarizable toxicants and the role of the net energy and entropy of adsorption in the adsorption of apolar toxicants is presented.

Wolfe (Chapter 5) provides us with an insight into the abiotic reaction conditions which occur in subsurface ecosystems and with this information is able to impose boundaries on the activity of selected chemical species in porous media and to narrow the ranges of reactivities to be studied in quantifying abiotic transformation processes. Data collected from the literature suggest that neutral (pH independent) hydrolysis is more important than either acid or base hydrolysis in groundwater systems. From the kinetics of abiotic hydrolysis of selected organic compounds in soils and aquifer systems it appears that adsorption affects the observed hydrolysis rate constant depending on the hydrolytic pathway. Abiotic redox reactions have been limited to reductive transformations and while the sediment is necessary for the reaction to occur the transformation does not occur in the sorbed phase.

3 Interactions of Toxic Organic Chemicals with Humic Substances

N. SENESI[1] and Y. CHEN[2]

3.1 Introduction

Humic substances (HS) are the most widespread and ubiquitous natural non-living organic materials occurring in all terrestrial and aquatic environments. HS occur not only as the major fraction of soil organic matter (OM), but also in marine, river and lake waters and sediments, sewage effluents of various nature, peat, coal and lignite. The amount of carbon on the earth occurring as humic substances (60×10^{11} tons) has been estimated to exceed about 10-fold of that occurring in living organisms (7×10^{11} tons).

A pollutant or toxic chemical may be defined as any substance that is foreign to the natural ecosystem which, either directly or indirectly, adversely affects the natural physical, chemical and biological equilibria and processes in the global environment or a portion of it. Many types of toxic organic chemicals reach the soil either directly in the form of herbicides, insecticides or fungicides, etc., or indirectly through municipal and industrial wastes and composts of various origin applied to the soil as fertilizers and/or amendments. Among the various ways toxic organic chemicals may reach aquatic systems – i.e. direct application, spray drift, aerial spraying, washing from the atmosphere by precipitation, erosion and run-off from agricultural land, discharge of effluent from factories, and sewage-run-off from agricultural land is considered the main source of gradual pollution, while discharge of effluent may result in more serious, but localized contamination. The most hazardous organic chemicals seem to be organochlorine pesticides and certain herbicides.

Humic substances due to their polyelectrolitic nature interact substantially with and commonly adsorb man-made organic chemicals. In addition, the ascertained presence of a great variety of chemical reactive functional groups in the molecular structure of humic substances, renders them able to chemically bind various organic compounds including pesticides, petrol, plastic derivatives, surfactants and detergents. Due to their functional groups and hydrophylic and hydrophobic sites, fulvic acid (FA) and humic acid (HA) fractions of humic substances may interact with toxic organic chemicals by various mechanisms which include physical and chemical adsorption, partitioning, solubilization,

---

[1] Istituto di Chimica Agraria, Università di Bari, Via Amendolo, no. 165/A, Bari 70126, Italy.
[2] The Seagram Center for Soil and Water Sciences, Faculty of Agriculture, P.O. Box 12, Rehovot 76100, Israel.

hydrolysis, photosensitization, and others. In particular, chemical bonds of various strength and stability are suggested to form between humic substances and toxic organic chemicals, ranging from weak, partially reversible, physical associations to strong, irreversible covalent bonds (Stevenson 1972; Hayes and Swift 1978). All these processes will have evident implications in the fate of toxic organic chemicals in the environment, affecting their degradation and detoxication processes, residue persistence and monitoring, mobilization and transport phenomena, phytotoxicity and bioavailability in soils, waters and sediments, and bioaccumulation in organisms.

The main purpose of this chapter is, therefore, to summarize and review some quantitative and qualitative aspects of the modes of interactions between the principal components of organic matter (humic substances) and some classes of toxic organic chemicals.

3.2 Soil and Aquatic Humic Substances

3.2.1 Introduction

During the decay process of organic materials, macromolecules of a mixed aliphatic and aromatic nature are formed. The term "humus" is widely accepted as synonymous to soil organic matter. It is defined as the total of the organic compounds in soil exclusive of undecayed plant and animal tissues, their partial decomposition products and the soil biomass. Other important definitions for this chapter are presented in Table 1.

The chemical and colloidal properties of soil organic matter can be studied only in the free state, that is, when freed of inorganic soil components. Thus, the first task of the researcher is to separate organic matter from the inorganic matrix of sand, silt, and clay. Alkali, usually 0.1 to 0.5 NaOH, has been a popular extractant of soil organic matter, but this reagent may alter the organic matter through hydrolysis and autoxidation. In recent years, milder but less efficient extractants have been used with variable success.

Other methods of extraction have been used by researchers, but since this is beyond the objectives of this chapter, extraction procedures will, in general, be ignored.

Approximately 60-70% of the total soil-organic carbon occurs in HS. Estimated levels of soil organic carbon on the earth surface occurring as HS are 30×10^{14} kg (Stevenson 1982). Dissolved aquatic HS constitute 40-60% of dissolved organic carbon (DOM) and are the largest fraction of natural organic matter (OM) in water (Thurman 1986). The concentration of HS varies for different natural waters, widely ranging from the highest values determined in colored rivers and wetlands, to the lowest ones measured in groundwaters and seawaters. The majority of the HS in natural waters are FA.

Average elemental composition and acidic functional group content of soil, dissolved aquatic, and sediment HA and FA are presented in Tables 2 and 3.

Table 1. Definitions of various components of soil organic matter (after Stevenson 1982)

Term	Definition
Organic residues	— Undecayed plant and animal tissues and their partial decomposition products.
Soil biomass	— Organic matter present as live microbial tissue.
Humus	— Total of the organic compounds in soil exclusive of undecayed plant and animal tissues, their "partial decomposition" products, and the soil biomass.
Soil organic matter	— Same as humus.
Humic substances	— A series of relatively high-molecular-weight, brown to black colored substances formed by secondary synthesis reactions. The term is used as a generic name to describe the colored material or its fractions obtained on the basis of solubility characteristics. These materials are distinctive to the soil (or sediment) environment in that they are dissimilar to the biopolymers of microorganisms and higher plant (including lignin).
Nonhumic substances	— Compounds belonging to known classes of biochemistry, such as amino acids, carbohydrates, fats, waxes, resins, organic acids, etc. Humus probably contains most, if not all, of the biochemical compounds synthesized by living organisms.
Humin	— The alkali insoluble fraction of soil organic matter humus.
Humic acid (HA)	— The dark-colored organic material which can be extracted from soil by various reagents and which is insoluble in dilute acid.
Fulvic acid (FA)	— The colored material which remains in solution after removal of humic acid by acidification.

Table 2. Average values of elemental composition (%) of humic substances from soil, lake sediment and various aquatic environments

Sample		C	H	O	N	P	S	Ash
Soil	FA	48.0	4.5	45.0	1.0	—	0.4	1.2
	HA	56.0	4.5	37.0	1.6	—	0.3	2.4
Lake sediment	FA	45.0	5.1	42.3	7.6	—	—	—
	HA	52.1	5.7	36.6	5.6	—	—	—
Lake water	FA	52.0	5.2	39.0	1.3	0.1	1.0	5.0
	HA	—	—	—	—	—	—	—
Ground water	FA	59.7	5.9	31.6	0.9	0.3	0.65	1.2
	HA	62.1	4.9	23.5	3.2	0.5	0.96	5.1
Seawater	FA	50.0	6.8	36.4	6.4	—	0.46	3.4
	HA	—	—	—	—	—	—	—
River water	FA	51.9	5.0	40.3	1.1	0.2	0.6	1.5
	HA	50.5	4.7	39.6	2.0	—	—	5.0
Wetland water	FA	51.0	4.3	40.2	0.7	0.2	0.4	2.0
	HA	51.2	4.4	40.9	0.6	0.1	0.6	2.0

Table 3. Average values of major functional groups (carboxyl and phenolic hydroxyl) content of soil and aquatic humic substances

Sample		Carboxyl (meq/g)	Phenolic (meq/g)
Soil	FA	5.2–11.2	0.3–5.7
	HA	1.5–5.7	2.1–5.7
Lake water	FA	5.5–6.2	0.3–0.5
Ground water	FA	5.1–5.5	1.4–1.6
Seawater	FA	5.5	–
River water	FA	5.5–6.0	1.5
	HA	4.0–4.5	2.0
Wetland water	FA	5.0–5.5	2.5
	HA	4.0–4.5	2.5

Carbon in aquatic HS is greater and oxygen is less than in soil HS. C/N ratios for aquatic FA (45–55:1) and HA (18–30:1) are considerably greater than C/N ratios in soil FA (average 20:1) and HA (average 10:1) and in HS from aquatic sediments. Aquatic HS appears, therefore, depleted of nitrogen as compared to adjacent soils and sediments. Phenolic OH content of aquatic HS (ranging from 1–2 meq g^{-1}) are considerably lower than that in soil HS. Soil HA and FA differ in elemental composition, molecular weight and functional group contents (Schnitzer 1978). The FA fraction exhibits lower molecular weight and lower levels of total and aromatic carbons. Oxygen and oxygen-containing functional group levels are higher in soil FA than HA (Schnitzer and Khan 1972). HA contains longer chain fatty-acid products than FA, thus suggesting a higher hydrophobicity than FA. ^{13}C NMR spectra indicate that approximately 65% of the carbon in aquatic FA is aliphatic and many of the COOH and OH groups are attached with aliphatic carbons. Aquatic FA similarly to soil FA, generally exhibits low molecular weight (500–2000 daltons), whereas aquatic HA ranges in molecular weight from 2000–5000 daltons, or greater, being therefore more polydisperse in nature and colloidal in size (Thurman 1986).

HA and FA cannot be regarded as a single chemical entity and described by a unique, chemically-defined molecular structure. Both HA and FA are operationally-defined by a model structure, containing the same basic structural units and the same types of reactive functional groups common to all the single, indefinitely-variable and unknown molecules (Stevenson 1982). The macromolecular structure is, therefore, constituted by "building blocks" or monomeric units of aromatic, phenolic, quinonic and heterocyclic nature, randomly-condensed and/or linked by aliphatic oxygen, nitrogen or sulphur bridges. The macromolecule will bear surface-chains of aliphatic, glucidic, amino acidic and lipidic nature, as well as chemically-active functional groups of various nature (i.e. carboxylic, phenolic and alcoholic OH, aldehydic and ketonic), which confer the polymer an overall acidic character.

For more information and details on soil HS, the reader is referred to comprehensive literature on the topic (Gieseking 1975; Hayes and Swift 1978; Schnitzer and Khan 1978; Stevenson 1982). Aquatic HS have been reviewed recently for various types of natural waters (Thurman 1986), groundwaters (Thurman 1985), streams (Malcolm 1985), lakewaters (Steinberg and Muenster 1985), estuarine environments (Mayer 1985), seawaters (Harvey and Boran 1985), and for lake sediments (Ishiwatari 1985) and marine sediments (Vandenbrouke et al. 1985).

3.2.2 Interactive Properties of Humic Substances

Humic substances are polydisperse materials exhibiting polyelectrolitic behavior in aqueous solution. Electron microscopy, viscosity and ultracentrifuge measurements have been widely used to obtain information on these properties (Schnitzer 1978; Hayes and Swift 1978; Stevenson 1982).

Soil, water dissolved, and sedimentary aquatic HA and FA were shown to be surface-active (Visser 1982; Tschapek and Wasowski 1976; Chen and Schnitzer 1978). This is an important property of HS that renders them particularly interactive toward hydrophobic compounds in soil and aquatic environments, as far as it can affect the structure and solubilization of some HS-interacting organic chemicals. Chen and Schnitzer (1978) showed that the surface tension of soil FA and HA solutions is concentration and pH dependent as a function of ionization of acidic functional groups. The amphopholic character and, therefore, the surface activity of HA and FA will increase at high pH values, when COOH and phenolic OH groups form more hydrophylic sites. Increasing pH and concentration of both HA and FA lower the surface tension of water, thus increasing soil wettability and affecting the interaction phenomena of HS with both hydrophobic and hydrophylic organic chemicals in solution. Surface-active properties of HS assume higher importance in interaction phenomena occurring in aquatic environments. Aquatic FA and HA seem, in fact, to be more surface active than their terrestrial correspondents.

HS also exhibit high concentrations of stable free radicals, probably of the semiquinone type, which are considerably reactive for the binding of certain organic molecules (Stevenson 1972). Senesi et al. (1977a,b) and Senesi and Schnitzer (1977, 1978) showed that the free radical content of HA, FA and FA fractions were pH and visible-light irradiation dependent. They found that the higher the pH, the greater is the spin content. Solid samples showed small increases in free radical concentrations, while HS in solution exhibited much higher increases. In a later review paper, Senesi and Steelink (1989) suggest that the pH dependence and photo-induced radical production might be the basis for many photodegradation reactions in soils and surface waters.

3.3 Toxic Organic Chemicals

Approximately one half of the industrially-produced organic chemicals reach the global environment via direct and indirect ways, such as agriculture practices, municipal and industrial wastes and landfill effluents (Weber 1978). These products include a variety of pesticides and their metabolites, aliphatic and aromatic organic derivatives of petrol and plastics, organic solvents and surfactants and detergents (Weber 1978). When these substances reach the natural environment, various degradation and transfer processes are initiated. Chemical properties of each specific organic compound, such as molecular structure, volatility, ionic charge and ionizability, polarizability, and water-solubility will determine which processes predominate. Nowadays, a prevalent opinion is that interaction processes, leading to activation-inactivation, physical sorption and/or chemical binding or partitioning are among the most widespread and important phenomena that toxic organic chemicals (TOCs) are subjected to in the global environment. Some general considerations and properties of major groups of TOCs of relevance to the environment and of importance to human health, will be briefly summarized in the following subsections.

3.3.1 Pesticides

Organic pesticides that are presently used belong to widely differing families of organic chemicals and may be grouped in various ways. In this chapter the classification that will be used is based on the interactive properties toward HS (Weber 1972). The following groups will be discussed: cationic, basic, acidic and anionic, and non ionic. Selected pesticides of various applications such as herbicides, insecticides, fungicides and germicides and their interactions with HS will be considered.

3.3.1.1 Cationic Compounds

Bipyridilium herbicides such as diquat and paraquat are the only important compounds of this group that have been thoroughly investigated in relation to interactions with aquatic and soil HS. They are available commercially as dibromide and dichloride salts, respectively, and are used as herbicides and dessiccants. These chemicals were shown to be toxic to humans (Vettorazzi 1977; Calderbank 1968; Wolfhart 1980). The solubility of cationic pesticides is generally high in aqueous solutions, where they dissociate readily to form divalent cations. Diquat and paraquat are non-volatile compounds and do not escape as vapors from aquatic or soil systems. They are known to readily photodecompose when exposed to sun or UV light, but are not photodecomposed when adsorbed onto particulate matter (Calderbank 1968). These compounds are able to form well-defined charge-transfer complexes with phenols and many other donor molecules (Haque and Lilley 1972).

The fungicide phenacridane chloride, the germicide thyamine, and the plant-growth regulator phosphon have also been studied somewhat for their interaction with soil organic matter (Weber 1972).

3.3.1.2 Basic Compounds

The most important and studied pesticides of this group are amitrole and several members of the family of *s*-triazines. Amitrole had been widely used as a herbicide, but its uses as a registered product for application on food crops were cancelled starting in 1971 because it was suspected of inducing thyroid tumors in rats (Carter 1975). Amitrole is soluble in water, showing a weak basic character ($pK_b = 10$) and behaves chemically as a typical aromatic amine.

S-Triazines that are currently used as selective or general herbicides are substituted diamino-*s*-triazines which have a chlorine, methoxy, methylthio, or azido group attached to carbon-3 ring atom. The presence of electron-rich nitrogen atoms confer to *s*-triazines the well-known electron donor ability i.e., weak basicity and the capacity to interact with electron acceptor molecules, giving rise to electron donor-acceptor (charge-transfer) complexes.

Symmetric-triazines have low solubilities in water, the 2-chloro-*s*-triazines being less soluble than the 2-methylthio and 2-methoxy analogues. Water solubility increases at pH values where strong protonation occurs, e.g. between pH 5.0 and 3.0 for 2-methoxy- and 2-methylthio-*s*-triazines, and at pH 2.0 or lower for 2-chloro-*s*-triazines. Structural modifications of the substituents significantly affect solubility at all pH levels. Increasing solubility is associated with increasing electron-donating capability of the substituents at carbon-2 and increasing size and branching of the N-alkyl groups at the 4- and 6-positions. The *s*-triazines, and especially the chloro-*s*-triazines are chemically hydrolized in aqueous systems. Chloro- and methylthio-*s*-triazines are partly photodecomposed in aqueous systems by UV and IR radiation, including sunlight, while methoxy-substituted compounds are not photodegradable. Most *s*-triazines are relatively volatile, so that they can be lost from aquatic and soil systems by volatilization processes (Weber 1972).

3.3.1.3 Acidic Compounds

This group of pesticides comprises different families of chemicals with herbicidal action including substituted phenols, chlorinated aliphatic acids, chlorophenoxyalkanoic acids, and substituted benzoic acids, which possess carboxyl or phenolic functional groups capable of ionizing in aqueous media to yield anionic species. These materials range in acid strength from strong acid (TCA) to relatively weak acids such as MCPB.

Chlorinated aliphatic acids show the highest water solubility and the strongest acidity among this group of chemicals, owing to the strong electronegative inductive effect of the chlorine atoms replacing the hydrogens in the aliphatic chain of these acids. The water solubilities of the phenoxyalkanoic acids are low as they have a considerable lipophilic component. Most commercial formulations of these herbicides, however, contain the compound in the soluble salt form, thus the anionic species predominate in neutral aqueous systems, while at low pH levels they are present in the molecular rather than the anionic form. These herbicides may undergo reactions of alkylcarboxylic acids, aromatic compounds and ethers. Dinitrophenols and pentachlorophenol (PCP) are generally of intermediate solubility in water, while they are highly water soluble as alkali salts which represent most of their common commercial formulations.

With the exception of picloram and phenols, acidic pesticides are considered non-volatile from aqueous and soil systems (Weber 1972). Some ester formulations of these compounds also behave as herbicides. They do not ionize in solution and are less water soluble than the acid or salt forms. They are eventually hydrolized to acid anions in aqueous and soil systems, but in the ester form are non-ionic and relatively volatile.

2,4-D and 2,4,5-T are among the most widely known and used phenoxyalkanoic acids. These two herbicides were used as defoliant in Vietnam. Teratogenic (fetus deforming) effects on rats and mice were reported for 2,4,5-T and the isooctyl ester of 2,4-D (2,4-DOE; Courtney et al. 1970; Crosby et al. 1971; Woolson et al. 1972), while mortality and physical abnormalities were shown to increase in chick embrios of gamebird eggs sprayed with 2,4-D at rates commonly used in field applications (Lutz-Ostertag and Lutz 1970). The most extensively used halogenated benzoic acid herbicides are chloramben, dicamba, and TBA. Phenols most used as herbicides are Dinoseb and DNOC, while ioxynil and bromoxynil, PCP, TCA and Dalapon are the major chlorinated aliphatic acids used as insecticides. Picloram is the only prominent member of pyridine derivatives that has been extensively studied and commercially developed as a herbicide. The uracil herbicides bromacil, isocil, and terbacil that are usually considered as non-ionic compounds are also classified sometimes as ionic herbicides, but with no specific acidic/basic character (Weber 1972).

3.3.1.4 Nonionic Pesticides

Pesticides of this category do not ionize significantly in aqueous systems and vary widely in their chemical composition and properties, i.e. in water solubility, polarity, tendency to volatilization and molecular volume.

Chlorinated hydrocarbon insecticides are among the most widely known and studied group of nonionic pesticides. DDT, in particular, has been studied more than any other pesticide. It has been imputed as detrimental to numerous wildlife species and to accumulate in the food chain. Several chlorinated hydrocarbons have been detected in various marine and terrestrial living organisms, food crops, surface waters and soils. Toxaphene, lindane, chlordane and heptachlor have been found in the biosphere in much smaller levels than DDT, aldrin, and dieldrin (Weber 1972). DDT content of phytoplankton in the sea has been shown to increase since 1955 even though the amount used has been declining since 1965. With the exception of lindane, all these compounds are insoluble in water. DDT is about ten times more insoluble than the other compounds of this family, thus it is considered to be immobile in soil systems. Endrin, dieldrin and aldrin show higher water solubility and are, therefore, slightly mobile in soils. The vapor pressure of chlorinated hydrocarbons varies widely: from low (DDT, endrin and dieldrin), to moderate (toxaphene and aldrin), to high (chlordane and lindane), and very high (heptachlor). Volatilization of DDT from soils and other surfaces is, therefore, almost insignificant.

Organophosphates are more toxic than chlorinated hydrocarbons, in particular to humans, but they exhibit lower persistence in soils and do not seem to accumulate in soil fauna or to concentrate in birds and fish. This behaviour is also related to an enhanced water solubility and higher vapor pressure of organophosphates. Malathion and parathion insecticides are known to be chemically drolyzed and biodegraded by microorganisms in soil systems. The most important organophosphates herbicide is glyphosate.

Phenylcarbamates, or carbanilates, generally exhibit low water solubilities, thus they are almost immobile in soil systems. Chlorpropham and propham are readily volatilized from soil systems, but terbutol and carbaryl are not. Among chemical reactions carbamates may undergo, there are ester- and amide-hydrolysis, N-dealkylation and hydroxylation. N-methylcarbamate insecticides commonly used in soils are carbaryl, methiocarb, aldicarb and carbofuran.

More than 25 different substituted urea herbicides are available commercially today. The most important are phenylureas (fenuron, monuron, diuron, fluometuron, and chlortoluron) and cycluron which has the aromatic nucleus replaced by a saturated hydrocarbon moiety. Benzthiazuron and methabenzthiazuron are more recent selective herbicides of the class, with the aromatic moiety replaced by a heterocyclic ring system. With the exception of fenuron, substituted ureas exhibit low water solubilities, which decrease with increasing molecular volume of the compound. The majority of phenylureas have relatively low vapor pressures and are, therefore, not very volatile. These compounds show electron donor properties and thus they are able to form charge-transfer complexes by interaction with suitable electron acceptor molecules. Hydrolysis, acylation and alkylation reactions are also possible with these compounds.

The most important substituted anilide herbicides are propanil and chloracetamide, while propachlor and alachlor, commercially introduced in the late 1960s, are classified also as chloracetamides (Jaworski 1975). Another commercially important herbicide classified as acetamide is CDAA. Its water solubility and vapor pressure are relatively high. It is thus relatively mobile and volatile in soil systems.

Substituted dinitroanilines are an important, relatively recent series of selective herbicides commercially introduced in agriculture in the 1960s. Trifluralin is the most prominent member of this series. Nitralin and benefin have also received widespread usage, while butralin and profluralin are relatively recent herbicides of this class. Dinitroanilines show very low water solubilities. Nitralin and benefin have low vapor pressures and are non-volatile, while trifluralin is relatively volatile. All these compounds have been shown to be relatively immobile in soil systems.

The phenylamide herbicide diphenamid is moderately water soluble and not volatile. It probably behaves much like the acetanilides in aqueous and soil systems.

Thiocarbamate and carbothioate herbicides generally exhibit low water solubility and high vapor pressures, thus being relatively mobile in soil systems. Surface losses are attributed to volatilization because of the herbicides high vapor pressures.

The most important nonionic benzonitrile herbicide is dichlobenil, which has a low solubility in water and low vapor pressure, thus it is relatively immobile in most soils.

3.3.2 Polynuclear Aromatic Hydrocarbons

Polynuclear aromatic hydrocarbons (PAH) are in general a hazardous class of widespread contaminants produced in large quantities from the combustion of fossil fuels, in chemical manufacturing, petroleum refining, metallurgical processes, and in some coal, oilshale and tar sand conversion systems. PAH are present in waste streams from these processes and, through various environmental pathways, they can accumulate in food chains leading to man.

PAH are neutral, nonpolar organic molecules consisting of two or more benzene rings arranged in various configurations with hydrophobicity increasing with molecular weight. Many members of this class of chemicals have been identified to exhibit toxic and hazardous properties. Some materials have been demonstrated to cause mutations and certain types of cancer (Pitts et al. 1977; TRW Systems and Energy 1976). The World Health Organization (WHO 1971) has, therefore, recommended limits for certain PAH in drinking waters and the U.S.E.P.A. has included 16 PAH products in its list of priority pollutants to be monitored in industrial effluents (Keith and Telliard 1979).

Although there is evidence that the environmental sources of PAH also include natural inputs such as combustion (e.g. forest fires, Youngblood and Blumer 1975), sediment diagenesis (Wakeham et al. 1980), geological phenomena (e.g. volcanos, tar pits, seepage from rock formation, Greiner et al. 1977) and biological conversion of biogenic precursors (Aizenshtat 1973), most of the PAH contamination of aquifers, soils, sediments and water bodies comes from anthropogenic sources (Cole 1975). Hence, the occurrence of PAH in natural environments is generally recognized as contamination from anthropogenic sources (Cole 1975). This is a cause for environmental concern because PAH can be hazardous at low concentration and some PAH are degraded relatively slowly. As they have a hydrophobic nature, adsorption is very important in determining their fate in surface and subsurface water-soil/sediment systems.

3.3.3 Phthalates

Most of the industrially produced *ortho*-phthalic anhydride is esterified with various alcohols to form phthalic acid diesters (PAE) which are used mainly as plasticizers. For PVC plastics, the PAE to resin ratio is usually 1:2. PAE plasticizers generally occur in monomeric forms, only loosely linked to the polymers, and therefore are easily extruded. PAE are also used as nonplasticizers as pesticide carriers and insect repellents, in dyes, in cosmetics, and in lubricants (Anonymous 1972). PAE are lipophilic or lyophobic liquids of medium viscosity and low vapor pressure. Di-2-ethylphtalate (DEHP) and di-octylphtalate (DOP) are among the most widely used plasticizers for plastic and synthetic rubber.

Although DEHP is a moderately viscous liquid, practically immiscible with water and has a very low vapor pressure, it and other PAE have been detected in animal tissues, milk, blood and various aqueous solutions. PAE products have also been reported as natural constituents of biological and biologically derived materials, including wood, lignin, fungal metabolites, humic compounds, petroleum and coal, occurring as intermediates in biochemical pathways and as common products of chemical and biological oxidations of many organic compounds (Schnitzer and Khan 1972; Mathur 1974). Yet, these chemicals are regarded mostly as pollutants from industrial sources, as their levels in the environment are high enough to threaten human health directly or indirectly through disturbance of the ecological balance (Mathur 1974). Man is commonly exposed to PAE through polluted water and food packing materials although at levels far below those which can cause acute toxicity (Dillingham and Autian 1973). Nevertheless, recent indications have been obtained that in some organisms PAE may act as teratogens and mutagens at lower levels (Authian 1973).

Many natural water bodies seem to be contaminated with PAE. Examples are water samples from the Ohio river, Gulf Stream, Mississippi river delta, the Gulf of Mexico, Lake Superior, and other rivers and lakes (Corcoran 1973; Hites 1973; Mathur 1974).

DEHP has been observed to accumulate in aquatic organisms and to concentrate through the food chain (Metcalf et al. 1973). PAE contamination of soils and river and stream sediments has also been reported (Cifrulak 1969).

3.4 Evidence of Interactions of TOC with Humic Substances

The fate and behavior of TOC in the environment is affected by many different processes. Most of them, including degradation, persistence and mobility, bioactivity, phytotoxicity, volatility and leachability, bear a direct relationship to the nature and content of HS in the environment.

Many physical, chemical and biochemical properties, climatic factors and geochemical effects will influence the behavior of TOC in the environment, but the highest correlations and general dependence was found with concentrations of HS.

3.4.1 Soil

Among various TOC, pesticides represent the group of compounds which are most commonly found in soils and frequently in interaction with the soil humus. Comparative studies have suggested that most pesticides have a greater affinity for organic surfaces than for mineral surfaces, thus organic matter, and in particular humic fractions, play a major role in the performance of soil-applied pesticides (Stevenson 1982). Nevertheless, in most soils HS are partly associated with clay minerals, thus firmly clay-bound fractions of HS (e.g. humin) do not behave as a separate entity in the interaction of soil constituents with pesticides (Stevenson 1982).

Diquat and paraquat interactions in soils were studied extensively. These compounds become partly inactivated in highly organic soils (Khan 1974a, 1978b). The phytotoxicity of paraquat to plants grown in media containing organic matter is reduced (Scott and Weber 1967). Sorption of diquat and paraquat on soil OM has been proposed to be the major factor responsible for the decrease in herbicide activity, although the chemicals are still biologically active toward plants and microorganisms (Weber 1972; Weber and Weed 1974). Tucker et al. (1967, 1969) suggested that paraquat and diquat in a muck soil are present in "tightly" bound fractions, not available to plants, and a "loosely" bound fraction which can potentially become available. Phenacridane chloride and thyamine are cationic compounds that were found to adsorb strongly on soil organic matter. Levels of adsorption were the highest for these two compounds followed by phosphon, diquat and paraquat (Weber 1972).

The biological activity and transport in soil of basic pesticides, such as the s-triazine and triazole herbicides, has been shown to be negatively correlated with the level of OM in soils due to adsorption phenomena (Hayes 1970; Weber 1972; Weber and Weed 1974). Chemical hydrolysis and sunlight photodecomposition of s-triazines are also influenced by the presence of soil OM (Armstrong et al. 1967; Jordan et al. 1970; Burkhard and Guth 1981).

Bioactivity and transport of acidic pesticides and their ester formulations were found to correlate with the OM content of soil, even though the adsorption level of these herbicides is much lower than that of cationic or basic pesticides (Weber 1972; Khan 1978). Persistence of these pesticides, measured in terms of biological activity, as well as residual toxicity, was shown to be the highest in soils containing high levels of OM (Weber 1972; Weber and Weed 1974; Weed and Weber 1974). Leachability of picloram, amiben and PCP was negatively correlated with OM (Eliasson et al. 1969; Weed and Weber 1974). The efficacy of PCP was observed to decrease with increase in OM due to adsorption.

Insecticidal activity, degradation, inactivation, leaching and volatilization of several chlorinated hydrocarbons, including aldrin, dieldrin, endrin, endosulfan, lindane, heptachlor, DDT, toxaphene and chlordane was observed to decrease as the OM content of the soil increased. The most prominent effect was found in moist soils (Bowman et al. 1965; Farmer et al. 1974; Khan 1978). In particular, retention and inactivation of DDT is quite generally correlated with the OM content of the soil and especially with the humified fraction (Khan 1978). Higher levels of OM also increase the persistence of DDT, lindane and aldrin which was higher in a muck soil than in a mineral soil (Lichtenstein 1959). Transport of DDT in forest soils has been associated with soil HA and FA fractions (Ballard 1971). Peterson et al. (1971) reported that soil OM was the principal means of deactivation of DDT as determined by a bioassay technique.

Insecticidal activity of organophosphates and their adsorption in soils was also correlated with OM content of soils (Weber and Weed 1974). The bioactivity of phorate decreased with increased OM content of soils (Kirk and Wilson 1960), while diazinon and parathion showed the same relationship in moist soils, but not in dry soils (Harris 1966). Fonofos was shown to be persistent for more than two years in an organic soil and its mobility and persistence in soils was suggested to be partly a function of adsorption on HS (Khan et al. 1976; Khan 1977). Phenylcarbamate herbicides showed a lower herbicide activity in fine-textured soils than in coarse-textured ones. This was related to the higher OM-content of the former soils (Weber et al. 1974). Phytotoxicity of chlorpropham was greatly reduced by OM added to the soil (Scott and Weber 1967). Vapor losses of propham and chlorpropham from moist soils decreased as OM % increased (Weed and Weber 1974).

The herbicidal activity, adsorption and transport of several phenylureas, including monuron, linuron, neburon, fluometuron and diuron in soils was shown to decrease as the OM of soils increased (Dubey et al. 1966; Obien et al. 1966; Douding and Freeman 1968; Nash 1968; Weber and Weed 1974). Addition of OM to sandy soil significantly reduced the herbicidal activity of fluometuron and fenuron in a growth chamber study and in field experiments (Weber 1972).

Substituted anilides (or chloracetamides) alachlor and propachlor were found to perform better in high organic soils and poorer in coarse textured soils, thus showing, in this respect, behavior opposite to that exhibited by many other chemicals (Weber 1972). This is generally attributed to their low adsorptivities and/or to the different nature of interaction of these chemicals in soils.

The herbicidal activity of substituted dinitroanilines, i.e. trifluralin and benefin, was reported to be reduced by soil OM (Lambert et al. 1965; Hollist and Foy 1971; Menges and Hubbard 1970; Weber 1971; Weber et al. 1974). The mobility of these herbicides in muck soils was much less than in mineral soils, thus suggesting a partial adsorption of these compounds by organic soil colloids (Weber 1972). In greenhouse experiments, orizalyn was shown to be inactivated by soil OM at a much lower level than trifluralin and nitralin due to much lower adsorption of orizalyn by soil OM (Weber 1972).

Movement and bioactivity of diphenamid was decreased as the OM content of the soil increased (Dubey et al. 1966). Similarly, additions of soil OM to model soil systems significantly reduced the herbicidal activity of diphenamid (Weber 1972).

Increasing soil OM content resulted in decreased bioactivity, movement, leaching and volatilization of relatively volatile pesticides, which include thiocarbamates, acetamides and benzonitriles such as EPTC, pebulate, cycloate,

CDEC, CDAA and dichlobenil (Ashton and Sheets 1959; Fang et al. 1961; Miller et al. 1966; Koren et al. 1968, 1969; Briggs and Dawson 1970; Weber and Weed 1974). Higher levels of OM resulted in lower initial toxicity of many herbicides and lower losses through volatilization. Thus these compounds were slightly less effective but persisted longer in finer textured soils (Weber 1972).

Movement and adsorption of aldicarb, carbofuran and pyrazone were shown to be positively and significantly correlated to soil OM content (Jamet and Piedallu 1975a,b).

Aqueous concentrations, transport and surface-associated chemical and biological degradation processes of PAH are highly dependent on adsorptive/desorptive equilibria with sorbents, e.g. HS, present in the system. In particular, liquid to solid phase partitioning can play a significant role in the adsorption phenomena on HS, as will be extensively discussed later in this chapter.

The isolation and identification of small amounts of PAE (0.03% of the dry weight) from soil HA and FA (Ogner and Schnitzer 1970; Matsuda and Schnitzer 1971; Khan and Schnitzer 1972) indicates that the lyophylic HS can interact with lyophobic PAE to form water-stable products. Water-soluble FA may therefore mediate mobilization, transport, and immobilization of PAE in soil solutions and aquatic environments.

3.4.2 Waters

The importance of dissolved organic matter (DOM) in determining the speciation and long-term environmental and health impact of hydrophobic TOC in aquatic systems has only recently been recognized. A substantial fraction of pollutant found in aqueous phases may in fact interact with dissolved HS. This could significantly affect the rate of dissolution, volatilization, transfer to sediments, biological uptake and bioaccumulation, or chemical degradation. The latter may be induced via hydrolysis or de-alkylation. The distribution and total mass of a pollutant in an ecosystem as well as its environmental behavior and fate would, therefore, depend, in part, on the extent of HS-TOC interaction. The decrease in bioavailability of TOC interacting with dissolved HS, and the resulting decrease in biological uptake and bioaccumulation, might thus mitigate the biological impact of hydrophobic contaminants. Interaction of TOC with HS would also be a cause for problems in the analytical identification and quantitation of dissolved TOCs. HS can also exert a competitive effect on the adsorption of volatile TOCs on activated carbon during treatment of drinking water and adversely affect the efficiency of oxidative water treatment processes (Peyton et al. 1987).

3.4.2.1 Water Solubility, Mobility, and Volatilization

Several recent studies have shown that solubilities of hydrophobic TOC is considerably greater in the presence of HS than in their absence (Chiou et al. 1986). In general, solubilities increase 2 to 3 times in the presence of aquatic HS at concentrations similar to those found in natural waters (Gjessing and Berglind 1981; Perdue 1983; Carter and Suffet 1982, 1983). Wershaw et al. (1969) found that the solubility of DDT in an 0.5% aqueous solution of sodium humate is at least 20 times greater than that in water (Fig. 1). Based on previous experiments by Ballard (1971), Khan (1980) suggested the HS could act as quantitatively important carriers of DDT in the organic layer of a forest soil.

Boehm and Quinn (1973) found that dissolved organic matter (DOM) in seawater and sewage increased the rate of dissolution of the normal alkanes, hexadecane and eicosane. Haas and Kaplan (1985) showed that in the presence of a HA solution, toluene solubility increased by 8% within the range 0–70 $mgCL^{-1}$. Recently, Chiou et al. (1986) determined water solubility enhancements by dissolved HA and FA from aquatic and soil origin for selected TOCs, i.e. DDT, PCB and lindane. Griffin and Chian (1980) found that the rate of volatilization of PCB and hexachlorobutadiene in water decreased when HS were present.

3.4.2.2 Catalytic Effects

Evidence for catalytic effects of HA and FA on the hydrolysis of chloro-s-triazines in soils and aquatic systems has been presented by many authors (e.g. Gwo-Chen and Felbeck 1972; Khan 1978a; Perdue and Wolfe 1982; Gamble and Khan 1985) and recently summarized (Choudry 1984) while very limited studies are available on other TOCs (Bronsted and Pedersen 1924).

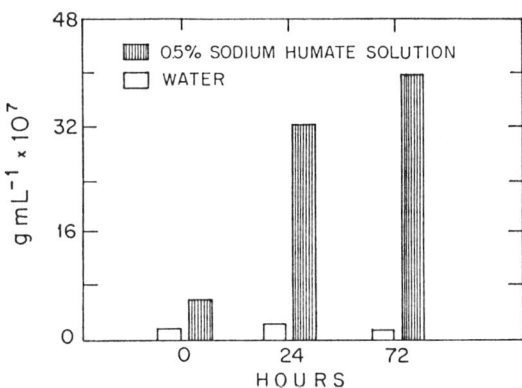

Fig. 1. Solubility of DDT in water and in sodium humate solution (after Wershaw et al. 1969)

Plimmer et al. (1968) demonstrated that free radical reactions caused N-dealkylation of the s-triazines and speculated that a similar mechanism may be responsible for N-dealkylation of the s-triazines in soils as well as by soil fungi and higher plants.

Chemical reductions induced by HS on TOCs such as nitroaromatic compounds to anilines and de-halogenation of chlorinated hydrocarbons might also be possible, but this research area is still relatively unexplored (Perdue 1987).

3.4.2.3 Photolysis Effects

Many pesticides and other common TOCs undergo light-energized transformations in aqueous solution or suspension. Even if the effective energy range of sunlight appears to be limited (290–450 nm), light within this wavelength span is sufficient to provide the required energy for the rupture of many types of chemical bonds, thus energizing a variety of common reactions including oxidation, reduction, hydrolysis, substitution, and isomerization (Crosby 1976a). Results of many investigations have indicated that photolysis rates of some chemicals increase in natural water samples and that some compounds which are completely sunlight inert in distilled water, photoreact rapidly in river or sea waters (e.g. Ross and Crosby 1973, 1975; Draper and Crosby 1976; Zepp et al. 1976, 1977; Mill et al. 1980). These results are generally attributed to photosensitizer interactions exerted by dissolved or suspended HS (Zepp et al. 1981, 1985). Dissolved FA and HA have been shown to sensitize the photolysis of atrazine and prometryn (Khan and Schnitzer 1978; Khan and Gamble 1983), ethylenthiourea (Ross and Crosby 1973), and 3,4-dichloroaniline (Miller et al. 1980).

Zepp et al. (1981, 1985) reported that the rates of photolysis of some TOCs such as dimethylfuran, disulfoton, aniline and pentadienes were increased in the presence of several commercial and soil-derived HS at similar rates to those observed in river waters containing comparable concentrations of HS.

Mudambi and Hassett (1987) found that photochemical action spectra of mirex (one of the major organochlorine contaminants detected in various biota and the waters of Lake Ontario) measured in distilled water, water containing HA and Lake Ontario water indicated the existence of specific interactions between mirex and DOM, which led to increased degradation of mirex to photomirex (8-monohydromirex), one of the three possible monohydro-decomposition products of mirex. Mill and Carter (1987) found that the photolysis of sodium tetraphenylboron (TPB) used to precipitate the major radioactive contaminant ^{137}Cs from radioactive wastewaters was enhanced by both HA and FA. Similar results were obtained using DOM isolated from stream water.

In a recent paper, Hoigné et al. (1987b) have summarized the behavior of aquatic HS as sources and sinks of photochemically produced transient reactants which are known to degrade certain classes of micropollutants on timescales of environmental interest (Haag and Hoigné 1985, 1986).

Peyton et al. (1987) obtained evidence that low concentrations of HS could actually catalyze the destruction of some organic compounds and their reaction by-products by photolytic ozonation. The results of these authors suggested that it might be possible to remove an organic pollutant and its by-product using oxy-radical processes in the presence of comparable amounts of HS. On the other hand, Hoigné et al. (1987) suggested that the presence of HS in water being treated by photolytic ozonation (ozone/UV) could cause lowered treatment efficiency by acting as a scavenger for the hydroxyl radicals which are formed, thereby protecting micropollutants from oxidation.

3.4.2.4 TOC Bioaccumulation

In aquatic systems, the availability, uptake and resulting bioconcentration of TOCs such as PAH, PCB, and polychlorinated dioxins are reduced for fish and aquatic invertebrates due to association of TOCs with HS thus resulting in a corresponding decrease in toxic effects (McCarthy 1987). The presence of HS was shown to inhibit the effect of several organic mutagens. The in vivo toxicity of DDT to *Daphnia magna* was reduced in the presence of HS while the toxicity of lindane was not affected because of its low affinity for binding to HS. All these effects appeared to be due to the inability of TOC associated with the humic macromolecule to cross biological membranes (McCarthy 1987).

Spacie et al. (1984) have shown that as little as 1 mg C L^{-1} of dissolved HS reduced the rate of uptake of benzo(a)pyrene by bluegill sunfish (*Leponis machrochinus*) by about 67%. McCarthy and Jimenez (1985a) observed that dissolved HS at concentration 20 mg C L^{-1} were able to reduce the bioaccumulation of benzo(a)pyrene by 90%, while they had little effect on uptake of naphtalene by bluegill sunfish. Levels of dissolved HS of 1.5 mg C L^{-1} and less have also been demonstrated to reduce bioaccumulation of benzo(a)pyrene and benzanthracene by *Daphnia magna* (Leversee et al. 1983; McCarthy 1983). The presence of increasing amounts of dissolved HS decreased the rate of uptake and the bioaccumulation of benzo(a)pyrene, benzanthracene and anthracene. The decrease in uptake was quantitatively related to the amount of the contaminant bound to HS. The reduction in bioaccumulation was greatest for the more hydrophobic compounds which have the greatest affinity for binding to HS and also the greatest potential for bioaccumulation (McCarthy 1987).

Landrum et al. (1985) imputed the reduced bioavailability to *Pantoporeia hoyi* of a number of PAHs, PCBs, DDT and DEHP to adsorption or binding by DOM in either simulated lake water or sediment interstitial water. The apparent biological uptake rate constant of each compound was found to be inversely proportional to the HA concentration and greatest for the more water insoluble compounds such as benzo(a)pyrene, while it was near zero for phenanthrene, the most water soluble compound that was studied (Landrum et al. 1987). The presence of dissolved HS may also compete with sediment or suspended particles for the binding of PAH and other TOCs, thus altering the fate and transport of TOC in aquatic systems (McCarthy and Jimenez 1985b; McCarthy et al. 1985).

3.4.2.5 TOC Analyses and Identification

Hassett and Anderson (1979) and Landrum and Geisy (1981) found that DOM could affect the identification and analysis of TOCs in water. Recovery of benzo(a)pyrene using cyclohexane extraction decreased with increasing HS concentration (Gjessing and Berlind 1981). Carlberg and Martinsen (1982) showed that HS in surface waters negatively influenced efficiency of liquid/liquid extractions and of the XAD-2 resin adsorption method for a number of alkanes, PAH, PAE and PCB. Thurman (1987) found that there was the potential to coisolate LAS surfactants with aquatic HS, which may thus interfere with the analytical determination of LAS in waters.

3.4.2.6 TOC Elimination

Activated carbon treatment is one of the fundamental technologies used for purification of waters contaminated by TOC. The presence of DOM complicates the engineering and operation of adsorption treatment processes because of its interactions both with activated carbon and with the target TOCs in solution. Recent studies have indicated that the presence of background OM in solution reduced adsorption capacities and rates of target pollutants, in particular, of volatile halogenated organics, such as 1,1-dichloroethane, trichloroethene, 1,1,1-trichloroethane, tetrachloroethene, p-dichlorobenzene (Baker et al. 1987; Narbaitz and Benedek 1987; Weber and Smith 1987; Zimmer and Sontheimer 1987) by interacting with the carbon surface in a way which affected the subsequent adsorption of trace organic compounds. The impact of HS on the adsorption capacity was higher for compounds with high adsorbability (Zimmer and Sontheimer 1987). Reduction in the carbon adsorbing capacity for PCB was observed in the presence of HA which was attributed to a combination of competitive interaction and complexation phenomena between HA and PCB (Pirbazari et al. 1987).

3.4.3 Sediments

Since most TOC contaminating water are adsorbed onto particulate matter, which tends continuously to flocculate and precipitate at the bottom of aquatic systems, TOC residues usually occur in higher concentrations in the sediment of any water body. The sorption behavior of a number of hydrophobic TOCs including some PAHs on sediments seems to be determined by the organic carbon content of the substrate and is often independent on factors such as substrate pH, cation exchange capacity, textural composition and clay mineralogy (Karickoff et al. 1979; Hasset et al. 1980b; Means et al. 1980).

Organic coating of sediment particles showed high affinity for organic contaminants (Karickoff et al. 1979; Means et al. 1980), comparable with that of dissolved HS even when the binding to the sediment appeared to be less rapid than comparable binding to dissolved HS (McCarthy and Jimenez 1985b).

3.5 Modes of Interaction of Humic Substances with Toxic Organic Pollutants

The significance of interactions between HS and TOCs to chemical and biological processes involving TOCs has been described in Sect. 3.4. In this section the most important interactions involving HS and TOC will be discussed from a mechanistic point of view. Emphasis will be given to adsorption mechanisms, solubilization effects, hydrolysis, catalysis and photosensitization. It should be noted at this point that the chemical properties and behavior of the dissolved-phase and solid-phase fractions of HS may be sufficiently different that these two fractions will interact differently with a given organic pollutant and that various chemical properties of TOCs will result in several interaction mechanisms that may frequently operate in combination.

3.5.1 Adsorption

Sorption represents probably the most important interaction phenomena exerted by HS on the fate of TOCs in the environment. Adsorption will control the quantity of free TOCs in solution and thus determine its persistence, mobility and bioavailability. The extent of adsorption will depend on the amount and properties of both HS and TOC. Once adsorbed on HA a TOC may be easily desorbed, desorbed with difficulty, or not at all. Thus sorption phenomena may vary from complete reversibility to total irreversibility.

The most important properties of a TOC molecule that will determine its mode of interaction with a HS macromolecule are the chemical character of the molecule, shape and configuration, acidity (pK_a) or basicity (pK_b), water solubility, polarity, molecular size, polarizability, and charge distribution (Bailey and White 1970).

3.5.1.1 Adsorption Isotherms

Construction and use of adsorption isotherms from equilibrium adsorption data has been employed by numerous researchers to describe adsorption of TOC on a solid matrix. An isotherm represents a relation between the amount of solute adsorbed per unit weight of solid adsorbent and the solute concentration in solution at equilibrium. Giles et al. (1960) investigated the relation between solute adsorption mechanism on solid surfaces and the shapes of the adsorption isotherms obtained, thus developing an empirical classification which recognized four main types of adsorption isotherms (Fig. 2). The L-type, or Langmuir isotherm, is the most common and occurs when the adsorbent has a moderately high affinity for the solute in the initial stages of the isotherm. As adsorption sites are filled, the solute molecules have increasing difficulty in finding vacant sites and the slope of the curve decreases. The S-type indicates that adsorbed molecules favor the retention of additional molecules. The C-type isotherms are obtained by solutes which penetrate into the solid more readily than does the solvent. A C-type isotherm can also be an L+H type at low solution concentrations. These curves are characterized by a constant partition of solute between the solution and substrate, and therefore the adsorption is always directly proportional to the solution concentration. The H-type isotherms represent very high affinity between the solute and the solid. These classes of isotherms have been referred to in the literature in many instances concerning TOCs adsorption on OM.

Adsorption of cationic compounds, including paraquat and diquat, phosphon, hyamine and chlordimeform on soil OM or HS resulted in L-shaped isotherms, characterized by a curvilinear response at all concentrations used, which seem to level off at a certain adsorption maximum (Weber 1972; Maqueda et al. 1983). This indicates specific adsorption on homogeneous sites. pH-dependent adsorption of s-triazines by organic soil colloids and HA also showed L-shaped isotherms, indicating that adsorbed species were in equilibrium with species in solution at each pH value (Weber et al. 1969a,b; Gaillardon et al. 1977). Some typical experimental adsorption isotherms are shown in Fig. 3.

In general, two mathematical equations are used to quantitatively describe TOC adsorption on organic matter adsorbent:

(a) the empirically derived Freundlich equation:

$$x/m = K_{oc}C^{1/n} \tag{1}$$

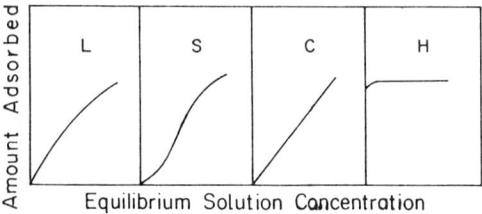

Fig. 2. Classification of adsorption isotherms according to Giles et al. (1960) (L = normal or "Langmuir" isotherms; S = "cooperative adsorption"; C = "constant partition"; H = "high affinity")

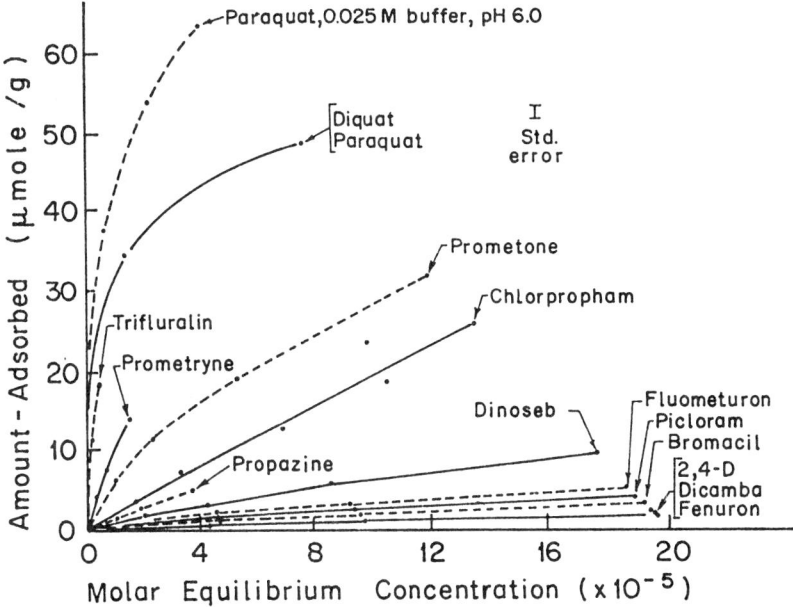

Fig. 3. Adsorption isotherms for selected organic herbicides from various chemical groups on soil organic matter: cationic (diquat, paraquat); polar, low water solubility (trifluralin); basic (prometryn, prometone, propazine); polar, moderate water solubility (chlorpropham, fluometuron); acidic (dinoseb, picloram, bromacil, 2,4-D, dicamba); and polar, high water solubility (fenuron); (after Weber 1972)

where x/m is the ratio of adsorbate to organic matter mass, C is the adsorbate concentration in solution at the equilibrium, and K_{oc} and n are constants. Normally, within a reasonable range of adsorbate concentrations, the logarithmic form of Eq. (1) is linear with 1/n being constant. The K_{oc} value may be considered as an index of the degree of adsorption of various adsorbates by different organic surfaces, assuming the determinations are made at the same concentration range. In general, adsorption of pesticides on soil OM fit the Freundlich equation reasonably well with an exponent $1/n = 1$ reduction to a linear equation. Khan (1973a, 1977) and Khan and Mazurkevith (1974) showed that the adsorption of linuron, 2,4-D, picloram, and fonofos on HA followed a Freundlich-type isotherm.

(b) The Langmuir equation

$$x/m = K_1 K_2 C / 1 + K_1 C \qquad (2)$$

where K_1 is a constant of the system dependent on temperature and K_2 is a monolayer capacity of adsorbate. The reciprocal of Eq. (2) provides a straight line with the intercept $1/K_1 K_2$. The adsorption of a number of pesticides on organic surfaces was found to fit the Langmuir-model isotherm. Li and Felbeck (1972) showed that the adsorption of atrazine on HA had the form of a Langmuir isotherm. Gupta et al. (1985) found that adsorption of terbutryne on HA fit both isotherms. Adsorption data for chlordimeform fit only the Langmuir equation (Maqueda et al. 1983). Burns et al. (1973b) observed that neither the Freundlich nor the Langmuir equations fit the data if the adsorption of a TOC on OM is predominantly due to an ion-exchange mechanism, e.g. the case of the adsorption of paraquat on HA. In this case, only the Rothmund-Kornfeld equation was found to fit the results satisfactorily.

In conclusion, carefully controlled adsorption measurements and shapes of related isotherms may provide information on the mechanism involved in the adsorption of TOC to HS.

3.5.1.2 Adsorption Mechanisms

Several types of mechanisms often operate simultaneously in the adsorptive interaction between HS and TOC. The following mechanisms have been proposed: ionic bonding (ion exchange), hydrogen bonding, Van der Waals attractions, ligand exchange, charge-transfer (electron donor-acceptor process), covalent binding (chemical or enzyme-mediated) and hydrophobic bonding.

Ionic Bonding (Ion Exchange). Adsorption by this mechanism applies only to the relatively small number of TOC which are cations in solution or can accept a proton, i.e. protonate, to become cationic. Adsorption via cation exchange or ionic bonding operates through ionized carboxylic and phenolic hydroxyl functional groups of the HS (Schnitzer and Khan 1972). Diquat and paraquat, being divalent cationic pesticides, can react with more than one negatively charged site on HS (e.g. two COO^- groups or a COO^- plus a phenolate ion). Infrared (Burns et al. 1973a,b; Khan 1974b) and potentiometric titrations data (Khan 1973a; Narine and Guy 1982) were used to demonstrate that the predominant mechanism for adsorption of bipyridilium herbicides by HS is ion exchange. Similar results were obtained from IR studies on the interaction between the cationic pesticide chlorodimeform and HA (Maqueda et al. 1983). Adsorption of other cationic pesticides, such as phosphon and phenacridane chloride through ionic bonds onto OM also have been reported (Weber 1972). Khan (1973a) observed that HA and FA retained paraquat and diquat at levels that were considerably lower than the exchange capacity of the HS. Thus, not all negative sites of HS seem to be positionally available to large organic cations, which will result in steric hindrance effects.

Less basic pesticides such as *s*-triazines may become cationic through protonation depending on the basicity of the herbicide and on the proton-supplying power of the system. The pH of the solution will, therefore, govern not only the degree of ionization of the acidic groups on the adsorbent (HS) but also the relative quantity of the *s*-triazine occurring in the cationic form. Evidence that maximum adsorption of *s*-triazines on organic soil colloids occurs at pH levels close to the pK_a of the herbicide is indicative of ion-exchange (Weber et al.

1969b). However, the pH at the HS surface may be of the order of two units lower than that of the liquid environment (Hayes 1970), thus surface-protonation of a basic molecule may occur even though the measured pH of the water-adsorbent system is greater than the pK_a of the adsorbate.

On the basis of an IR study of some *s*-triazine + HA systems Sullivan and Felbeck (1968) concluded that ionic bonding could take place between a protonated secondary amino group of the *s*-triazine and a carboxylate anion on the HA. Successive studies, mainly conducted by IR spectroscopy, confirmed previous results and also gave evidence of possible involvement of the acidic phenol-OH of HA in the proton exchange of the *s*-triazine molecule (Turski and Steinbrich 1971; Carringer et al. 1975; Senesi and Testini 1980, 1982, 1983a; Kalouskova 1986; Senesi et al. 1987b).

Differential thermal analysis (DTA) curves measured by Senesi and Testini (1982, 1983) showed an increased thermal stability of the HA-*s*-triazine complexes in comparison to unreacted HA, thus confirming that ionic binding took place between the interacting products. The basicity and reactivity of triazine compounds were related to the nature of the substituent in the 2-position, as well as with the alkyl groups in the 4- and 6-positions (Weber 1967, 1970). The higher reactivity of simazine with respect to atrazine and prometryn was related to the smaller steric hindrance of the reactive N-H group of this herbicide (Turski and Steinbrich 1971). It should be noted, however, that other possible mechanisms have been suggested for the adsorption of *s*-triazines on HS, including hydrogen-bonding and charge-transfer which will be discussed later in this section.

Amitrole, another weakly basic pesticide, has been shown to be adsorbed by HA through ionic bonding (Senesi et al. 1986b). The insecticide dimefox (tetramethylphosphorodiamidic fluoride) was shown to be strongly adsorbed by HA through a mechanism including its protonation in acidic solution and subsequent adsorption by ion exchange (Grice et al. 1973).

Hydrogen Bonding. The presence of oxygen and nitrogen containing functional groups as well as hydroxylated and amino groups on HS strongly suggests that H-bonding will represent an important adsorption mechanism for TOCs containing similar complementary groups. The organic compound, however, will be in competition with water molecules for such adsorption sites.

Hayes (1970) stressed the participation of a H-bonding mechanism in the interaction of *s*-triazine with OM. A large body of evidence for this type of bonding was obtained from IR and DTA studies (Turski and Steinbrich 1971; Senesi and Testini 1982, 1983; Kalouskova 1986; Senesi et al. 1987). It was shown that multiple sites are available on both HS and herbicide molecules for this type of bonding, in particular $C=O$ groups of HS and secondary amino groups of *s*-triazines (Sullivan and Felbeck 1968). From the heat of formation of the HA-atrazine complex it was assumed that the formation of one or more H-bonds are involved (Sullivan and Felbeck 1968; Li and Felbeck 1972).

Hydrogen-bonding plays an important role as an adsorption mechanism for substituted ureas, and phenylcarbamates (Gaillardon et al. 1980; Senesi and

Testini 1980, 1983a,b) as well as other nonionic polar TOC which possess functional groups that can form H-bonding with HS-sites (i.e. alachlor and cycloate (Senesi et al. 1976b), metolachlor (Kozak 1983), malathion (Khan 1973b), bromacil (Weber 1972), and dialkylphthalates (Khan 1980).

Acidic or anionic pesticides, such as chlorophenoxyalkanoic acids and esters, asulam and dicamba can be adsorbed by H-bonding onto HS at pH values below their pK_a in nonionized forms through their COOH and COOR groups and other analogous groups (Khan 1973b; Carringer et al. 1975; Senesi et al. 1984, 1986, 1987a).

Van der Waals attractions. Van der Waals forces, although very weak, operate in all adsorbent-adsorbate interactions, and result from short range dipole-dipole or dipole-induced dipole or induced dipole-induced dipole attractions of several types. Although Van der Waals interactions are universally acting forces, they assume particular importance in the adsorption of nonionic and non-polar molecules or portion of molecules on similar sites of the adsorbent humic molecule (Burchill et al. 1981). These forces are additive, thus their contribution increases with the size of the molecule and with its capacity to adapt to the adsorbent surface. Van der Waals attractions were often invoked in case of difficulties in explaining adsorption of a TOC onto HS, but the experimental evidence was not always convincing. Van der Waals forces were considered to be involved in the physical adsorption of carbaryl and parathion on soil OM (Leenheer and Aldrichs 1971) and alachlor and cycloate (Senesi et al. 1986b) while they were suggested to represent the principal adsorption mechanism for picloram and 2,4-D by HS (Khan 1973a; Kozak 1983). Due to the large size of bipyridilium cations, adsorption by short-range Van der Waals forces can feasibly contribute to adsorption (Burns et al 1973b). Van der Waals forces have also been claimed to be involved in the interactions of HS with PCBs (Strek and Weber 1982), thiocarbamates, carbothioates and acetanilides (Weber 1972), benzonitrile and DDT (Pierce et al. 1971).

Ligand Exchange. Adsorption by this mechanism involves the replacement of water of hydration or other weak ligands partially holding transition metal ions bound to HS functional groups by suitable adsorbent molecules (Hayes 1970; Hayes and Thompson 1968). This type of mechanism might be involved in the binding of s-triazines on incompletely coordinated transition metals of HA. On the basis of IR studies Khan and Mazurkevitch (1974) found that it was very unlikely that complex formation with cations is a possible mechanism of linuron adsorption on HA. Nearpass (1976) suggested that bridging of a pesticide in the anionic form to polyvalent cations associated with OM might be possible.

Electron Donor-Acceptor Interaction (Charge-Transfer). The presence of groups possessing an electron-deficient acceptor (i.e. quinones) and an electron-rich donor (nitrogen or activated aromatic rings) in HS and the existence of TOCs possessing the same ability renders the possibility of an interaction based on the formation of electron donor-acceptor or charge-transfer systems between suitable TOC and HS structural moieties.

The feasible formation of charge-transfer complexes between paraquat and HS was postulated by Burns et al. (1973a), but UV spectroscopy failed to provide evidence for such a mechanism in an aqueous system. Successively, Khan (1973c, 1974b) provided IR evidence for the charge-transfer complex formation between the bipyridilium herbicides and HS on the basis of shifts in the out-of-plane C-H bending vibration frequencies of paraquat and diquat observed upon their interaction with HS. Recently, IR studies by Maqueda et al. (1983) suggested that charge-transfer may represent a possible additional or alternative mechanism for the interaction of chlordimeform with HA.

The formation of charge-transfer complexes was postulated by Hayes (1970) as a possible mechanism involved in the adsorption of methylthiotriazines onto OM. As in the case of the bipyridilium herbicides, IR evidence was obtained by Müller-Wegener (1977) and Senesi and collaborators (Senesi 1981; Senesi and Testini 1983a,b; Senesi et al. 1987b) indicating that s-triazines might be bound to HA in the form of electron donor-acceptor complexes. A shift of the triazine CH wagging vibration absorption band towards lower frequencies was in fact observed after interaction of the s-triazines with HA.

Electron spin resonance (ESR) studies by Senesi (1981), Senesi and Testini (1983a,b) and Senesi et al. (1987b) confirmed these findings showing that free radical concentrations increased in the products of interaction between a number of s-triazines and HA of different origin and nature. This effect was explained assuming that electron-deficient quinone-like structures in the HA molecule are able to remove electrons from the electron-rich donating amine or heterocyclic nitrogen atoms of the triazine molecule. Such an electron transfer is assumed to occur by single-electron donor-acceptor processes through the formation of semiquinone free radical intermediates, which can be stabilized and increase their lifetimes by enhanced conjugation possibilities offered to free electrons by the increased molecular complexity of the electron donor-acceptor systems, thus giving rise to the observed enhancement of the ESR signal (Senesi 1981; Senesi and Testini 1983a,b, 1984). Donor-acceptor systems, such as HA-s-triazine, may also give rise to charge-transfer complexes under the effect of light, which may introduce an unpairing of the electrons involved, producing an increase in the ESR signal (photo-induced transfer). These interpretations are supported by similar effects observed in several different organic electron donor-acceptor systems in the dark or in the light (Foster 1969; Lagerkrantz and Yhland 1962, 1963).

Müller-Wegener (1977) observed that levels of s-triazines bound to HA were inversely proportional to the levels of phenolic OH in HA. Senesi and Testini (1982) found a reverse relationship between OH and COOH functional groups content of three soil HA and free radical concentrations of their interaction products with s-triazines. A successive study on the interaction of HA of different nature and origin with s-triazine herbicides (Senesi et al. 1987b) showed that a synthetic HA characterized by low acidic and high quinonic functional group content was the most efficient in *s*-triazine adsorption and showed the highest relative increase in free radical concentrations upon interaction with *s*-triazines. The results suggested that the lower the capacity of HA molecules to form ionic and hydrogen bonds, the higher is their tendency to give rise to electron donor-acceptor systems in their interaction with *s*-triazines (Senesi and Testini 1982). Furthermore, the structure and properties of *s*-triazine chemicals were shown to affect to some extent the efficiency in forming electron donor-acceptor systems with HA. Prometone was shown to be able to produce the highest increases of free radical concentration upon interaction with HA (Senesi and Testini 1982; Senesi et al. 1987b). Thus, the presence of a methoxyl group on the 2-position and of an isopropyl group on each amino-group in 4- and 6-positions of the triazine ring apparently renders prometone the most efficient *s*-triazine in forming electron donor-acceptor systems with HA. Senesi and Testini (1982) concluded, therefore, that the basicity of the *s*-triazines and hence their tendency to form ionic bonds with HA is not the main factor governing adsorption and that electron donor-acceptor processes do play a more important role in the adsorption of *s*-triazines on HA than previously thought.

Briggs (1969) suggested that phenylurea herbicides could be bound by their deactivated ring to activated sites of soil OM through charge-transfer bonding, but that this is in contrast with the well-known electron donor nature of these compounds. Senesi and Testini (1983a,b) and Senesi et al. (1987b) showed by ESR that a prominent role is played in the interaction of HA of different nature and origin with a number of substituted urea herbicides by electron donor-acceptor processes involving organic free radicals which lead to the formation of charge-transfer complexes. Substituted ureas are, in fact, expected to act as electron donors from the nitrogen (or oxygen) atoms to electron acceptor sites on quinone or similar units in HA molecules. This electron-transfer would, therefore, give rise to observed enhancement in the ESR signal of the interaction products (Senesi et al. 1987b).

The observed positive trend between the bioactivity of the urea herbicides and the free radical concentrations measured in the HA-urea interaction products (Senesi and Testini 1983b) suggested that the same molecular parameters influence soil adsorption and phytotoxicity of these herbicides. Thus, the most phytotoxic ureas, which are the most potent inhibitors of the Hill reaction, appear to be strongly interactive by electron donor-acceptor mechanism with HA. These findings led to the hypothesis that important parallels, based on similar electron donor-acceptor mechanisms, exist between the chemical reactions that these herbicides may undergo with HS in soil and waters and the biological processes involved in their herbicidal action in plants, i.e. photosynthesis inhibition by interference with single-electron transfers in the Hill reaction in chloroplasts (Moreland and Hilton 1976).

The chemical structure and properties of the substituted urea herbicides seem to influence the extent of formation of electron donor-acceptor systems with HA. For instance, the highest increase of free radical concentrations in the products of interaction of fenuron with HA may be attributed to the absence of deactivating chlorine atoms on the phenyl ring of fenuron which is expected to reinforce the electron donor tendency of this urea in comparison with other ureas studied which contain one or two chlorine atoms on the ring (Senesi et al. 1987b). In conclusion, electron donor-acceptor mechanism appear to play a major role in the adsorption phenomena of substituted ureas on HA.

ESR studies recently conducted (Senesi et al. 1986b) on amitrole interactions with HA from various natural and synthetic sources showed in some cases moderate increases of free radical concentrations in the HA-amitrole interaction products. This was attributed to the well-known electron donor capacity of amitrole which can, similar to s-triazines and substituted ureas, interact with electron acceptor quinone moieties of HA molecules through single electron donor-acceptor mechanism of adsorption.

The importance of charge-transfer interactions in HA chemistry under the conditions prevailing in soil and water systems has been emphasized by Lindqvist (1982, 1983). The charge-transfer acceptor and donor properties of a HA in aqueous solutions were established by UV spectrophotometric studies of the interactions, respectively, with donor dihydroxobenzene and acceptor p-benzoquinone at varying pH levels. The capability of electron donation of HS to chloranil which is a known electron acceptor was recently confirmed by direct measurement of binding constants using UV spectroscopy (Melcer et al. 1987). The authors concluded that a charge-transfer mechanism could be postulated for the chloranil-HS interactions as well as for similar systems.

Finally, specific π-π (charge-transfer) interactions between the condensed ring structure of PAH and similar, complementary portions of the HA molecule, have been recently suggested for the binding of PAH to OM by Gauthier et al. (1987a) in their fluorescence quenching studies of interactions between PAH and HS. PCBs were also suggested to bind to HS by charge-transfer complexing (Strek and Weber 1982).

Covalent Binding. Formation of covalent bonds which lead to stable, mostly irreversible incorporation of TOCs into HS, or more likely of their toxic degraded reaction intermediates or products (e.g. anilines and phenols) have been shown to occur, often mediated by chemical, photochemical, or enzymatic catalysts.

Acylanilides, phenylcarbamates, phenylamide, phenylureas and analogous herbicides are known to be biodegraded in soil with the release of free chloro-aniline residues. These residues were shown to be prevalently immobilized by chemical binding to the soil OM without the intervention of microbial activity (Hsu and Bartha 1974, 1976). The chemical attachment of chloroanilines to HS was proposed to occur by at least two mechanisms involving carbonyl, quinone and carboxyl groups of HS and leading to hydrolyzable (probably anil, a Schiff base, and anilinoquinone) and to nonhydrolyzable (probably heterocyclic rings or ether) bound forms. In his study on the reaction of several ring-substituted anilines with HS in aqueous solution, Parris (1980) proposed that the primary amines interact with HS in a pathway consisting of an initial rapid, reversible formation of imine linkages with the HS carbonyl, followed by a slow, irreversible 1,4 nucleophilic addition to quinone-like rings which is then followed by tautomerization and oxidation to yield an amino-substituted quinone. The slow reaction with HS was also given by secondary amines.

A structural analogue of chloroaniline, the carcinogenic 3,3'-dichlorobenzidine, widely used in the production of dyes and pigments, has also been demonstrated to form covalent linkages with soil HS by a mechanism similar to that suggested for the chloroanilines (Hsu and Bartha 1974, 1976; Katan et al. 1976).

Pesticides that do not originally contain aromatic amines may be bound to HS in a similar fashion if they can be microbially transformed to aromatic amines. This was demonstrated for organophosphate insecticides such as parathion and methylparathion (Katan et al. 1976; Lichtenstein et al. 1977), dinitroaniline herbicides, and nitroaniline fungicides (Wang and Broadbent 1973; Van Alfen and Kosuge 1976).

Enzyme-mediated, oxidative cross-coupling mechanisms leading to stable covalent incorporation of various aniline and phenols arising during the biodegradation of various pesticides (i.e. PCB) into HS has also been ascertained and will be discussed in the next section.

Recently, the binding mechanism of a number of water dissolved chlorophenoxyalkanoic acid and ester herbicides by aqueous suspensions of HA originating from soils and other sources has been studied by ESR spectroscopy (Senesi et al. 1984, 1986a, 1987a). ESR measurements showed a considerable quenching of the original HS free radical concentrations and a broadening of the resonance linewidths in the products of interaction between the herbicide and HA. These results suggested that homolytic cross-coupling reactions, leading to incorporation through the formation of strong covalent bonds, should have occurred between indigenous free radicals and highly reactive phenoxy and/or aryloxy radicals originated from the herbicide (Senesi et al. 1984). Chlorophenoxyalkanoic compounds are in fact known to generate these kinds of free radicals as intermediates in the course of their chemical, photochemical and/or enzymatic oxidative degradation (Bollag et al. 1980). The presence of indigenous inorganic catalysts (i.e. cupric or ferric ions) and residual enzymatic (phenoloxidase) activity, maintaining the ability to mediate chemically or biologically the oxidative degradation of these herbicides has been ascertained for HS. It has also been demonstrated that phenoxy-type radicals may be photochemically generated from phenoxy-alkanoic herbicides in the initial oxidation step of the non-biological degradation process which they undergo in aqueous solution and in the presence of light and air (Slawinska et al. 1975; Crosby 1976b; Zepp et al. 1981).

HA of different origin and nature showed different levels of interacting power in the homolytic reaction with chlorophenoxy residues, assuming that the interacting ability is proportional to the extent of lowering of the residual free radical concentration in the interaction products. The average order of reactivity of HA according to origin is: synthetic > peat > soil > coal > compost (Senesi et al. 1986a). In addition, the reactivity of HA towards chlorophenoxy herbicides was found to be negatively correlated with the carboxyl content and COOH to phenolic OH ratio of HA (Senesi et al. 1986a, 1987a). The molecular structure of the chlorophenoxy units was shown to affect the reactivity for the coupling with HA. Most likely, the presence of increasing number of chlorine atoms on the phenoxy ring interferes to a certain extent with cross-linking to the humic macromolecule (Senesi et al. 1986a).

The increased adsorption of s-triazines on HA at high temperatures (McGlamery and Slife 1966) was interpreted by Hayes (1970) as an indication of the formation of covalent bonds. The significant increase in free radical contents together with the enrichment with nitrogen measured by Hayes et al. (1975) when HA and FA were extracted or treated after extraction with ethylenediamine were tentatively ascribed to the formation of covalent bonds by condensation of the primary amine with HA-carbonyl groups and/or by reaction of the amine with HA-quinone units in the presence of oxygen. Senesi and Testini (1982, 1983a) and Senesi et al. (1987b) suggested that the enrichment of free radical concentrations and the enlargement of the ESR signal linewidths were due to stabilization of free electrons onto the extended aromatic structure attained in consequence of the covalent binding between amino groups of s-triazine with carbonyl or quinone groups of the HA.

Enzyme-Mediated Bonding. An early study by Wolf and Martin (1976) showed that melanic fungi incorporated significant amounts of the ring portion of 2,4-D into self-produced HS. Similarly, Mathur and Morley (1978) observed that the insecticide methoxychlor was incorporated into model HA during their chemical synthesis from hydroquinone. Both research groups ascertained the binding of the pesticides, or their transformation products, into HS-like compounds but the nature of the chemical binding was not determined.

Enzymatically-mediated oxidative coupling reactions, which are universally recognized to be important in the synthesis of HS, may also be responsible for the incorporation of many agricultural and industrial TOC, and more likely of their reactive degradation intermediates (e.g. phenols and anilines), into soil OM and thus are important in determining the fate of many TOCs in soils.

Oxidative-coupling enzymes are metal-containing and belong to the groups of monophenol monooxygenases and peroxidases, that may catalyze the oxidation and covalent linking of phenolic compounds and aromatic amines together or to HS. The most important class of monooxygenases in soil are laccases which contain copper and require oxygen for their activity, while most peroxidases contain a heme group and require H_2O_2 for their activity.

Phenolic compounds are thought to be coupled by laccases and peroxidases by a mechanism involving free radicals (Sjoblad and Bollag 1981). The initial step of the enzymatic catalysis involves the production of an aryloxy radical from phenol by removal of an electron and hydrogen ion from the hydroxyl group as shown in Eq. (3).

OH   (−e),(−H⁺)/(ENZYME) → aryloxy radicals

PHENOL   ← ARYLOXY RADICALS →

Formation of aryloxy radicals from phenol

Free radical intermediates may then be converted to stable products by self-coupling or cross-coupling with other radical species, i.e. indigenous humic free radicals, becoming incorporated into the HS macromolecule.

Bollag (1983, 1987) provided clear evidence that covalent binding of substituted phenols and aromatic amines to humic monomers and to HS occurred readily in the presence of various phenoloxidases. An extracellular laccase isolated from the fungus *Rhizoctonia praticola* was shown to be able to mediate cross-coupling between phenolic constituents of HS and 2,4-dichlorophenol formed during the decomposition of 2,4-D, thus leading to the incorporation of this xenobiotic into soil OM (Bollag et al. 1980). The fungal enzyme from *R. praticola* was also able to catalyze the oxidative coupling of pentachlorophenol (PCP) and syringic acid, a representative of carboxylic phenols occurring in HS structures. The catalytic effect of laccase was suggested to be through the formation of a radical, but it was difficult to determine whether subsequent coupling reactions were of biochemical or pure chemical nature (Bollag et al. 1980). Liu and Bollag (1985) showed also that a laccase from the fungus *Trametes versicolor* could catalyze the copolymerization of syringic acid and 2,6-xylenol, a major pollutant in streams and other water resources, which is known to be toxic to fish and other organisms.

It has also been shown that various chloro- and alkyl-substituted anilines which represent the aromatic base of a large number of TOCs readily reacted with phenolic humic constituents in the presence of a phenoloxidase isolated from *R. practicola*, while no reaction occurred when only the anilines were incubated with the fungal laccase (Bollag et al. 1983). Evidence was obtained that anilines were bound via an imine linkage or through a N-C single bond to a quinone originating from various humic constituents (Bollag et al. 1983; Bollag 1987). Suflita and Bollag (1980, 1981) succeeded in extracting phenoloxidase-like catalysts from soil that were capable of polymerizing phenolic substances in a manner analogous to the fungal laccase. No evidence was available for the enzymatic binding and incorporation of triazine or phenylurea residues in HS (Bollag and Loll 1983).

Stott et al. (1983) showed that chlorocatechols, known intermediates in the decomposition of 2,4-D and 2,4,5-T and other pesticides were readily incorporated by enzymatic polymerization into HA polymers when reacted with purified horseradish peroxidase. Recently, results presented by Berry and Boyd (1984) suggested that the degree to which anilines and phenols become bound to soil HS through horseradish peroxidase-mediated oxidative coupling reactions might be affected by substituent groups on the aromatic ring. The authors found that in some cases, e.g. trifluralin, a dinitroaniline herbicide, these reactions may be inhibited due to the presence of NO_2 which is a strong electron-withdrawing group. In contrast, reactivity would be facilitated by electron-donating groups such as OCH_3, which are common substituents on HS. In another study on the peroxidase-mediated oxidative coupling operative in soil between mono-substituted anilines and soil OM, Berry and Boyd (1985) showed that the reactivity of aniline, nitroaniline and chloroaniline was greatly enhanced in the presence of a highly reactive electron donor such as ferulic acid, a lignin-derived phenol containing the acrylic group, which is a potential humic phenolic constituent. The authors (Berry and Boyd 1985) suggested, as the most plausible mechanism to explain their results a secondary chemical reaction between anilines and intermediates or products released during the enzymatic oxidation of phenolic electron donors. These observations, together with previous suggestions of Bollag et al. (1983) of a secondary mechanism for the laccase-mediated cross-coupling of phenols and chloroanilines, indicated that aniline binding to soil humic components results from a chemical reaction which might be only indirectly controlled by the presence of oxidative coupling enzymes in soils (Berry and Boyd 1985).

In conclusion, because practically all aromatic TOCs that release phenols or anilines in the course of their degradation could bind HS through enzymatic catalysis, methods employing enzyme-catalyzed polymerization reactions as processes for minimizing the presence, by partial removal of TOCs in aquatic and terrestrial environments might be utilized in pollution control (Bollag 1987).

Hydrophobic Adsorption. Hydrophobic adsorption has been proposed as a mechanism for retention of nonpolar TOCs or TOCs having predominant nonpolar to polar regions, by hydrophobic surfaces of HS. This kind of adsorption originated from a weak solute-solvent interaction, i.e. low solubility, or hydrophobic nature, of the solute. Water molecules will not be good competitors with nonpolar molecules for adsorption on HS hydrophobic surfaces. Walker and Crawford (1968) referred to the "squeezing out" of nonpolar solute molecules from aqueous solutions and its accumulation at any solid interphase where competition with the solvent is negligible.

Hydrophobic active sites of HS include aliphatic side chains or lipid portions and lignin-derived moieties with high carbon content and a small number of polar groups (Schnitzer and Khan 1972; Walker and Crawford 1968). Hydrophobic retention needs not be an active adsorption but can also be regarded as a partitioning between a solvent and a non-specific surface, as will be discussed in the next section. Adsorption by this mechanism would be independent of pH (Hance 1965; Walker and Crawford 1968) but would increase with methylation of the HS that blocks hydrophylic groups (Hance 1965).

Adsorption by OM and HS through this mechanism has been suggested for DDT and other organochlorine insecticides (Pierce et al. 1971), diakylphthalates (Matsuda and Schnitzer 1971; Khan and Schnitzer 1972), leptophos, methazole, norflurazon, oxadiazonon, butralin and profluralin (Carringer et al. 1975), metolachlor (Kozak 1983), chlorinated hydrocarbons (Pierce et al. 1971), picloram and dicamba (Khan 1973a), 2,4-D (Khan 1973b), parathion (Leenheer and Aldrichs 1971), phenylcarbamates (Briggs 1969), substituted anilines (Weber 1972) and PCBs (Strek and Weber 1982).

Hydrophobic association has also been proposed as a possible mechanism for adsorption of the s-triazines (Walker and Crawford 1968) and the phenylureas (Hance 1965; Khan and Mazurkewitch 1974) by soil OM, but for these pesticides most of the evidence favors higher bond strength with the HS.

3.5.2 Partitioning

Reversible physical adsorption of hydrophobic pollutants with dissolved-phase and solid-phase HS is a well-established and fundamental interaction, affecting equilibrium distribution and rate of an organic pollutant between soil or sediment, water, biomass and air. There is continuous discussion in the literature as to whether physical association of hydrophobic TOC with sediment and soil involves a process of adsorption or partitioning (Chiou et al. 1983; Mingelgrin and Gerstl 1983; Chiou et al. 1984). We shall focus in this section on partitioning and criticisms to this approach.

According to Weed and Weber (1974) the process of accumulation of nonpolar molecules at hydrophobic surfaces in soil OM, i.e. aliphatic side chains on HA and FA, is called unproperly "hydrophobic bonding" but is actually a partitioning between a solvent and non specific surfaces rather than an active adsorption process. Earlier studies by Lambert (1967) expressed sorption of neutral organic pesticides by soils in terms of partition coefficients between soil OM and water, neglecting mineral soil constituents. Lambert (1967) was the first to propose that the role of soil OM was similar to that of an organic solvent in solvent extraction and that the partitioning of an organic compound between soil OM and water should correlate with its partitioning between water and an immiscible organic solvent. Successively, Swoboda and Thomas (1968) suggested that partition uptake by soil was mainly due to partitioning in the soil OM. Recently, Chiou et al. (1979, 1983, 1985) suggested that the controlling sorptive mechanism of nonionic organic compounds from water consists primarily of solute partition, rather than adsorption into the soil humus. This concept was principally based on results from sorption studies performed for a number of chlorinated hydrocarbons, benzene derivatives and PCBs on specific soils. The results showed, in fact, that sorption isotherms were linear over a wide range of aqueous concentrations relative to solute solubilities, and that soil uptake of organic solutes exhibited a small heat effect with a lack of apparent solute competition. Nevertheless, the authors (Chiou et al. 1983) considered that the solute partition coefficient, K_{om}, might be affected by the network and polarity of soil HS and by variations of HS properties with different soil conditions.

In brief, partitioning interaction is modeled as an equilibrium reaction, similarly to the partitioning of a solute between two immiscible solvents (Yalkowski and Valvani 1979; Chiou et al. 1979; Karickoff 1980; Schwarzenbach and Westall 1981).

In other words, HS both in the solid- and dissolved-phase, are treated as a non-aqueous solvent into which the organic pollutant can partition from water. The distribution of organic pollutant between aqueous solution and organic carbon component of soils and sediments may be described, therefore by the use of partitioning equilibrium constants (K_{oc} or K_{om}). The fraction of pollutant in the organic phase, x_{oc}, will be given by:

$$x_{oc} = \rho_{oc} K_{oc}/(1 + \rho_{oc} K_{oc}) \qquad (4)$$

where ρ_{oc} is the humic organic carbon-to-water weight ratio and K_{oc} is the partition coefficient (Perdue 1987). For example, according to this equation in a water sample with a dissolved organic carbon content of 100 mg L^{-1} ($\rho_{oc} = 10^{-4}$), a pollutant with a K_{oc} greater than 10^4 will exist primarily in association with the HS. For a more complete discussion of the treatment of partitioning equilibria in the environment, the reader is referred to Karickhoff (1981).

An important advantage of this approach is the fact that K_{oc} values can be closely correlated with octanol/water partition coefficients, K_{ow}, and water solubilities, thus facilitating the estimation of K_{oc} values that have not been experimentally determined (Chiou et al. 1979).

Earlier studies by Briggs (1969) showed that alkyl-N-phenylcarbamate sorption on soils could be depicted as an accumulation at hydrophobic sites at the OM/water interface in a way similar to surface active agents and that Hansch's π constants (Fujita et al. 1964), derived from partition distribution between 1-octanol and water, expressed this behavior better than other parameters (i.e. parachor and water solubility). Later, Briggs (1973, 1981) found that soil adsorption from aqueous solutions of a number of phenylurea and anilide herbicides could also be predicted from their octanol/water partition coefficients. Excellent linear correlations between K_{oc} and K_{ow} were found for a variety of nonpolar organic compounds, including various PAHs and halogenated alkenes and benzenes, and various soils and sediments that were investigated for sorption (Karickhoff et al. 1979; Schwarzenbach and Westall 1981). Nys and Rekker (1974) presented a method by which the K_{ow} of PAH could be calculated on the basis of molecular structure. The method involved summation of hydrophobic fragmental constants (or f-values) for all groups in a molecule of a specific compound. The relationship between molecular structure and K_{ow} is expressed as: $\log K_{ow} = \sum_{i=1}^{n} a_i f_i$, where n is the number of carbon atoms or fragments in a compound, and a_i are numerical factors indicating the incidence of a given atom or fragment in the structure. The f-values developed by Nys and Rekker (1974) are statistically derived numbers based on experimental measurements of K_{ow}. Yalkowski and Valvani (1979) used this method to compute $\log K_{ow}$ values for 32 PAHs of wide ranging molecular weight and showed that the calculated $\log K_{ow}$ values were quantitatively correlated to the PAH solubility data of MacKay and Shiu (1978). Chiou et al. (1977) and Means et al. (1980) also arrived at the conclusion that the log of PAH aqueous solubility, calculated from molecular and physical properties, is a good estimation of organic-water partitioning, and, therefore, it should be a reliable means of estimating $\log K_{oc}$. In a recent review on the information available on adsorption of PAHs in water/soil or sediment systems, Dzombak and Luthy (1984), concluded that estimation of K_{oc} based on K_{ow} values computed by means of hydrophobic fragmental constants, represents a useful approach for the evaluation of adsorption of lower molecular weight PAH, which have the highest mobility in groundwater. The authors (Dzombak and Luthy 1984) suggested that estimates of K_{oc} based on aqueous solubility seemed more reliable than estimates based on K_{ow} for the higher molecular weight PAHs. However, the use of large-group fragmental constants is indicated as a promising technique to predict approximations of log K_{ow} for higher molecular weight PAHs (Dzombak and Luthy 1984). Nevertheless, the authors propose to exercise caution in applying the empirical equations, due to the lack of fundamental understanding of the interactions of compounds such as PAHs with soil OM. The authors (Dzombak and Luthy 1984) emphasized the need for having detailed knowledge of the composition of soil OM in order to accurately predict adsorption characteristics for a particular compound. They concluded that until more information is available on HS, it will not be possible to clearly identify specific mechanisms of interaction of dissolved organic compounds with soil OM.

3.5.2.1 Critical Evaluation of the Partition Model

The evidence presented in the literature on the dominance of a partition mechanism in the process of adsorption of nonionic TOC on soil OM appears insufficient to prove its general application and to exclude a physical adsorption model based on weak chemical forces of interaction (Mingelgrin and Gerstl 1983; Chiou et al. 1984).

Mingelgrin and Gerstl (1983) in a comprehensive paper have critically reconsidered the universal applicability of the partitioning model to nonionic compounds. The authors question the model which was based mainly on the assumption that accurate estimation of the distribution coefficients, K_{oc} or K_{om}, were possible from solvent-water partition coefficients, K_{ow} or from aqueous solubility data. In particular, Mingelgrin and Gerstl (1983) have reevaluated available data and theoretical assumptions leading to the partitioning model. Thermodynamic arguments used to differentiate between partition and adsorption (Chiou et al. 1979) were over simplified and cannot be used to distinguish between adsorption or partition. This is mainly due to the fact that enthalpy (ΔH) and entropy (ΔS) values may vary in magnitude and sign. The observed linearity of adsorption isotherms and absence of competitive effects (Chiou et al. 1984) are not evidence for partition, because such behavior is also consistent with a physical adsorption model (Briggs 1969; Mingelgrin and Gerstl 1983). Since soil OM is not uniform in all soils it cannot be treated as a well-defined organophylic phase. The appreciable variation of reported K_{om} values for many nonionic compounds between soils with change in soil OM composition is a strong argument limiting the universality of the partition model (Mingelgrin and Gerstl 1983; Chiou et al. 1984). Recent data (Felsot and Wilson 1980; Hasset et al. 1980a) demonstrate that deviations of an order of magnitude or more from the calculated fit between K_{om}

and solubility or K_{ow} exist, thus the use of these parameters to predict K_{om} should be cautiously treated.

Mingelgrin and Gerstl (1983) also proved that partition coefficients may be inconvenient parameters to predict adsorption even when they correlate well with K_{om} and when K_{om} is approximately constant within a group of soils. In addition, the molecular structure and conformation of the solute is a property which may affect adsorption onto a solid surface and partition into an organic lipid phase differently, thus hindering the expected correlation between K_{om} and K_{ow} (Mingelgrin and Gerstl 1983).

In summary, correlations between solubility, liquid-liquid partition and soil uptake have been shown to be insufficient proof of a partition process and do not allow predictions applicable to all diverse groups of various TOC and soils (Mingelgrin and Gerstl 1983). The complexity of uptake processes on soil surfaces cannot be simply defined as adsorption or partition, but rather it should be viewed as summation of the many possible mechanisms which are determined by the structural and chemical parameters of the adsorbates and adsorbants. Estimations of partitioning uptake based on K_{om} correlations with K_{ow} or solubility are acceptable as long as the limitations of these correlations are taken into consideration.

3.5.3 Solubilization Effects

Gschwend and Wu (1985) observed that partition coefficients, K_{oc}, for a group of model hydrophobic organic compounds (PCBs) remained constant over a wide range of solid-to-solution ratios, if precautions were taken to eliminate or account for organic macromolecules (i.e. dissolved HS) which remains in the aqueous phase during laboratory sorption tests. Thus, any description of the equilibrium environmental speciation of hydrophobic organic compounds in natural waters should include not only dissolved and sorbed to suspended sediment fractions, but also a component sorbed to non-settling microparticles or organic macromolecules (DOM).

Results of a study by McCarthy and Jimenez (1985b) demonstrated that dissolved HS could reversibly bind some PAH with an affinity comparable to that of sediment HS and that binding to dissolved HS appeared to be more rapid than comparable binding to sediment. The presence of dissolved HS might be expected therefore to increase the amount of hydrophobic contaminant that will bind to suspended particles or sediment, thus altering the fate and transport of TOC in aqueous systems.

The extent of binding of many classes of TOCs, including PAHs, PCBs, PAEs, DDT and lindane by dissolved HS was shown to increase with an increase in its octanol/water partition coefficient, K_{ow}, or with a decrease in its water solubility (Carter and Suffet 1983; Landrum et al. 1984; McCarthy and Jimenez 1985b; Chiou et al. 1986). Chiou et al. (1986) recently reported that the apparent water solubility of a number of selected water-insoluble TOC, including DDT, PCBs, lindane and 1,2,3-trichlorobenzene, increased linearly with concentration of dissolved HA and FA from soil and aquatic origins and showed no competitive effects between solutes. Chiou et al. (1986) described these results in terms of a partition-like interaction of the solute with the "microscopic organic environment" of the high-molecular-weight dissolved HS. However, the effectiveness of such an interaction appeared to be largely controlled by the molecular size and polarity of the HS, i.e., the cosolute macromolecule must be sufficiently large and possess a sizeable intramolecular nonpolar environment to promote solute partitioning (Chiou et al. 1986). The calculated partition coefficients, K_{om}, for given solutes with soil HA were about four times higher than the values with soil FA and five to seven times higher than the values with aquatic HA and FA. A lower concentration of a small-sized organic cosolute is not expected, therefore, to produce significant solubility enhancements regardless of its polarity (Chiou et al. 1986). This assumption appears to be in contrast with the other interaction possibilities offered to the solute (TOC) by the cosolute (DOM) examined in other sections of this publication.

A large number of recent publications have emphasized the importance of the source, and of molecular and structural properties of DOM as factors controlling extent of HS interactions with TOC in water systems. Landrum et al. (1984) and Morehead et al. (1986) found that the partition coefficients for several TOCs, including DDT, some PAHs, PCBs, and a phthalate, bound to HS from different sources did not depend on pollutant concentration but were inversely proportional to the concentration of HS in solution and were related to the source of the DOM. The partition coefficient for natural water DOM was about one order of magnitude lower than that determined for a commercial HA at similar concentrations. Carter and Suffet (1982) found that a significant fraction of the dissolved DDT found in natural waters might be bound to dissolved HS. The association constants or partition coefficients measured covered a wide range of values showing that the extent of binding for three samples isolated from natural waters was less than that for HA isolated from soil and sediment. The data of Carter and Suffet (1982) showed also that there was more bound DDT at lower pH levels, and that the presence of calcium and high ionic strength also caused an increase in the association constant. These effects indicated that as the charge is neutralized the humic macromolecule becomes less hydrophylic and thus would bind hydrophobic compounds more effectively.

In a successive study on the interactions between dissolved HA and FA and a number of compounds, including PAH, PAE, DDT and lindane, Carter and Suffet (1983) found that different HS bind compounds to a significantly different extent. HA appeared to bind compounds to a greater extent than FA, but there were large differences between various HA and FA. While the extent of binding was imputed to the nature of HS, various attempts that were made by the authors (Carter and Suffet 1983) to determine what characteristics of the HS resulted in the differences in binding, were unsuccessful. Thus, they concluded that with the data currently available it was not possible to predict the strength at which a particular sample of HS would bind a pollutant without actually measuring the binding constants. These were shown to be strongly related to K_{ow} for a particular compound. Thus it was predicted that compounds with log K_{ow} lower than 4 (such as lindane) will probably not be bound to an appreciable extent, while compounds with a high log K_{ow} values (DDT and DEHP) may be bound to a significant extent.

Gauthier et al. (1986, 1987a,b) have shown that partition coefficients (K_{oc}) for the binding of pyrene to 14 different HA and FA originating from estuarine and off-shore sediments, soil or commercial sources, varied by as much as a factor of 10 depending on the chemical and/or structural characteristics of the HS. The binding of pyrene to dissolved HA and FA gave results highly dependent on their degree of aromaticity, as was suggested by the good correlation of this parameter with the magnitude of K_{oc} values. Other important variables affecting K_{oc} values for pyrene were shown to be the elemental and functional group composition of the HS. Thus, binding of TOC to DOM appeared to involve, at least to some extent, specific interactions between TOC and DOM. Because binding seemed to involve specific interactions, the authors concluded that additional parameters accounting for compositional and structural variations of the HS should be considered in any attempts to model pollutant-DOM distribution coefficients used to predict the transport and fate of hydrophobic TOC in the aquatic systems.

In conclusion, a number of models have been proposed in attempts to predict partition coefficients of TOC to sediment and soil OM and to DOM on the basis of aqueous solubility or K_{ow} of solutes. These models consider almost exclusively the quantity of carbon in the sorbent but ignore chemical properties and structure of the DOM. It seems that a much more detailed parameterization of the sorbent is required to the establishment of future environmental fate models for hydrophobic TOCs.

3.5.4 Hydrolysis Effects

The catalytic effects of HS on dechloro-hydroxylation of the chloro-*s*-triazine herbicides in soils and aqueous solutions have been extensively studied although their mechanism is not completely understood. Armstrong et al. (1967) first suggested that the formation of H-bonding between the ring- or side-chain nitrogens of the *s*-triazine and HA-surface acid groups will cause further electron-withdrawing from the electron-deficient carbon atom already surrounded by electronegative nitrogen and chlorine atoms, thus enabling the weak nucleophile water to replace the chlorine atom and thereby increasing the rate of hydrolysis. This mechanism explains the correlation established between soil-catalyzed hydrolysis of chloro-*s*-triazines and the amount adsorbed by soil OM (Burkhard and Guth 1981). Successively, Li and Felbeck (1972) measured a first order kinetic reaction with respect to herbicide concentration for the hydrolysis of atrazine at pH 4 in an aqueous suspension of HA. The half-life of atrazine varied non-linearly with concentration of HA. The formation of H-bonding between HA and atrazine was suggested as being responsible for the observed decrease of the activation energy barrier for atrazine hydrolysis. Because no correlation was found between the rate of atrazine hydrolysis at pH 4 and the number of carboxyl groups or total acidity of different HA, the authors (Li and Felbeck 1972) concluded that the catalytic effect of HA could depend not only on the number of effective acid groups but also on the arrangement of these groups on the HA molecule. Khan (1978a) found that the hydrolysis of atrazine in aqueous FA solution at pH $\leqslant$ 7 also followed first order kinetics with respect to the herbicide concentration. Increase in FA concentration resulted in a higher hydrolysis rate constant and a shortened half-life, but had no effect on the activation energy, which, however, increased with increase in pH of the reaction mixture. The author (Khan 1978) speculated that a change in pH of the FA solution would change the types and concentrations of acidic functional groups involved in the hydrolysis of atrazine, which in turn might affect the mechanism of hydrolysis as indicated by the change in the activation energy.

Weber et al. (1969) provided evidence that undissociated carboxyl groups of soil OM were the only catalytically active sites, leading directly to catalyzed hydrolysis of the H-bonded atrazine, while phenolic groups had no catalytic effects. More recent studies by Gamble and Khan (1985), based on the knowledge of the types, numbers and pK_a values of acidic functional groups in a quantitatively characterized FA, confirmed that hydrogen ions and undissociated carboxyl groups were the only catalytic agents for atrazine hydrolysis. No evidence of catalysis was observed by weakly acidic functional groups having very high pK, such as phenolic OH, and carboxylate ions (Gamble and Khan 1985).

Evidence for catalytic effects of HS on the rate of hydrolysis of other TOCs is extremely limited. Perdue and Wolfe (1982) demonstrated that aquatic HS inhibited the base-catalyzed hydrolysis of the n-octyl ester of 2,4-D (2,4-DOE). The rate of hydrolysis of 2,4-DOE at pH 9 to 10 decreased by a factor equal to the fraction of the ester associated with the HS. These observations were consistent with an unreactive humic-bound 2,4-DOE in equilibrium with reactive aqueous-phase 2,4-DOE. Thus, association between dissolved HA and 2,4-DOE inhibited the base-catalyzed hydrolysis reaction. Further support to these conclusions is furnished by lower alkaline hydrolysis rates observed by Macalady and Wolfe (1984, 1985, 1987) for sediment OM-associated organophosphorothioate esters compared to the rate in the aqueous phase. The same authors also showed that neutral (pH-independent) hydrolysis rates of several classes of organic compounds were unaffected by association with sediment HA.

Recently, Perdue (1983) proposed a general mechanism for the effects of HS on the hydrolysis kinetics of hydrophobic organic pollutants. The overall catalysis model for HS provided by Perdue (1983) is derived by a combination of equations that separately describe partitioning equilibria, general acid-base catalysis and micellar catalysis. The resulting model predicted that the overall effect of HS in modifying hydrolysis reaction rates of TOC may be almost totally attributed to partitioning equilibria and micellar catalysis, with only a minor effect due to general acid-base catalysis. General acid-base catalysis by HS is predicted by the model to be relatively unimportant, and to remain insignificant even in the presence of rather high concentrations of HS (e.g. 200 mg L^{-1}) when other processes such as partitioning or association equilibria may become significant for hydrophobic pollutants. Application of the model to experimental data obtained for the base-catalyzed hydrolysis of 2,4-DOE by Perdue and Wolfe (1982) and the acid-catalyzed hydrolysis of atrazine by Li and Felbeck (1972) indicated that HS may alter the reactivities of bound substrates in a way similar to that of anionic surfactants (inhibiting base-catalyzed and accelerating acid-catalyzed reactions). These effects were attributed to electrostatic stabilization of the transition state for the acid catalysis in which the substrate becomes more positively charged, and to destabilization of the transition state for base-catalyzed hydrolysis in which the substrate becomes more negatively charged (Perdue 1983). The overall catalysis model was also used by Perdue (1983) in a predictive fashion. It was estimated that in natural waters the base-catalyzed hydrolysis rate of parathion, a weakly HS-associated pollutant, will not be significantly affected by HS, while for more strongly associated pollutants, such as DDT, the effect of HS would clearly be potentially significant in this reaction. In conditions where much higher concentrations of HS are possible, i.e. in sewage sludge or in sediment and soil interstitial water, the impact of HS on TOC hydrolysis kinetics was predicted to be larger. Hydrolysis reactions of both parathion and DDT were expected, therefore, to be strongly retarded in these environments.

Finally, the inhibition effect exerted by HA on hydrolytic enzymes in soils indicated by some authors (Malini de Almeida et al. 1980; Mulvaney and Bremner 1978) may be regarded as an additional mechanism by which HA may indirectly influence hydrolysis reactions.

3.5.5 Photosensitization

HS are one of the primary sunlight adsorbing chemical species in freshwater and seawater, thus representing one of the major photochemically active materials in natural water bodies (Zepp et al. 1985). A significant portion of the solar radiation adsorbed by aquatic HS results in the formation of electronically excited molecules (HS*) that are capable of greatly accelerating or even determining a number of light-induced transformations which TOCs may undergo in natural waters (Slawinski et al. 1978a,b; Wolff et al. 1981; Zepp et al. 1981, 1985). In surface waters aquatic HS can act both as sensitizers or precursors for the production of singlet oxygen ($^1O_2$), humic-derived peroxy radicals (ROO·), hydrogen peroxide and solvated electrons (e^-_{aq}), and as the scavenger which controls their lifetimes (Hoignè et al. 1987).

A proposed mechanism taking place when an excited sensitizer (HS*) interacts with an energy acceptor can be described by the key energy-transfer steps depicted in the following scheme:

$$HS \xrightarrow{h\nu} {}^1HS^* \to {}^3HS^* \qquad (5)$$
$$^3HS^* \to HS + heat \qquad (6)$$
$$^3HS^* + TOC \to TOC^* + HS \qquad (7)$$
$$TOC^* \to photoproducts \qquad (8)$$
$$^3HS^* + O_2 \to HS + {}^1O_2 \qquad (9)$$
$$^1O_2 + TOC \to TOC\text{-}O_2 \qquad (10)$$

Light absorption promotes the photosensitizer molecules (HS) to their first excited states $^1HS^*$ which are very short-lived and undergo in part to excited triplet states $^3HS^*$, Eq. (5), which are in turn considerably longer-lived. Such triplets may in part decay to the ground state, Eq. (6), or transfer energy to the substrate (TOC) forming its triplet state (TOC*), Eq. (7), which then gives its photoproducts, Eq. (8), or transfer energy to groundstate triplet oxygen producing excited singlet molecular oxygen $^1O_2$, Eq. (9), which is a powerful oxidant and may in turn decay back to its groundstate or react rapidly with an acceptor (TOC) thus producing its photooxidation products, Eq. (10). Different types of aquatic DOM exhibit different quantum efficiencies of singlet oxygen production. Summarizing studies using a variety of different HS, Hoigné et al. (1987) concluded that HS with higher specific light absorption exhibit somewhat lower quantum efficiencies. However, no significant relationship with molecular weight fraction was found (Haag and Hoigné 1986). The occurrence of singlet oxygen was suggested to be of importance for the elimination of dissociated forms of phenolic compounds (e.g. chlorinated phenols), for cyclic dienes and for sulfur compounds (Hoigné et al. 1987).

Hydroxyl radicals, which are of importance for the elimination of refractive micropollutants (Hoigné et al. 1987), are predominantly consumed by fast scavenging reactions of the DOM present in natural waters (Haag and Hoigné 1985). Different types of aquatic HS were shown to exhibit comparable rate constants for trapping the hydroxyl radicals (Hoigné et al. 1987). Peroxy radical photooxidants, a mixture of different HS-derived species, were shown to be important for the elimination of alkylphenols, which are typical compounds classified as antioxidants.

Direct photo-ionization or photo-induced electron transfer from marine and terrestrial HS to a variety of polyaromatic electron acceptors have been documented by time-resolved and steady-state laser flash kinetic spectroscopy studies under conditions which facilitate extrapolation to the environment (Zepp et al. 1985). Thus, another common photoprocess could be the generation of a solvated electron, followed by the production of the very reactive superoxide radical anion, which finally produces hydrogen peroxide. Because the formation rate of solvated electron from photolysis of DOM is extremely low, it was considered to be relevant only for the elimination of highly refractive compounds (Hoigné et al. 1987). Polychlorinated organic compounds, for example, may undergo dehalogenation upon reaction with solvated electrons.

HS can also photosensitize reactions involving hydrogen atom transfer, which are likely to involve triplet state intermediates (Zepp et al. 1985). Zepp et al. (1981) showed that both commercial and aquatic HA were able to photosensitize reactions involving hydrogen transfer from the nitrogen of aniline to the sensitizer at much higher rates than those observed in the aniline photoreaction in distilled water. In both photoreacting systems azobenzene was found to be a major product resulting from coupling of the anilino-free radical intermediates (Zepp et al. 1981).

Quantitative kinetic data obtained by Zepp et al. (1981, 1985) showed that photosensitized oxygenations of 2,5-dimethylfuran and the insecticide disulfoton in air-saturated natural water samples containing aquatic HS and in distilled water containing soil-extracted or commercial HA and FA, were at least one order of magnitude faster than those in distilled water. The same photooxygenated products were obtained in both natural water samples and solutions of soil-derived HS. The authors also observed that *cis*-1,3-pentadiene gave no detectable isomerization when distilled water solutions of the chemical were exposed to sunlight for one day. With the river water and HA solutions, isomerization proceeded to a photostationary mixture of the trans and cis isomers having the same composition observed when a known triplet energy donor for this reaction was employed as photosensitizer. Up to half the triplet states of HS were estimated to have energies sufficiently high to photosensitize reactions of various TOCs, including polycyclic aromatic hydrocarbons, nitroaromatic compounds, polyenes and diketones (Zepp et al. 1985).

Slawinska et al. (1975) proposed that HS could act as a photosensitizer of some previously-bound substances, that might undergo detoxication stimulated by light and oxygen:

$$HS + TOC \rightarrow HS\text{-}TOC \xrightarrow{h\nu} \text{photoproducts} \qquad (11)$$

through a mechanism analogous to that followed by biologically acting photosensitizers like riboflavin. Zepp et al. (1985) suggested that this mechanism, referred to as "static photosensitization", may act with substrates much more hydrophobic than dimethylfuran or the pentadienes, i.e., DDT, which was the substrate used in the study of Carter and Suffet (1982).

Khan and Schnitzer (1978) found that dissolved FA initially retarded the photolysis of atrazine in water, but photolysis was more extensive in the presence of FA. FA was able, in fact, to photosensitize the further degradation of the initially formed 2-hydroxy-atrazine to N-dealkylated products (Fig. 4) (Khan and Schnitzer 1978). UV irradiation (254 nm) experiments conducted with prometryn showed a first order reaction in distilled water and HA suspension at pH 3, while second-order reaction rate kinetics were observed in the presence of dissolved HA at pH 6 and 8 and FA at pH 3,6 and 8 (Khan and Gamble 1983). An additional, phytotoxic dealkylated product, 4-amino-6-(isopropylamino)-1-triazine, was formed when the photolysis of prometryn was performed in aqueous solutions of HA or FA (Khan and Gamble 1983). The authors suggested that hydroxy radicals generated by UV irradiation of dissolved HS were responsible for the dealkylation of the 2-hydrogen analogue of prometryn.

ESR studies have suggested that visible and UV light irradiation of HS may enhance the indigenous free radical contents of HS (Senesi and Schnitzer 1977,

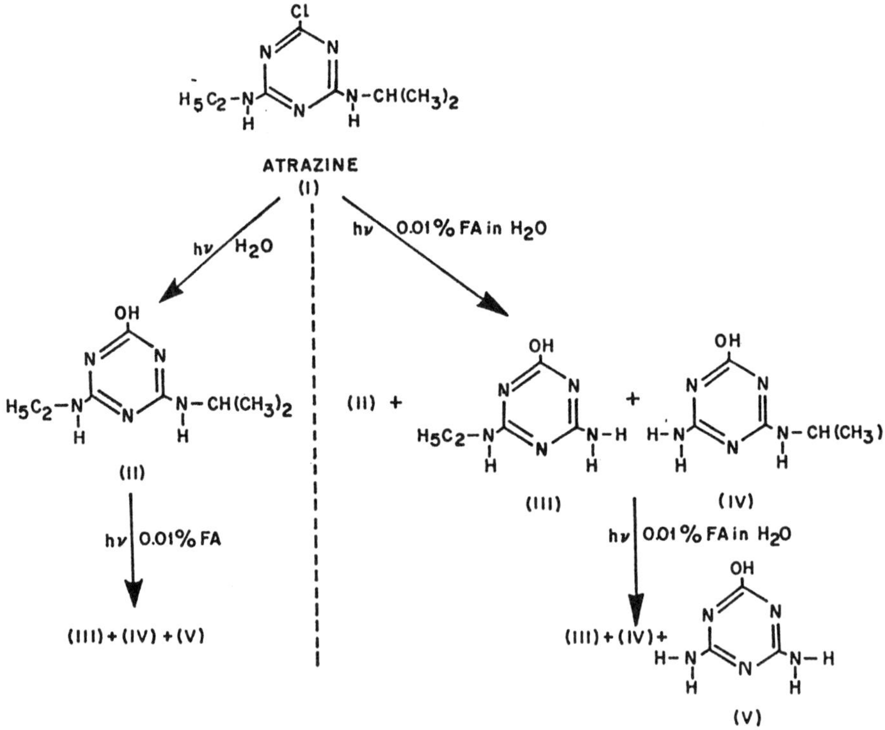

Fig. 4. Pathway of photolytic degradation of atrazine in the absence and presence of FA (after Khan and Schnitzer 1978)

1978; Senesi et al. 1977a) which are highly susceptible to free-radical mediated interaction of HS with TOCs. ESR monitored free radical increase observed in many donor-acceptor systems, such as HA-s-triazine and HA-urea herbicides has been suggested also to be imputable to the unpairing of electrons originating from the formation of charge-transfer complexes under the effect of light as well (Senesi and Testini 1982, 1983a, b, 1984).

Mill and Carter (1987) measured first order kinetics for the photolysis of TPB and its decomposition product diphenylborinic acid (DPBA) in the presence of HA and FA or DOM isolated from stream water. The sensitized enhanced reaction rates for TPB were lower than for DPBA, with HA having a somewhat greater effect. The authors (Mill and Carter 1987) also observed that dissolved oxygen might compete with the organoborates in energy transfer reactions with the excited DOM and thus act as a quenching agent.

Miller et al. (1980) found that solutions of a commercial HA were significantly less active than aquatic HS in the photosensitization of 3,4-dichloroaniline, a metabolite of several important herbicides. This was probably imputed to trapping of the reactive 3,4-DCA intermediates by the HA.

In addition, HS may feasibly be able to accelerate or induce by photosensitization a number of photoreactions that other TOCs undergo in the presence of artificial sensitizers, such as riboflavin, methylene blue, rose bengal, acetone, benzophenone, etc. Obviously, the occurrence of these compounds is unlikely to occur under natural environmental conditions.

3.6 Conclusions and Recommendations

Humic substances interact with TOCs in different ways. Adsorption is probably the most important mode. The reversibility or irreversibility of the adsorption processes is of major importance. The question whether the bound residues are to be considered definitely inactivated was the focus of research and will continue to be in the near future. Have the adsorbed materials lost their toxic identity forever and have they become common components incorporated in the humic polymer, or are they only momentarily inactivated in reversibly bound-forms thus representing a possible source of contamination by a time-delayed release of toxic units? Experimental evidence of at least partial re-mobilization of TOC residues have been furnished: (a) dichloroaniline, a degradation product of the herbicide propanil which is applied to rice fields at an early stage of plant growth and then stabilized in bound forms by HS, was microbially and chemically remobilized and made available to the rice plant during the period of grain ripening; (b) intact methoxychlor was released from the pesticide containing model HS subjected to partial decomposition by the soil fungus *Marasmius oreades*; (c) earthworms and oat plants were able to adsorb solvent-unextractable residues of methylparathion; (d) soil-bound dinitroaniline herbicides were adsorbed by soybeans; and (e) soil-bound methabenzthiazuron residues were biologically available to maize plants. Thus, a better understanding of TOCs adsorption by HS may allow a better evaluation of their influence on the final fate of TOC in the environment.

Other interactions are catalytic effects in hydrolysis and dealkylation reactions of some TOCs and photosensitization effects in various photodegradation reactions of a number of TOCs. These processes lead to the formation of degradation products or intermediates having a mobility, toxicity and persistence which are different form those of products obtained in the absence of HS. More studies are needed in this field to test beneficial or adverse effects of the HS intervention.

Solubilization effects of TOCs produced by interaction with HS may influence the speciation of TOCs in natural systems, in particular their partition between the solid and fine-particulate or colloidal adsorbents. Thus, a deeper study of these processes and factors which control them appears extremely necessary.

The existence of interactions between TOCs and natural HS were demonstrated to give rise to a problem in the analytical, qualitative and quantitative determination of TOCs and in water purification processes from TOCs. Thus, it appears necessary to develop new analytical and process methods which may take into consideration the various aspects of interaction and bring to their solution.

New methods employing enzyme-catalyzed polymerization of TOCs into HS may be extremely useful for removing or minimizing TOC concentrations in aquatic and terrestrial environments. As well, HS may feasibly act as naturally occurring photosensitizer to induce or accelerate a number of photodegradation processes which TOC are known to undergo in the presence of added artificial sensitizers.

The source of HS and their diverse macromolecule size and chemical properties are extremely important in determining the mode and extent of interaction with TOCs. The importance of improving our understanding of the interacting HS and the nature of their interaction with TOCs is recognized but needs further research by a more extended application of advanced techniques, including NMR, for the identification of the structure of TOC-bound residues; and ESR, for the investigation of chemical, enzymatic and photochemical HS-TOC interactions involving free radicals species as starting reagents and/or intermediates, or products of reaction. Fluorescence spectrometry, which allows the study of a number of chemical and functional modifications which occur upon interaction between HS and TOCs in situ, without separation of the interacted TOC molecules from uninteracted ones, provides an important yet scarcely unexploited means for the investigation of TOC and DOM interactions.

This review indicates that interdisciplinary research in which soil scientists, water scientists, analytical chemists and environmentalists will provide joint efforts is essential to elaborate interaction mechanisms between TOCs and DOM, and for the establishment of predictive measures for TOC pollution problems.

References

Aizenshtat Z (1973) Perylene and its geochemical significance. Geochim Cosmochim Acta 37:559–597

Alfen NK Van, Kosuge T (1976) Metabolism of the fungicide 2, 6-dichloro-4-nitroaniline in soil. J Agric Food Chem 24:584–588

Anonymous (1972) Phthalate effect on health still not clear. Chem Eng News 50:14–15

Armstrong DE, Chesters G, Harris RF (1967) Atrazine hydrolysis in soil. Soil Sci Soc Am Proc 31:61–66

Ashton FM, Sheets TJ (1959) The relationship of soil adsorption of EPTC to oats injury in various soil types. Weeds 7:88–90

Autian J (1973) Toxicity and health threats of phthalate esters: review of the literature. Environ Health Perspect 4:3–26

Bailey GW, White JL (1970) Factors influencing the adsorption, desorption, and movement of pesticides in soil. Residue Rev 32:29–92

Baker R, Yohe T, Suffet IH (1987) The competitive effect of humic substances on the adsorption of volatile halogenated organics in drinking water. Am Chem Soc-Div Environ Chem 27 (1):363–364

Ballard TM (1971) Role of humic carrier substances in DDT movement through forest soil. Soil Sci Soc Am Proc 25:145–147

Berry DF, Boyd SA (1984) Oxidative coupling of phenols and anilines by peroxidase: structure-activity relationships. Soil Sci Soc Am J 48:565–569

Berry DF, Boyd SA (1985) Reaction rates between phenolic humus constituents and anilines during cross-coupling. Soil Biol Biochem 17:631–636
Boehm PD, Quinn J (1973) Solubilization of hydrocarbons by the dissolved organic matter in seawater. Geochim Cosmochim Acta 37:2459–2477
Bollag JM (1983) Cross-coupling of humus constituents and xenobiotic substances. In: Christman RF, Gjessing ET (eds) Aquatic and Terrestrial Humic Materials. Ann Arbor Sci Publ, Ann Arbor, MI, pp 127–141
Bollag JM (1987) Enzymatic detoxication through binding of pollutants to humic substances. Am Chem Soc-Div Environ Chem 27 (1):289–290
Bollag JM, Loll MJ (1983) Incorporation of xenobiotics into soil humus. Experientia (Basel) 39:1221–1231
Bollag JM, Liu S-Y, Minard RD (1980) Cross-coupling of phenolic humus constituents and 2,4-dichlorophenol. Soil Sci Soc Am J 44:52–56
Bollag JM, Minard RD, Liu S-Y (1983) Cross-linkage between anilines and phenolic humus constituents. Environ Sci Technol 17:72–80
Bowman MC, Schechter MS, Carter RL (1965) Behavior of chlorinated insecticides in a broad spectrum of soil types. J Agric Food Chem 13:360–365
Briggs GG (1969) Molecular structure of herbicides and their sorption by soils. Nature 223:1288
Briggs GG (1973) A simple relationship between soil sorption of organic chemicals and their octanol/water partition coefficients. Proc 7th Brit Insecticide Fungicide Conf 11:475–478
Briggs GG (1981) Adsorption of pesticides by some Australian soils. Aust J Soil Res 19:61–68
Briggs GG, Dawson JE (1970) Hydrolysis of 2,6-dichlorobenzonitrile in soils. J Agric Food Chem 18:97–99
Bronsted JN, Pedersen K (1924) Die Katalytische Zersetzung des Nitramids und ihre physikalische-chemische Bedentung. Z Phys Chem 108:185–235
Burchill S, Hayes MHB, Greenland DJ (1981) Adsorption of organic molecules. In: Greenland DJ, Hayes MHB (eds) The Chemistry of Soil Processes, Chapt 6, Wiley, New York; pp 221–400
Burkhard N, Guth JA (1981) Chemical hydrolysis of 2-chloro-4, 6-bis(alkylamino)-1,3,5-triazine herbicides and their breakdown in soil under the influence of adsorption. Pestic Sci 12:45–52
Burns IG, Hayes MHB, Stacey M (1973a) Spectroscopic studies on the mechanisms of adsorption of paraquat by humic acid and model compounds. Pestic Sci 4:201–209
Burns IG, Hayes MHB, Stacey M (1973b) Some physico-chemical interactions of paraquat with soil organic materials and model compounds II. Adsorption and desorption equilibria in aqueous suspensions. Weed Res 13:79–90
Calderbank A (1968) The bipyridilium herbicides. Adv Pest Control Res 8:127–190
Carlberg GE, Martinsen K (1982) Adsorption/complexation of organic micro-pollutants and comparison of two analytical methods for analysing organic pollutants in humus waters. Sci Total Environ 25:245–254
Carringer RD, Weber JB, Monaco TJ (1975) Adsorption-desorption of selected pesticides by organic matter and montmorillonite. J Agric Food Chem 23:569–572
Carter CW, Suffet IH (1982) Binding of DDT to dissolved humic materials. Environ Sci Technol 16:735–740
Carter CW, Suffet IH (1983) Interactions between dissolved humic and fulvic acids and pollutants in aquatic environments. ACS 1983
Carter MC (1975) Amitrole. In: Kearney PC, Kaufman (eds) Herbicides: Chemistry, Degradation and Mode of Action. Dekker, New York, pp 377–398
Chen Y, Schnitzer M (1978) The surface tension of aqueous solutions of soil humic substances. Soil Sci 125:7–15
Chiou CT, Freed VH, Schmedding DW, Kohnert RG (1977) Partition coefficients and bioaccumulation of selected organic chemicals. Environ Sci Technol 11:475–478
Chiou CT, Peter LJ, Freed VH (1979) A physical concept of soil-water equilibria for nonionic organic compounds. Science 206:831–832
Chiou CT, Porter PE, Schmedding DW (1983) Partition equilibria of nonionic organic compounds between soil organic matter and water. Environ Sci Technol 17:227–231

Chiou CT, Porter PE, Shoup TD (1984) Comment on "Partition equilibria of nonionic organic compounds between soil organic matter and water". Environ Sci Technol 18:295–297

Chiou CT, Shoup TD, Porter PE (1985) Mechanistic roles of soil humus and minerals in the sorption of nonionic organic compounds from aqueous and organic solutions. Org Geochem 8:9–14

Chiou CT, Malcolm RL, Brinton TI, Kile DE (1986) Water solubility enhancement of some organic pollutants and pesticides by dissolved humic and fulvic acids. Environ Sci Technol 20:502–508

Choudry GG (1984) Humic substances: interactions with environmental chemicals (excluding sorptive interactions). In: Humic Substances, Vol 7. Current Topics in Environ and Toxicol Chem, Gordon and Breach, New York, pp 143–169

Cifrulak CF (1969) Spectroscopic evidence of phthalates in soil organic matter. Soil Sci 107:63–69

Cole JA (1975) Groundwater pollution in Europe. Water Information Center, Port Washington, New York

Corcoran EF (1973) Gas chromatographic detection of phenolic acid esters. Environ Health Perspect 3:13–15

Courtney KD, Gaylor DW, Hogan MD, Falk HL, Bates RR, Mitchell I (1970) Teratogenic evaluation of 1,4,5-T. Science 168:864–866

Crosby DG (1976a) Non-biological degradation of herbicides in soil. In Audus LJ (ed) Herbicides: Physiology, Biochemistry, and Ecology, Vol 2, 2nd edn. Academic Press, London pp 65–97

Crosby DG (1976b) Herbicide photodecomposition. In: Kearney PC, Kaufmann DD (eds) Herbicides: Chemistry, Degradation and Mode of Action, Vol 2. Dekker, New York, Chapt 18, pp 835–890

Crosby DG, Wong AS, Plimmer JR, Woolson EA (1971) Photodecomposition of chlorinated dibenzo-p-dioxins. Science 173:748–749

Dillingham EO, Autian J (1973) Teratogenity, mutagenicity, and cellular toxicity of phthalate esters. Environ Health Perspect 3:81–89

Douding RL, Freeman JF (1968) Residual phytotoxicity of fluomethuron in soils. Weed Sci 16:226–229

Draper WM, Crosby DG (1976) Measurement of photochemical oxidants in agricultural field water. ACS 172 Meet, San Francisco-CA, September 1976

Dubey HD, Sigafus RE, Freeman JF (1966) Effect of soil properties on the persistence of linuron and diphenamid in soils. Agron J 58:228–231

Dzombak DA, Luthy RG (1984) Estimating adsorption of polyciclic aromatic hydrocarbons on soils. Soil Sci 137:292–308

Eliasson L, Hallman V, Tolf E (1969) Leaching of picloram from different soils. Sver Skogsvardsforb Tidskr 67:491–501

Fang SG, Theisen P, Freed VH (1961) Effect of water evaporation, temperature and rates of application on the retention of ethyl-N-N-di-n-propyl thiocarbamate in various soils. Weeds 9:569–574

Farmer WJ, Spencer WF, Shepherd RA, Cliath MM (1974) Effect of flooding and organic matter applications on DDT residues in soil. J Environ Qual 3:343–346

Felsot A, Wilson J (1980) Adsorption of carbofuran and movement on soil thin layers. Bull Environ Contam Toxicol 24:778–782

Foster R (1969) Organic Charge-Transfer Complexes. Academic Press, London, p 470

Fujita T, Iwasa J, Hansch C (1964) A new substituent constant, derived from partition coefficients. Am Chem Soc J 86:5175–5180

Gaillardon P, Calvet R, Tercé M (1977) Adsorption et désorption de la terbutryne par une montmorillonite-Ca et des acides humiques seul on en melanges. Weed Res 17:41–48

Gaillardon P, Calvet R, Gaudry JC (1980) Adsorption de quelques phenylurées herbicides par des acides humiques. Weed Res 20:201–204

Gamble DS, Khan SU (1985) Atrazine hydrolysis in soils: catalysis by the acidic functional groups of fulvic acid. Can J Soil Sci 65:435–443

Gauthier TD, Shane EC, Guerin WF, Seitz WR, Grant CL (1986) Fluorescence quenching method for determining equilibrium constants for polycyclic aromatic hydrocarbons binding to dissolved humic materials. Environ Sci Technol 20:1162–1166

Gauthier TD, Booth KA, Grant CL, Seitz WR (1987a) Fluorescence quenching studies of interactions between polynuclear aromatic hydrocarbons and humic materials. Am Chem Soc-Div Environ Chem 27(1):246–248

Gauthier TD, Seitz WR, Grant CL (1987b) Effects of structural and compositional variations of dissolved humic materials on pyrene K_{oc} values. Environ Sci Technol 21:243–248

Gieseking JE (1975) Soil Components, Vol 1. Organic Components. Springer, Berlin Heidelberg New York, p 534

Giles CH, MacEwan TH, Nakhwa SN, Smith D (1960) Studies in adsorption. Part XI. A system of classification of solution adsorption isotherms and its use in diagnosis of adsorption mechanisms and in measurement of specific surface areas of solids. J Chem Soc 3973–3993

Gjessing ET, Berglind L (1981) Adsorption of PAH to aquatic humus. Arch Hydrobiol 92:24–30

Greiner AC, Spyckerelle C, Albrecht P, Ourisson G (1977) Hydrocarbures aromatiques d'origin geologique. V. Derives mono-et diaromatiques du hopane. J Chem Res 3828–3836

Grice RE, Hayes MHB, Lundie PR (1973) Adsorption of organo-phosphorus compounds by soil constituents and by soils. Proc 7th Brit Insecticide and Fungicide Conf 11:73–81

Griffin RA, Chian ESK (1980) Attenuation of water soluble polychlorinated biphenyls by earth materials. EPA Publ 600/2–80–027, 1980

Gschwend PM, Wu S (1985) On the constancy of sediment-water partition coefficient of hydrophobic organic pollutants. Environ Sci Technol 19:90–96

Gupta RK, Raman S, Raman KV (1985) Adsorption of terbutryne on humic acid. J Indian Soc Soil Sci 33:255–259

Gwo-Chen Li K, Felbeck GT Jr (1972) Atrazine hydrolysis as catalyzed by humic acids. Soil Sci 114:201–209

Haag WR, Hoigné J (1985) Photosensitized oxidation in natural water via OH radicals. Chemosphere 14:1659–1671

Haag WR, Hoigné J (1986) Single oxygen in surface water. 3. Photochemical formation and steady-state concentrations in various types of waters. Environ Sci Technol 20:341–348

Haas CN, Kaplan BM (1985) Toluene-humic acid association equilibria: isopiestic measurements. Environ Sci Technol 19:643–645

Haque R, Lilley S (1972) Infrared spectroscopic studies of charge-transfer complexes of diquat and paraquat. J Agric Food Chem 20:57–58

Hance RJ (1965) Observations on the relationship between the adsorption of diuron and the nature of the adsorbent. Weed Res 5:108–114

Harris CR (1966) Influence of soil type on the activity of insecticides in soil. J Econ Entomol 59:1221–1225

Harvey GR, Boran DA (1985) Geochemistry of humic substances in seawater. In: Aiken GR, McKnight DM, Wershaw RL, MacCarthy P (eds) Humic Substances in Soil, Sediment and Water. Wiley, New York, pp 233–248

Hassett JJ, Anderson MA (1979) Association of hydrophobic organic compounds with dissolved organic matter in aquatic systems. Environ Sci Technol 13:1526–1529

Hassett JJ, Means JL, Banwart WL, Wood SG (1980a) Sorption properties of sediments and energy-related pollutants. EPA Report 600/3–80–041

Hassett JJ, Means JC, Banwart WL, Wood SG, Ali S, Khan A (1980b) Sorption of dibenzothiophene by soils and sediments. J Environ Qual 9:184–186

Hayes MHB (1970) Adsorption of triazine herbicides on soil organic matter, including a short review on soil organic matter chemistry. Res Rev 32:131–174

Hayes MHB, Swift RS (1978) The chemistry of soil organic colloids. In: Greenland DJ, Hayes MHB (eds) The Chemistry of Soil Constituents. Wiley, New York, pp 179–320

Hayes MHB, Swift RS, Wardle RE, Brownn JK (1975) Humic materials from an organic soil: a comparison of extractants and of properties of extracts. Geoderma 13:231–245

Hayes MHB, Thompson JM (1968) Adsorption of s-triazine herbicides by soil organic matter preparations. In FAO/IAEA (eds) Isotopes and Radiation in Soil Organic Matter Studies, Proc Symp FAO/IAEA, Vienna, pp

Hites R (1973) Phthalates in the Charles and the Menimeck rivers. Environ Health Perspect 3:17–21

Hoigné J, Bader H, Nowell LH (1987b) Rate constants of OH radical scavenging by humic substances: role in ozonation and in a few photochemical processes for the elimination of micropollutants. Am Chem Soc-Div Environ Chem 27 (1):208–211

Hoigné J, Faust BC, Haag WR, Zepp RG (1987a) Aquatic humic substances as sources and sinks of photochemically produced transient reactants. Am Chem Soc-Div Environ Chem 27 (1):221–224

Hollist RL, Foy CL (1971) Trifluralin interaction with soil constituents. Weed Sci 19:11–16

Hsu TS, Bartha R (1974) Interaction of pesticides-derived chloroaniline residues with soil organic matter. Soil Sci 116:444–452

Hsu TS, Bartha R (1976) Hydrolysable and nonhydrolysable 2,4-dichloroaniline-humus complexes and their respective rates of biodegradation. J Agric Food Chem 24:118–122

Ishiwatari R (1985) Geochemistry of humic substances in lake sediments. In: Aiken GR, McKnight DM, Wershaw RL, MacCarthy P (eds) Humic Substances in Soil, Sediment and Water. Wiley, New York, pp 147–180

Jamet P, Piedallu MA (1975a) Mouvement du carbofuran dans differents types de sols. Phytiatrie-Phytopharmacie 24:279–296

Jamet P, Piedallu MA (1975b) Étude de l'adsorption et de la desorption de la pyrazone (amino-5-chloro-4-phenyl-2,2 N pyridazinone-3) par different types de sols. Weed Res 15:113–121

Jaworski EG (1975) Chloroacetamides. In: Kearney PC, Kaufman DD (eds) Herbicides: Chemistry, Degradation and Mode of Action. Dekker, New York, pp 349–376

Jordan LS, Farmer WJ, Goodin JR, Day BE (1970) Nonbiological detoxication of the s-triazine herbicides. Res Rev 32:267–285

Kalouskova N (1986) Kinetics and mechanisms of interaction of simazine with humic acids. J Environ Sci Health B21:251–267

Karickoff SW (1980) Sorption kinetics of hydrophobic pollutants in natural sediments. In: Baker RA (ed) Contaminants and Sediments Vol 2. Ann Arbor, Sci Publ, Ann Arbor MI, pp 193–205

Karickoff SW (1981) Semi-empirical estimation of sorption of hydrophobic pollutants on natural sediments and soils. Chemosphere 10:833–846

Karickoff SW, Brown DS, Scott TA (1979) Sorption of hydrophobic pollutants on natural sediments. Water Res 13:241–248

Katan J, Fuhreman TW, Lichtenstein EP (1976) Binding of ^{14}C-parathion in soil: a reassessment of pesticide persistence. Science 193:891–894

Keith LH, Telliard WA (1979) Priority pollutants. A perspective view. Environ Sci Technol 13:416–423

Khan SU (1973a) Equilibrium and kinetic studies of the adsorption of 2,4-D and picloram on humic acid. Can J Soil Sci 53:429–434

Khan SU (1973b) Interaction of humic acid with chlorinated phenoxyacetic and benzoic acids. Environ Lett 4:141–148

Khan SU (1973c) Interaction of humic substances with bipyridilium herbicides. Can J Soil Sci 53:199–204

Khan SU (1974a) Humic substances reactions involving bipyridilium herbicides in soil and aquatic environments. Res Rev 52:1–26

Khan SU (1974b) Adsorption of bipyridilium herbicides by humic acids. J Environ Qual 3:202–206

Khan SU (1977) Adsorption of dyfonate (o-ethyl-s-phenylethyl-phosphonodithioate) on humic acid. Can J Soil Sci 57:9–13

Khan SU (1978a) Kinetics of hydrolysis of atrazine in aqueous fulvic acid solution. Pestic Sci 9:39–43

Khan SU (1978b) The interaction of organic matter with pesticides. In: Schnitzer M, Khan SU (eds) Soil Organic Matter. Elsevier, Amsterdam, pp 137–170

Khan SU (1980) Determining the role of humic substances in the fate of pesticides in the environment. J Environ Sci Health B15:1071–1090

Khan SU, Gamble DS (1983) Ultraviolet irradiation of an aqueous solution of prometryn in the presence of humic materials. J Agric Food Chem 31:1099–1104
Khan SU, Mazurkevich R (1974) Adsorption of linuron on humic acid. Soil Sci 118:339–343
Khan SU, Schnitzer M (1972) The retention of hydrophobic organic compounds by humic acid. Geochim Cosmochim Acta 36:745–754
Khan SU, Schnitzer M (1978) UV irradiation of atrazine in aqueous fulvic acid solutions. J Environ Sci Health 3:299–310
Khan SU, Hamilton HA, Hague EC (1976) Fonofos residues in an organic soil and vegetable crops following treatment of the soil with the insecticide. Pestic Sci 7:553–558
Kirk RE, Wilson MC (1960) The effect of soil type and moisture on germination and growth from wheat seed treated with phorate. J Econ Entomol 53:771–774
Koren E, Foy CL, Ashton FM (1968) Phytotoxicity and persistence of four thiocarbamates in five soil types. Weed Sci 16:172–175
Koren E, Foy CL, Ashton FM (1969) Adsorption, volatility and migration of the thiocarbamates herbicides in soil. Weed Sci 17:148–153
Kozak J (1983) Adsorption of prometryn and metholachlor by selected soil organic matter fractions. Soil Sci 136:94–101
Lagerkrantz C, Yhland M (1962) Photo-induced electron spin resonance in solutions of some electron donor-acceptor complexes. Acta Chem Scand 16:1043–1045
Lagerkrantz C, Yhland M (1963) Photo-induced free radical reactions in the solution of some tars and humic acid. Acta Chem Scand 17:1299–1306
Lambert SM (1967) Functional relationship between sorption in soil and chemical structure. J Agric Food Chem 15:572–576
Lambert SM, Porter PE, Schieferstein RH (1965) Movement and sorption of chemicals applied to the soil. Weeds 13:185–190
Landrum PF, Giesy JP (1981) Anomalous break through of benzo (a)pyrene during concentration with Amberlite XAD-4 resin from aqueous solutions. In: Keith L.H. (ed) Advances in the Identification and Analysis of Organic Pollutants, Vol 1. Ann Arbor Sci Publ, Ann Arbor, MI, pp 345–355
Landrum PF, Nihart SR, Eadie BJ, Gardner WS (1984) Reverse-phase separation method for determining pollutant binding to Aldrich humic acid and dissolved organic carbon of natural waters. Environ Sci Technol 18:187–192
Landrum PF, Reinhold MD, Nihart SR, Eadie BJ (1985) Predicting the bioavailability of organic xenobiotics to *Pontoporeia hoyi* in the presence of humic and fulvic materials and natural dissolved organic matter. Environ Toxicol Chem 4:459–467
Landrum PF, Eadie BJ, Nihart SR, Reinhold MD (1987) Confirmation of the reverse-phase measure of xenobiotic partitioning to dissolved organic matter by toxicokinetic studies. Am Chem Soc-Div Environ Chem 27 (1):291–295
Leeneer JA, Aldrichs JL (1971) A kinetic and equilibrium study of the adsorption of carbaryl and parathion upon soil organic matter surfaces. Soil Sci Soc Am Proc 35:700–705
Leversee GJ, Landrum PF, Giesy JP, Fannin T (1983) Humic acids reduce bio-accumulation of some polycyclic aromatic hydrocarbons. Can J Fish Aquat Sci 40:63–69
Li GC, Felbeck GT Jr (1972) A study of the mechanism of atrazine adsorption by humic acid from muck soil. Soil Sci 113:140–148
Lichtenstein EP (1959) Adsorption of some chlorinated hydrocarbon insecticides from soils into various crops. J Agric Food Chem 7:430–433
Lichtenstein EP, Katan J, Anderegg BN (1977) Binding of "persistent" and "non-persistent" ^{14}C-labeled insecticides in an agricultural soil. J Agric Food Chem 25:43–47
Lindqvist I (1982) Charge-transfer interaction of humic acids with donor molecules in aqueous solutions. Swed J Agric Res 12:105–109
Lindqvist I (1983) The interaction between a humic acid and a charge-transfer acceptor molecule. Swed J Agric Res 13:201–203
Liu S-Y, Bollag JM (1985) Enzymatic binding of the pollutant 2, 6-xylenol to a humus constituent. Water Air Soil Pollut 25:97–106

Lutz-Ostertag Y, Lutz H (1970) Action de l'herbicide 2,4-D sur le developpement embryonaire et la fecondite due gibier a plumes. C R Acad Sci Paris, Series D271:2418–2421

Macalady DL, Wolfe NL (1984) Abiotic hydrolysis of sorbed pesticides. In Krueger RF, Seiber JN (eds) Treatment and Disposal of Pesticides Wastes. ACS Symp Series N 259:221–224

Macalady DL, Wolfe NL (1985) Effects of sediment sorption and abiotic hydrolysis. 1. Organophosphorothioate esters. J Agric Food Chem 33:167–173

Macalady DL, Wolfe NL (1987) Influences of aquatic humic substances on the abiotic hydrolysis of organic contaminants: a critical review. Am Chem Soc-Div Environ Chem 27(1):12–15

MacKay D, Shiu WJ (1978) Determination of the solubility behavior of some polycyclic aromatic hydrocarbons in water. Anal Chem 50:997–1000

Malcolm RL (1985) Geochemistry of stream fulvic and humic substances. In: Aiken GR, McKnight DM, Wershaw RL, MacCarthy P (eds) Humic Substances in Soil, Sediment and Water. Wiley, New York, pp 181–210

Malini de Almeida R, Pospisil F, Vockova K, Kutacek M (1980) Effect of humic acids on the inhibition of pea choline esterase and choline acyltransferase with malathion. Biol Plant 22:167–175

Maqueda C, Perez Rodriguez JL, Martin F, Hermosin MC (1983) A study of the interaction between chlordimeform and humic acid from a typic chromoxevert soil. Soil Sci 136:75–81

Mathur SP (1974) Phthalate esters in the environment: pollutants or natural products? J Environ Qual 3:189–197

Mathur SP, Morley HV (1978) Incorporation of methoxychlor-^{14}-C in model humic acids prepared from hydroquinone. Bull Environ Contam Toxicol 20:268–274

Matsuda K, Schnitzer M (1971) Reactions between fulvic acid, a soil humic material, and dialkylphthalates. Bull Environ Contam Toxicol 6:200–204

Mayer LM (1985) Geochemistry of humic substances in estuarine environments. In Aiken GR, McKnight DM, Wershaw RL, MacCarthy P (eds) Humic Substances in Soils, Sediments and Water. Wiley, New York, pp 211–232

McCarthy JF (1983) Role of particulate organic matter in decreasing accumulation of polynuclear aromatic hydrocarbon by *Daphnia Magna*. Arch Environ Contam Toxicol 12:559–568

McCarthy JF (1987) Humic substances reduce bioavailability and toxicity of contaminants. Am Chem Soc-Div Environ Chem 27(1):286–288

McCarthy JF, Jimenez BD (1985b) Interaction between polycyclic aromatic hydrocarbons and dissolved humic materials: binding and dissociation. Environ Sci Technol 19:1072–1076

McCarthy JF, Jimenez BD (1985a) Reduction in bioavailability to bluegills of aromatic polycyclic hydrocarbons bound to dissolved humic material. Environ Toxicol Chem 4:511–521

McCarthy JF, Jimenez BD, Barbee T (1985) Effect of dissolved humic materials on accumulation of polycyclic aromatic hydrocarbons: structure-activity relationship. Aquatic Toxicol 7:15–24

McGlamery MD, SLife FW (1966) The adsorption and desorption of atrazine as affected by pH, temperature and concentration. Weeds 14:237–239

Means JC, Wood SG, Hassett JJ, Banwart WL (1980) Sorption of polynuclear aromatic hydrocarbons by sediments and soils. Environ Sci Technol 14:1524–1528

Melcer MC, Zalewski MS, Hassett JP, Brisk MA (1987) Nature of the binding interactions between humic substances and hydrophobic molecules. Am Chem Soc-Div Environ Chem 27(1):414–416

Menges RM, Hubbard JL (1970) Phytotoxicity of bensulide and trifluralin in several soils. Weed Sci 18:244–247

Metcalf RL, Booth GM, Schuth CK, Hansen DJ, Lu PY (1973) Uptake and fate of di-2-ethylhexyl phthalate in aquatic organisms and in a model ecosystem. Environ Health Perspect 4:27–34

Mill GL, Carter DJ (1987) Photochemical degradation of tetraphenylboron sensitized by dissolved organic matter in natural water. Am Chem Soc-Div Environ Chem 27(1):146–148

Mill T, Hendry D, Richardson H (1980) Free radical oxidants in natural water. Science 207:886–887

Miller CW, Demoranville IE, Charig AJ (1966) Persistence of dichlobenil in nonberry bogs. Weeds 14:296–298

Miller GC, Zisook R, Zepp R (1980) Photolysis of 3,4-dichloroaniline in natural waters. J Agric Food Chem 28:1053–1056

Mingelgrin U, Gerstl Z (1983) Reevaluation of partitioning as a mechanism of nonionic chemicals adsorption in soils. J Environ Qual 12:1–11

Morehead NR, Eadie BJ, Lake B, Landrum PF, Berner D (1986) The sorption of PAH onto dissolved organic matter in Lake Michigan waters. Chemosphere 15:403–412

Moreland DE, Hilton JL (1976) Action on photosynthesis systems. In: Audus LI (ed) Herbicides: Physiology, Biochemistry and Ecology, Vol 2, Academic Press, New York, 2nd edit, pp 493–521

Mudambi AR, Hassett JP (1987) Photochemical activity of mirex associated with dissolved organic matter. Am Chem Soc-Div Environ Chem 27(1):201–203

Müller-Wegener U (1977) Uber die Bindung von s-Triazinen an Huminsäuren. Geoderma 19:227–235

Mulvaney FL, Bremner JM (1978) Use of p-benzoquinone for retardation of urea hydrolysis in soils. Soil Biol Biochem 10:297–302

Narbaitz RM, Benedek A (1987) The removal of 1,1,2-trichloroethane from highly colored river water by activated carbon. Am Chem Soc-Div Environ Chem 27 (1):359–362

Narine DR, Guy RD (1982) Binding of diquat and paraquat to humic acid in aquatic environments. Soil Sci 133:356–363

Nash RG (1968) Plant uptake of ^{14}C-diuron in modified soil. Agron J 60:177–179

Nearpass DC (1976) Adsorption of picloram by humic acids and humin. Soil Sci 121:272–277

Nys GG, Rekker RG (1974) The concept of hydrophobic fragmental constants (f-values). 2. Extension of its applicability to the calculation of lipophylicities of aromatic and heteroaromatic structures. Eur J Med Chem Chim Ther 9:361–375

Obien SR, Suchisa RH, Younge Or (1966) The effects of soil factors on the phytotoxicity of neburon to oats. Weeds 14:105–109

Ogner G, Schnitzer M (1970) Humic substances: fulvic acid-dialkylphthalate complexes and their role in pollution. Science 170:317–318

Parris GE (1980) Covalent binding of aromatic amines to humates. 1. Reactions with carbonyls and quinones. Environ Sci Technol 14:1099–1106

Perdue EM (1983) Association of organic pollutants with humic substances: partitioning equilibria and hydrolysis kinetics. In: Christman RF, Gjessing ET (eds) Aquatic and Terrestrial Humic Materials. Ann Arbor Sci Publ, Ann Arbor-MI, pp 441–460

Perdue EM (1987) Overview of the effects of humic substances on pollutant transformations. Am Chem Soc-Div Environ Chem 27(1):448–451

Perdue EM, Wolfe NL (1982) Modification of pollutant hydrolysis kinetics in the presence of humic substances. Environ Sci Technol 16:847–852

Peterson JR, Adams RS Jr, Cutkamp LK (1971) Soil properties influencing DDT bioactivity. Soil Sci Soc Am Proc 35:72–78

Peyton GR, Gee CS, Bandy J, Maloney SW (1987) Catalytic/competition effects of humic substances on photolytic ozonation of organic compounds, Am Chem Soc-Div Environ Chem 27(1):212–214

Pierce RH, Olney CE, Felbeck GT (1971) Pesticide adsorption in soils and sediments. Environ Lett 1:157–172

Pirbazari M, Stevens M, Ravindran V (1987) Effect of complexation of micro-pollutants with humic substances on activated carbon adsorption. Am Chem Soc-Div Environ Chem 27(1):434–435

Pitts JN Jr, Grosjean D, Mischke TM (1977) Mutagenic activity of airborne particulate organic pollutants. Toxicol Lett 1:65–70

Plimmer JR, Kearney PC, Rowlands JR (1968) Free radical oxidation of s-triazines mechanism of N-dealkylation. Am Chem Soc Annual Meeting, Atlantic City

Ross RD, Crosby DG (1973) Photolysis of ethylenethiourea. J Agric Food Chem 21:335–337

Ross RD, Crosby DG (1975) The photooxidation of aldrin in water. Chemosphere 4:277–282

Scott DC, Weber JB (1967) Herbicide phytotoxicity as influenced by adsorption. Soil Sci 104:151–158

Schnitzer M (1978) Humic substances chemistry and reactions. In: Schnitzer M, Khan SU (eds) Soil Organic Matter, Elsevier, Amsterdam, pp 1–64

Schnitzer M, Khan SU (1972) Humic Substances in the Environment, Dekker, New York, p 327

Schwarzenbach RD, Westall J (1981) Transport of non-polar organic compounds from surface water to groundwater. Laboratory sorption studies. Environ Sci Technol 15:1360–1367

Senesi N (1981) Free radicals in electron donor-acceptor reactions between a soil humic acid and photosynthesis inhibitor herbicides. Z Pflanzen Bodenkd 144:580–586

Senesi N, Schnitzer M (1977) Effects of pH, reaction time, chemical reduction and irradiation on ESR spectra of fulvic acid. Soil Sci 123:224–234

Senesi N, Schnitzer M (1978) Free radicals in humic substances. In: Krumbein WE (ed) Environmental Biogeochemistry and Geomicrobiology, Vol II, The Terrestrial Environment. Ann Arbor Sci Publ, Ann Arbor, MI, pp 467–481

Senesi N, Steelink C (1989) Application of ESR spectroscopy to the study of humic substances and their interactions with organic xenobiotics and metal ions. In: Hayes MHB, MacCarthy P, Malcolm RL, Swift RS (eds) Humic Substances: In Search of Structure. Wiley, New York (in press)

Senesi N, Testini C (1980) Adsorption of some nitrogenated herbicides by soil humic acids. Soil Sci 10:314–320

Senesi N, Testini C (1982) Physico-chemical investigations of interaction mechanisms between s-triazine herbicides and soil humic acids. Geoderma 28:129–146

Senesi N, Testini C (1983a) The environmental fate of herbicides: the role of humic substances. Ecological Bull Stockolm 35:477–490

Senesi N, Testini C (1983b) Spectroscopic investigations of electron donor-acceptor processes involving organic free radicals in the adsorption of substituted urea herbicides by humic acids. Pestic Sci 14:79–89

Senesi N, Testini C (1984) Theoretical aspects and experimental evidence of the capacity of humic substances to bind herbicides by charge-transfer mechanisms (electron donor-acceptor processes). Chemosphere 13:461–468

Senesi N, Chen Y, Schnitzer M (1977a) The role of free radicals in the oxidation and reduction of fulvic acid. Soil Biol Biochem 9:397–403

Senesi N, Chen Y, Schnitzer M (1977b) Aggregation-dispersion phenomena in humic substances. In IAEA (ed) Soil Organic Matter Studies, IAEA, Vienna, Vol II, pp 143–155

Senesi N, Testini C, Metta D (1984) Binding of chlorophenoxyalkanoic herbicides from aqueous solution by soil humic acids. Proc Int Conf Environmental Contamination, London 1984, CEP Cons Ltd, Edinburgh, pp 96–101

Senesi N, Miano TM, Testini C (1986a) Role of humic substances in the environmental chemistry of chlorinated phenoxyalkanoic acids and esters. In: Pawlowski L, Alaerts G, Lacy WJ (eds) Chemistry for Protection of the Environment 1985, Studies in Environmental Science 29. Elsevier, Amsterdam, pp 183–196

Senesi N, Padovano G, Loffredo E, Testini (1986b) Interactions of amitrole, alachlor and cycloate with humic acids. Proc 2nd Int Conf Environmental Contamination, Amsterdam 1986, pp 169–171

Senesi N, Miano TM, Testini C (1987a) Incorporation of water dissolved chlorophenoxyalkanoic herbicides by humic acids of various origin and nature. In: Giovannozzi-Sermanni G, Nannipieri P (eds) Current Perspectives in Environmental Biogeochemistry, CNR-IPRA, Rome, pp 295–308

Senesi N, Testini C, Miano TM (1987b) Interaction mechanisms between humic acids of different origin and nature and electron donor herbicides: a comparative IR and ESR study. Org Geochem 11:25–30

Sjoblad RD, Bollag JM (1981) Oxidative coupling of aromatic compounds by enzymes from soil microorganism. Soil Biochem 5:113–152

Slawinska D, Slawinski J, Sarna T (1975) The effect of light on the ESR spectra of humic acids. J Soil Sci 26:93–99

Slawinski J, Puzyna W, Slawinska D (1978a) Chemiluminescence in the photooxidation of humic acids. Photochem Photobiol 28:75–81

Slawinski J, Puzyna W, Slawinska D (1978b) Chemiluminescence during photooxidation of melanins and soil humic acids arising from a singlet oxygen mechanism. Photochem Photobiol 28:459–463

Spacie A, Landrum PF, Leversee GJ (1984) Uptake, depuration and biotransformation of anthracene and benzo(a)pyrene in Bluegill sunfish. Ecotoxicol Environ Safety 5:330–341

Steinberg C, Muenster U (1985) Geochemistry and ecological role of humic substances in lake water. In: Aiken GR, McKnight DM, Wershaw RL, MacCarthy P (eds) Humic Substances in Soil, Sediment and Water. Wiley, New York, pp 105–146

Stevenson FJ (1972) Organic matter reactions involving herbicides in soil. J Environ Qual 1:333–343

Stevenson FJ (1982) Humus Chemistry: Genesis, Composition, Reactions. Wiley, New York, p 443

Stott DE, Martin JP, Focht DD, Haider K (1983) Biodegradation, stabilization in humus, and incorporation into soil biomass of 2,4-D and chlorocatechol carbons. Soil Sci Soc Am J 47:66–70

Strek HJ, Weber JB (1982) Behavior of polychlorinated biphenyls (PCBs) in soils and plants. Environ Pollut A 28:291–312
Suflita JM, Bollag JM (1980) Oxidative coupling activity in soil extracts. Soil Biol Biochem 12:177–183
Suflita JM, Bollag JM (1981) Polymerization of phenolic compounds by a soil-enzyme complex. Soil Sci Soc Am J 45:297–302
Sullivan JD, Felbeck GT (1968) A study of the interaction of s-triazine herbicides with humic acids from three different soils. Soil Sci 106:42–50
Swoboda AR, Thomas GT (1968) Movement of parathion in soil columns. J Agric Food Chem 16:923–927
Thurman EM (1985) Humic substances in groundwater. In: Aiken GR, McKnight DM, Wershaw RL, MacCarthy P (eds) Humic Substances in Soil, Sediment and Water. Wiley, New York, pp 87–103
Thurman EM (1986) Aquatic Humic Substances. In Organic Geochemistry of Natural Waters, Chapt 10. Nijhoff-Junk, Dordrecht, pp 273–361
Thurman EM (1987) Linear alkylbenzene sulfonates in groundwater. Potential for co-isolation with humic substances. Am Chem Soc Div-Environ Chem 27(1):195–197
TRW Systems and Energy, Inc (1976) Carcinogens Relating to Coal Conversion Processes. US Energy Research and Development Administration, Oak Ridge, TN
Tschapek M, Wasowski C (1976) The surface activity of humic acid. Geochim Cosmochim Acta 40:1343–1345
Tucker BV, Pack DE, Ospenson JN (1967) Adsorption of bipyridylium herbicides in soil. J Agric Food Chem 15:1005–1008
Tucker BV, Pack DE, Ospenson JN, Omid A, Thomas WD Jr (1969) Paraquat soil bonding and plant response. Weed Sci 17:448–451
Turski R, Steinbrich A (1971) Studies on the possibilities of binding herbicides of triazine derivates by humic acids. Polish J Soil Sci 4:120–124
Vandenbrouke M, Pelet R, Debyser Y (1985) Geochemistry of humic substances in marine sediments. In Aiken GR, McKnight DM, Wershaw RL, MacCarthy P (eds) Humic Substances in Soil, Sediment and Water. Wiley, New York, pp 249–274
Vettorazzi G (1977) State of the art of the toxicological evaluation carried out by the joint FAO/WHO Expert Committee on Pesticide Residues 3. Miscellaneous pesticides used in agriculture and public health. Res Rev 66:138–184
Visser SA (1982) Surface active phenomena by humic substances of aquatic origin. Rev Franc Sci Eau 1:285–296
Wakeham SG, Schaffner C, Giger W (1980) Polycyclic aromatic hydrocarbons in recent lake sediments. II. Compounds derived from biogenic precursors during early diagenesis. Geochim Cosmochim Acta 44:415–429
Walker A, Crawford DV (1968) The role of organic matter in adsorption of the triazine herbicides by soil. In FAO/IAEA (eds) Isotopes and Radiation in Soil Organic Matter Studies, IAEA, Vienna, pp 91–108
Wang CH, Broadbent FE (1973) Effect of soil treatments on losses of two chloronitrobenzene fungicides. J Environ Qual 2:511–515
Weber JB (1967) Spectrophotometrically determined ionisation constants of 1,3-alkylamino-s-triazines and the relationships of molecular structure and basicity. Spectrochim Acta 23A:458–461
Weber JB (1970) Mechanism of adsorption of s-triazines by clay colloids and factors affecting plant availability. Res Rev 32:93–130
Weber JB (1971) Behavior of organic pesticides in soils. Proc Soil Sci North Carolina 14:74–118
Weber JB (1972) Interaction of organic pesticides with particulate matter in aquatic and soil systems. Adv Chem Series 111:55–120
Weber JB (1978) Fate of organics in sludges applied to the land. Abs 5th Nat Conf Acceptable Sludges Disposal Techniques, Orlando-Fla, Inform Transfer Inc 31:117–124
Weber JB, Weed SB (1974) Effect of soil on the biological activity of pesticides. J Series North Carolina State Univ Agric Experim Station, Raleigh-NC 4087:223–256
Weber JB, Weed SB, Best JA (1969a) Displacement of diquat from clay and its phytotoxicity. J Agric Food Chem 17:1075–1076
Weber JB, Weed SB, Ward TM (1969b) Adsorption of s-triazines by soil organic matter. Weed Sci 17:417–421

Weber JB, Weed SB, Waldrup TW (1974) Effect of soil constituents on herbicide activity in modified-soil field plots. Weed Sci 22:454–459

Weber WJ Jr, Smith EH (1987) The effects of background dissolved organic matter on adsorption processes. Am Chem Soc-Div Environ Chem 27(1):342–345

Weed SB, Weber JB (1974) Pesticide-organic matter interactions. J Series North Carolina State Univ Agric Experim Station, Raleigh-NC 3840:39–66

Wershaw RL, Burcar PJ, Goldberg MC (1969) Interaction of pesticides with natural organic matter. Environ Sci Technol 3:271–273

Wolff CJM, Halmans MTH, Van der Heijde HB (1981) The formation of singlet oxygen in surface waters. Chemosphere 10:59–62

Wolf DC, Martin JP (1976) Decomposition of fungal mycelia and humic-type polymers containing Carbon-14 from ring and side-chain labeled 2,4-D and chlorpropham. Soil Sci Soc Am Proc 40:700–704

Wolfhart D (1980) Controls needed on killer herbicide. Pac Island 51:7–10

Woolson EA, Thomas RF, Ensor PDJ (1972) Survey of polychlorodibenzo-p-dioxin content in selected pesticides. J Agric Food Chem 20:351–354

World Health Organization (1971) International Standards for Drinking Water 3rd edn, Geneve, Switzerland

Yalkowski SH, Valvani SC (1979) Solubilities and partitioning. 2. Relationships between aqueous solubilities, partition coefficients, and molecular surface areas of rigid aromatic hydrocarbons. J Chem Eng Data 24:127–129

Youngblood WW, Blumer M (1975) Polycyclic aromatic hydrocarbons in the environment: homologous series in soils and recent marine sediments. Geochim Cosmochim Acta 39:1303–1314

Zepp RG, Wolfe NL, Gordon JA, Fincher RG (1976) Light-induced transformations of methoxychlor in aquatic systems. J Agric Food Chem 24:727–733

Zepp RG, Wolfe NL, Baughman GL, Hollis RC (1977) Singlet oxygen in natural waters. Nature 267:421–423

Zepp RG, Baughman GL, Schlotzhauer PF (1981) Comparison of photochemical behavior of various humic substances in water: sunlight-induced reactions of aquatic pollutants photosensitized by humic substances. Chemosphere 10:109–117

Zepp RG, Schlotzhauer PF, Sink RM (1985) Photosensitized transformations involving electronic energy transfer in natural waters: role of humic substances. Environ Sci Technol 19:74–81

Zimmer G, Sontheimer H (1987) Activated carbon adsorption of organic pollutants in the presence of humic substances. Am Chem Soc-Div Environ Chem 27(1):346–348

4 Surface Interactions of Toxic Organic Chemicals with Minerals

U. MINGELGRIN[1] and R. PROST[2]

4.1 Introduction

The organic fraction of the solid phase is often assumed to dominate the interactions of contaminants in the unsaturated zone. This assumption may have helped to obscure the important role mineral surfaces play in determining the fate of organic chemicals in the unsaturated zone. While the role of the organic fraction is referred to in various places throughout this book, this chapter will discuss the interaction of organic pollutants with the mineral fraction.

Under many conditions, mineral surfaces may dominate the fate of toxic organic chemicals. At sufficiently large depths below the root zone, the nutrients and organic matter contents are often low and the gas exchange is slow. Biotic activity is expected to be retarded under such conditions and the low organic matter content at these depths allows the mineral surfaces to dominate the abiotic interactions of the organic toxicants. In arid zones, long periods of dryness minimize the importance of biotic interactions and the low organic matter content, even at the surface, maximizes the role of the mineral fraction in the abiotic interactions. When a significant quantity of organic matter is present, it often interacts with the mineral fraction. Clay-organic matter complexes have been studied by many authors (e.g. Greenland 1965). By interacting with the organic fraction, the mineral fraction affects the interaction between the contaminant molecule and the native organic matter. Since in typical soils and subsoils the clay content is an order of magnitude greater than that of the organic matter content, one may expect at least some contribution of the mineral fraction to the interaction of polar xenobiotics (even weakly polar ones) in those parts of the unsaturated zone in which the organic matter content is appreciable. This is true despite the existence of much of the organic fraction in the unsaturated zone as a coating over mineral surfaces. Finally, since most of the cation exchange capacity is usually contributed by the mineral fraction, the fate of cationic organic pollutants may often be controlled by that fraction.

The most obvious surface interaction is adsorption. Adsorption will affect both the mobility and the microbial transformations of adsorbate molecules. Yet, the more interesting interactions of surfaces with organic molecules are the

---

[1] Institute of Soils and Water, Volcani Center, ARO, P.O. Box 6, Bet Dagan, 50250, Israel.
[2] Station de Science du Sol, INRA, Versailles, France.

surface enhanced transformations. As a rule, adsorption is the first step in surface transformations and is analogous to the formation of a transition complex in homogeneous reactions. The more polar the organic molecule is, the more likely it is to strongly interact with 'hydrophilic' minerals (e.g. clays) which contribute the bulk of the mineral surface area in the unsaturated zone. As discussed below, strong interactions with the surface do not necessarily mean considerable adsorption, due for example, to competition with water. On the other hand, many weakly polar organic substances display a strong tendency to adsorb on mineral surfaces from aqueous solutions and to undergo heterogeneously catalyzed transformations at these surfaces.

The strength of an electrostatic interaction between an adsorbate molecule and a site on the surface is expected to be directly related to the charge density at that site. Exchangeable cations or dissociated groups at the surface should therefore be the sites of strongest interaction for charged, polar or polarizable adsorbates. The above applies to any surface, whether mineral or organic, which contains exchangeable cations or dissociated groups. The organic surfaces in the unsaturated zone are rich in charged sites as demonstrated by the cation exchange capacity of the organic matter in the pH range prevalent in that zone. The major difference between the organic and mineral surfaces is the presence in the organic fraction of 'hydrophobic' sites, (the relevance of which will be discussed below) in addition to the hydrophilic sites. The role of the exchangeable cations and their hydration status in the surface interactions of polar and polarizable adsorbates will be described in this chapter as will the role of the net enthalpy and entropy of the adsorption process in the uptake of organic molecules. The interactions of organic molecules with mineral surfaces may strongly depend on parameters such as the moisture content and pH. This dependence will also be discussed below.

4.2 Relevant Properties of the Mineral Components of the Solid Phase

Clay minerals are probably the most important inorganic components of the solid phase as far as the fate of organic pollutants in the unsaturated zone is concerned. These minerals dominate the clay size fraction of soils and subsoils and at times the coarser size fractions as well. In some cases more than half of the bulk of the solid phase consists of clay minerals. Clay minerals prevalent in soils were treated extensively in the past (e.g. Dixon and Weed 1977). They are characterized by a relatively high surface area and surface charge density and are thus efficient partners for electrostatic interactions with adsorbate molecules. Clays, being in general negatively charged, have exchangeable cations at their surface. Most clays have a laminar structure, and at a sufficiently low pH, they may be positively charged at their broken edges while maintaining a negative charge at the planar surfaces. Many clay minerals are expandable, affording interlayer adsorption sites, thus further expanding the active surface area.

Isomorphic substitution is the origin of the pH independent charge of clay minerals. The sequence of octahedral and tetrahedral layers in the clay lattice together with the type, location and extent of the isomorphic substitution determine the properties of the surface of the mineral and its swelling capacity. There are two basic sequences of layers: alternate layers of silicon oxide tetrahedra and oxy-hydroxyaluminum octahedra (1:1) and units composed of an octahedral layer surrounded by two tetrahedral layers (2:1). While the more common 2:1 minerals are characterized by an oxygen sheet over most of their surface (excluding broken edges), half of the planar surface of the 1:1 minerals consists of a hydroxyl sheet. Many of the 2:1 clays swell upon wetting. In the swelling process, solvent molecules penetrate between the 2:1 units (namely between two adjacent tetrahedral layers). The swelling increases the spacing between these units and exposes the large interlayer surface area to interaction with solute molecules. The negative charge of the clay minerals is neutralized by exchangeable cations located at the clay surface. Both adsorption and abiotic transformations are strongly affected by the nature of the exchangeable cations.

Smectites and in particular montmorillonite are very common in soils. Montmorillonite is a 2:1 clay mineral with a cation exchange capacity (CEC) of approximately 1 mole(+) kg^{-1}. It is expandable, namely, it swells in water, and has a large surface area (about 800,000 m^2 kg^{-1}). Illite is a very slightly expandable clay mineral having a 2:1 crystal form. K$^+$ ions fixed between the 2:1 plates prevent swelling and reduce the CEC of this mineral to 0.4 mole(+) kg^{-1} or less. Similarly, the surface area of illite is below 100,000 m^2 kg^{-1}. Illite is a dominant mineral in the silt fraction of loessial soils. As can be seen from its relatively low CEC and surface area, it is likely to be less effective than montmorillonite in adsorbing or enhancing surface reactions of organic molecules. Vermiculite is another common 2:1 mineral. Its swelling is limited to up to 0.5 nm or less and its interlayer spaces are usually saturated with Mg^{+2} and to a lesser extent Ca^{+2}. The CEC of vermiculite (including the interlayer Mg or Ca) is very high, in some cases exceeding 1.5 mole(+) kg^{-1}. Its surface area is similar to that of montmorillonite. The limited swelling of this mineral impedes the penetration of sufficiently bulky organic molecules into its interlayer spaces. The CEC of 2:1 clay minerals originates predominantly from isomorphic substitution which may take place, depending on the mineral, either in the octahedral or in the tetrahedral layers. Whether this substitution occurs in the octahedral or tetrahedral layer will affect the strength of the electric field at the surface and hence the interaction of organic adsorbates with the surface.

Among the 1:1 clay minerals, kaolinite and its hydrated form halloysite are probably the most common. This mineral is not expandable and accordingly has a small CEC (less than 0.15 mole (+) kg^{-1}) and available surface area (below 30,000 m^2 kg^{-1}). Dissociation of H$^+$ at the broken edges may contribute a significant fraction of the CEC. Since that dissociation is pH dependent, the CEC of kaolinite is strongly pH dependent as well. The above data suggests that the surface activity of kaolinite should strongly depend on the pH and be in general rather low. Yet, due to the fact that half of its planar surface consists of a hydroxyl sheet and to its relatively high surface charge density, kaolinite enhances some surface reactions which may not take place on 2:1 clay surfaces.

Clay minerals having a fibrous morphology also exist in the unsaturated zone. These are alumino-Mg-silicates with Al/Mg ratios varying from around unity (palygorskite or attapulgite) to almost nil (sepiolite). The fibrous crystals of these clays contain channels (0.38×0.63 or 0.38×0.94 nm respectively). The presence of channels and the exposed O and OH groups at the surface give these 2:1 clays interesting surface properties as far as adsorption and surface reactions are concerned. The surface area of the fibrous clays is not well defined because of the limited accessibility of the inner channels to various indicator molecules. Estimates ranged up to 392,000 m^2 kg^{-1} for sepiolite and about half that for palygorskite. Yet, measurements at 600° C gave a surface area of 335,000 m^2 kg^{-1} for the latter mineral (Zelazny and Calhoun 1977). CEC values reported for the fibrous clay minerals range from 0.45 down to 0.05 mole(+) kg^{-1}.

Numerous other crystalline alumino-silicate and related minerals are found in the unsaturated zone. Yet, they are less important as far as surface interactions are concerned because of their low surface area, chemically inactive surfaces or scarcity.

The unsaturated zone in arid areas (or in more humid areas if the appropriate parent material is present) contains significant quantities of the relatively soluble $CaCO_3$ or, less frequently, $CaSO_4$. These minerals may occur either in crystalline or in amorphous forms. Calcareous soils may contain large amounts of the carbonate and some agricultural soils may contain more than 50% $CaCO_3$. The importance of $CaCO_3$ is mainly in its buffering of the pH of the liquid phase to values over 7.5. In addition, the surface of the mineral itself may be the site of specific surface interactions (e.g. Mingelgrin and Yaron 1974).

Iron and other metal oxides are also found in the unsaturated zone in various crystalline and amorphous forms. The most common crystalline iron oxide is goethite, but other iron oxides are frequently present (Schwertman and Taylor 1977). The surface of iron oxides is often hydroxylated either structurally or through hydration of Fe atoms and the hydroxylated system adsorbs in many cases further layers of water. The presence of the hydroxyls and of the hydration water molecules on the oxide surfaces makes them rather active sites for surface interactions. The ability of iron (and some other metal) containing systems to undergo redox reactions, further increases the role of these metal oxides in the abiotic transformations which occur in the unsaturated zone. Iron (and other transition metal) oxides possess pH dependent CEC and AEC (anion exchange capacity). Thus, depending on the soil pH, anions, cations and various nonionic species may interact with these oxides. The point of zero charge of the metal oxides is in the pH range of 5.4 to 10.4, but is most frequently between pH 7.5 and 9.5. The wide range of physical forms in which iron oxides may exist in the unsaturated zone results in a correspondingly wide range of surface properties.

When these oxides appear as colloidal particles rather than as coating, their surface area is typically between 10,000–180,000 m² kg⁻¹ and their charge at pH levels considerably below the point of zero charge 0.02–0.37 mole(+) kg⁻¹. Iron (as well as other transition metals) ions tend to form complexes with many ligands. This fact is one more reason to suggest that oxide surfaces may be important in the adsorption and the surface enhanced transformation of organic pollutants. The ability of metal oxides to adsorb protons gives them a buffering capacity and, at the same time, iron oxides may serve as poising agents for redox potentials around 0. V, possibly through microbially mediated processes. While in the temperate zone iron and other metal oxides usually constitute only a small fraction of the solid phase of the soil and often of the subsoil as well, under tropical conditions, significant quantities of metal oxides are frequently present. In oxisols, 20–80% of the solid phase were reported to be iron as Fe_2O_3 (e.g. Soil Conservation Service 1975).

Crystalline iron oxides have, as a rule, low surface areas and charge densities as compared to swelling clays. The contribution of these metal oxides to the surface interactions of organic pollutants is therefore probably less important than that of the clays. It is harder to estimate the contribution of the amorphous metal oxides, which often exist as a coating on other mineral surfaces. In one study (Gorbunov et al. 1961), amorphous iron oxide adsorbed 109 times more phosphate than the crystalline minerals. The catalytic potential of iron oxides is demonstrated in Fig.1.

Not only transition metal oxides, but other minerals as well often appear in amorphous forms. The amorphous materials may coat crystals and thus, aside from their intrinsic contribution to adsorption and abiotic transformations, the amorphous materials modify the contribution of crystalline substances to surface

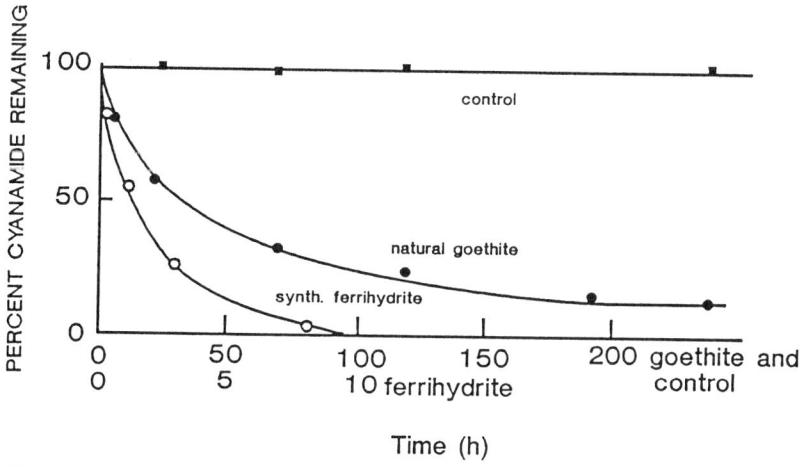

Fig. 1. Transformation of cyanamide at 18° C in the presence of natural geothite and a synthetic ferrihydrate (Amberger and Vilsmeier 1978)

interactions. The importance of the amorphous mineral fraction may, therefore, be considerably larger than suggested by its content in the system of interest. Oxides of silicon and iron are common in the amorphous fraction. The amorphous iron oxides are, as suggested above, often present as a coating on clay surfaces and may be more active in affecting surface reactions then the crystalline oxides. Allophane is an amorphous clay fraction which is found in many soils. It often appears in the colloidal size fraction, displaying a surface area and a CEC close to that of montmorillonite. The CEC of the allophanes is, however, strongly dependent on the pH. Some anion exchange capacity was also observed in allophanes.

4.3 Interrelation Between Components of the Solid Phase

The actual surface of the solid phase of the unsaturated zone is inhomogeneous and is characterized by the multi-component association between humic substances, clays, metal oxides, $CaCO_3$ and other minerals. In some cases, up to 90% of the soil organic matter was found to be associated with the mineral fraction of the soil (Greenland 1965). On the other hand, there is evidence that much of the surface of clay minerals, specifically in the interlayer spaces of smectites, is not covered by organic matter (Ahlrichs 1972). Iron oxide coating of clay minerals is also probably restricted mostly to the clay's external surfaces. The interlayer surfaces of smectites and of other expandable clays may therefore often be exposed to interaction with organic contaminants. On the other hand, intercalation of hydroxyaluminum species with clays was reported (e.g. Ahlrichs 1972). In many cases, cationic aluminum hydroxides or oxides coat clay surfaces thus reducing the clay's CEC. This effect is most important in acid soils since as the pH increases, the hydroxyaluminum oligomers lose their positive charge and their effect on the CEC is thus reduced. Because of the strong effect that the CEC and the type of the exchangeable cation dominating the exchange complex have on surface interactions, coating of the surface by cationic oligomers may have a considerable influence on both adsorption and abiotic transformations. Aluminum and iron, as well as other polyvalent cations, interact strongly with humic acids. Due to this interaction, polyvalent cations may serve as bridges between clay surfaces and humic materials. The humic materials may either coat the clay surface or be associated with it in mixed aggregates. Humic substances may also interact with oxides of aluminum or of other polyvalent cations, thereby possibly reducing the extent to which such oxides block active sites at clay surfaces.

Surfaces of coated or ill defined particles may contribute most of the surface area in the unsaturated zone. Despite this, much of the more detailed work on surface interactions was conducted with pure mineral or organic constituents. In many cases interactions between the different components of the solid phase poison the surface for adsorption or heterogeneous catalysis, since the more active sites at the surface are those most likely to interact with a potential coating agent or with neighboring particles. For example, coating clay surfaces with organic matter, may bind exchangeable cations which are the site of many surface enhanced transformations (e.g. Mingelgrin et al. 1977). Coating may also physically block access to active sites which are themselves not coated, for instance, in the interlayer surfaces of clays. A larger surface area of the amorphous coating agent is exposed than would have been if the coating material was present in a particulate form. Coating may, therefore, enhance certain surface interactions, just as it retards others.

The heterogeneity of the surfaces in the unsaturated zone may affect surface interactions in yet another way. In order for a surface reaction to occur, a substrate molecule must adsorb at a suitable site. The presence of numerous types of adsorption sites increases the likelihood that the molecule will adsorb at sites other than the reactive ones. A surface reaction which is enhanced by an isolated component may, therefore, not take place to a significant extent in the field even if an exposed surface of that component is present there.

4.4 The Exchange Complex

The solid phase of the unsaturated zone contains charged components which endow it with both cation exchange and, to a lesser extent, anion exchange capacities. Counterions located on or near the surface of the charged components balance this charge. Exchangeable ions, both hydrated and anhydrous, are the site of many important surface interactions. These ions, having a localized electric charge, create a strong electric field about them. This field brings about polarization and in some cases dissociation of adsorbed species and in particular the cations' hydration water. The polarization and dissociation of water molecules in the hydration shell of exchangeable cations enhances in turn many surface reactions some of which will be discussed below. The electric field emanating from the surface of a charged mineral will not only affect the fate of adsorbate molecules bound to specific sites at the surface, but it will also affect abiotic transformations in that volume of the liquid phase which is close enough to the surface to be under the influence of the field. This volume is defined as the diffuse electric double layer region. The surface charge density and the nature of the exchangeable cations in the exchange complex will strongly influence the transformations which may take place in that region.

Much of the cation exchange capacity of the unsaturated zone arises from isomorphic substitution in the clay minerals, and is thus pH independent. The broken edges of clay crystals may display either a cation or an anion exchange capacity depending on the pH. Largely pH dependent cation and anion exchange capacities are also displayed by the metal (especially iron) oxides. The dominant cations in the exchange complex of most solid phases in arid regions are Ca^{+2}, Na^+, Mg^{+2} and to a lesser extent, K^+. In humid regions, Na^+ disappears and Al^{+3} (and at a sufficiently low pH also H^+) join the other three cations in dominating the exchange complex. Less frequently, NH_4^+ may also be an important exchangeable cation.

Many ions display specific adsorption on charged surfaces and thus adsorb more strongly then expected from their charge. This is especially true for many organic ions and for the transition metal cations with their available d or higher orbitals. The capacity of these cations to participate in specific interactions with the surface, correspond to a potential for specific interactions with adsorbates. Sludge application or irrigation with effluents, for example, may introduce heavy metal ions to the unsaturated zone both in free or complexed forms. The high selectivity of many charged surfaces for transition metal cations may bring about the cations' accumulation at the surface. Due to the capacity of heavy metal ions to act as complexants and catalysts for many chemical reactions, the accumulation of these ions at the surface may enhance the adsorption and abiotic degradation of organic pollutants. While the availability of d and higher orbitals may control the surface activity of transition metal cations, in the case of most exchangeable cations, their charge and radius are probably the factors which have the strongest influence on the cations' efficiency in promoting surface interactions.

Heavy metals often possess slightly soluble hydroxides or salts, and the charged surfaces may serve as an efficient sink for the ions of these metals which might have otherwise precipitated in various insoluble forms. Adsorbed ions are more likely to participate in reactions then are precipitated species due to the higher accessibility to potential reactants of adsorbed ions which are located at the surface than of ions embedded in the bulk of solid particles. Although adsorption may reduce the precipitation of transition metal ions as hydroxides or other insoluble salts, polyvalent metals may exist at the surface in various hydroxy species, the nature of which is determined by conditions, such as pH. Some of these adsorbed hydroxymetal species are charged oligomers or polymers which may contain sites with a capacity to enhance surface reactions. Other hydroxymetal species may, on the other hand, bind or coat active sites on the original surface. Cationic hydroxy-Al and hydroxy-Fe^{+3} species are probably the most common hydroxymetal species found at mineral surfaces.

4.5 Interrelations Between the Phases

The interaction between water and mineral surfaces may affect the behavior of organic pollutants in a number of ways. Water molecules adsorb strongly on clay and other mineral surfaces. As a result, water may compete for adsorption sites with some solutes and enhance the adsorption (e.g. through water bridges) of others. Adsorbed water, and in particular hydration water of adsorbed ions, participates in numerous surface enhanced reactions. The interaction of water molecules with solid surfaces also affects the distribution and transport of solutes. The restricted mobility of water (and hence solutes) through capillary pores, for example, is a result of this interaction. The moisture content, together with the chemical composition of the liquid phase, control the extent of swelling or dispersion of expandable clays. This swelling, in turn, will influence the accessibility of the interlayer surfaces to organic solutes. The accessibility of reaction sites and the rate of transport of reactants and products may control both the nature and the kinetics of surface enhanced transformations.

The properties of the liquid phase in the vicinity of the surface of charged solids are different from those of the bulk liquid phase, due to the aforementioned electric field emanating from such solids. In the volume of the liquid phase in which the strength of the electric field is not negligible (the diffuse electric double layer region) the concentration of the counter ions is significantly larger then in the bulk solution. The thickness of the double layer is a function of the valence of the exchangeable cations and of the ionic concentration in the liquid phase. The higher this ionic concentration, the thinner the double layer. Under conditions prevalent in soils at field capacity or at higher moisture contents, the double layer may in many cases be considerably thicker than 10 Å (e.g. Van Olphen 1966).

Near negatively charged surfaces, the cations' concentration steeply increases as the distance to the surface decreases. The concentration of anions decreases as the surface is approached. The electric field in the diffuse double layer region is thus controlled by the negative charge of the solid, which often originates from a charge imbalance inside the solid lattice and by the positive, nearly point charges of the exchangeable cations. Under the influence of that field, the water molecules in the diffuse double layer region experience a significant increase in their polarization and dissociation (e.g. Mortland 1970). It is evident from the above discussion that the pH in the diffuse double layer region near negatively charged surfaces is lower than in the bulk solution. This fact should strongly influence the rates of pH controlled transformations, such as many hydrolysis reactions, which occur in the diffuse double layer region.

The electron charge distribution within polarizable species (such as water molecules) is modified by the electric field near charged surfaces. This field may, therefore, significantly increase the dipole moments and the dissociation constants of polarizable species which are located sufficiently close to the surface. Thus, the surface charge can affect both the composition of the solution and the properties of the solutes, making the tendency of many organic molecules to undergo transformations higher near the surface then in the bulk solution. The effect of charged surfaces on reactions in the diffuse double layer region should be distinguished from the enhancement by the surface of reactions which require direct contact with the solid surface or even adsorption at specific sites on the surface. While the former reactions occur in a liquid phase, the latter reactions occur at a solid surface and could in many cases take place in the absence of a liquid phase.

As mentioned above, the pH of the liquid phase is strongly affected by the distance from a charged surface. Thus, the pH of the bulk solution may be several units higher than that of the water layer closest to a clay surface (e.g. Bailey et al. 1968). The mean pH value encountered by an organic molecule may, therefore, be a function of the moisture content of the system. When a liquid phase is absent, surface acidity (the ability of the surface to act as either a Lewis or a Bronsted acid) rather than the pH is often the parameter of interest as far as surface interactions are concerned. In many cases at least some of the surface acidity is due to the greater ease of dissociation of adsorbed water molecules as compared to free water molecules. Yet, the dissociation of water contributes as many hydroxyls as it does protons. The many surface interactions in which adsorbed water molecules participate (such as surface hydrolysis of esters, e.g. Mingelgrin et al. 1977), may therefore, involve both acidic (proton) and basic (hydroxyl) species and the designation of 'surface acidity' as the parameter controlling these interactions may at times be misleading. The tendency of the adsorbed water molecules to dissociate may in the case of such interactions be a more relevant parameter than is the surface acidity.

The moisture content in the soil or subsoil regulates the rates of gas exchange and of respiration and may thus have a strong effect on the redox potential. Surface enhanced redox reactions may therefore be strongly affected by the moisture content of the system even when water molecules do not participate directly in the reaction.

4.6 Surface Enhanced Transformations

Helling et al. (1971) reviewed the chemical transformations of pesticides in soil and classified them into 'reactions catalyzed by soils' and 'reactions not catalyzed by soil'. This classification, as pointed out by the authors themselves, is often arbitrary. For example, the rates of acid and base-catalyzed hydrolysis reactions in the liquid phase may be actually determined by constituents of the solid phase which buffer the pH. Abiotic transformations in the unsaturated zone may be classified according to the environment in which they occur. Numerous transformations take place in the homogeneous phases, especially in the liquid phase. Other transformations occur in the interface between phases. These include heterogeneously catalyzed reactions and reactions which occur in that part of the liquid phase which is under the influence of the electric field existing near charged surfaces. Crosby (1970), in a short review on abiotic transformations in the soil, presented some interesting examples of both surface enhanced reactions and reactions which take place in the bulk liquid phase.

Heterogeneous catalysis which requires adsorption at specific sites on the surface, is similar in many respects to enzymatic catalysis. Sites at the surface which possess a catalytic capacity may therefore be considered as primitive analogs of enzymes. Just as in the case of enzymatic reactions, small changes in the nature of the surface or in the environment around it may strongly affect the efficiency of the surface as a heterogeneous catalyst. Examples of this effect will be given below.

An interaction of an adsorbed molecule with the surface alters the electron charge distribution in the molecule. Such an interaction may, therefore, lower the energy barrier for some transformations and thus enhance them. Yet, while many organic substances break down faster when they are adsorbed on certain solids than when they are in solution, the degradation of others is inhibited by adsorption. A higher persistence of organic pollutants in the adsorbed state than in the free state was reported by a number of authors (e.g. Hurle and Walker 1980; Macalady and Wolfe 1985).

In some cases, the surface is modified as a result of the reaction it enhances. The reaction product may, for example, remain bound to the surface. Consequently, the term 'catalysis' does not strictly apply to all cases in which adsorption accelerates the rate of a reaction. Enhancement of abiotic reactions by adsorption will, however, be referred to below as 'surface catalysis' since this has been the practice of many authors in the past.

Among the organic biocides which were reported to break down by surface catalysis are trifluralin (Probst and Tepe 1969), diazinon (Konrad et al. 1967) ciodrin (Konrad and Chesters 1969), azido-triazines (Barnsley and Gabbott 1966) and many others (e.g. Morrill et al. 1982). Spencer et al. (1980) reported the conversion of parathion to paraoxon on dust particles in the presence of atmospheric oxidants and Hance (1970), concluded that adsorption catalyzes the hydrolysis of chlorotriazines in soils. Catalytic reactions on clay surfaces were reported by numerous authors and a few reviews on that subject were published (e.g. Mortland 1970; Theng 1974). McAuliffe and Coleman (1955) demonstrated the catalytic effect of acid clays on the hydrolysis of ethyl acetate. Mortland and Raman (1967) showed that Cu(II), which catalyzes the hydrolysis of organophosphate esters in solution, retained its catalytic capacity when it existed as an exchangeable cation on montmorillonite. While Mg-montmorillonite exhibited a much weaker catalytic effect than the Cu-clay, the Al- and Zn-clays did not catalyze the hydrolysis at all. Cu-beidellite and nontronite exhibited a lower catalytic capacity then Cu-montmorillonite and a Cu-organic soil did not enhance the hydrolysis of organophosphate esters. The results of the above study support the aforementioned suggestion that, as in the case of enzymes, the catalytic activity of a surface is often strongly influenced by small changes in the properties of the surface, even when the active site (the exchangeable Cu(II) ion) remains the same. Air-dried Cu-montmorillonite also catalyzed the degradation of urea to ammonium (Mortland 1966).

Rosenfield and Van Valkenburg (1965), suggested that the degradation of the organophosphate ester ronnel when adsorbed on bentonite is catalyzed by aluminum. The mechanism of the surface catalyzed degradation of organophosphate esters on clays was studied in detail (e.g. Mingelgrin et al. 1977; Mingelgrin and Saltzman 1979). This degradation displayed a strong dependence on the nature of the exchangeable cation and on the type of adsorbent.

Brown and White (1969) discussed the effect of the substituents of s-triazines on the transformation of these compounds at montmorillonitic clay surfaces. The methoxy and methoxy-thio substituted s-triazines hydrolyzed less rapidly than their Cl analogs. While the protonation and adsorption of the triazines were directly related to their basicity, the rate of degradation of these compounds was inversely related to their basicity. The above authors also observed that after a certain portion of the adsorbed molecules decomposed, no further degradation took place. Similar reaction kinetics were reported, as will be discussed below, for a number of other surface enhanced transformations.

Russell and co-workers (1968a) reported the catalytic hydroxylation (the substitution of the chlorine atom by a hydroxyl) of atrazine by H-montmorillonite. It was suggested by these authors that the degradation product remained adsorbed on the clay as the protonated hydroxy analog probably in its keto form. Skipper et al. (1978) showed that while Al- and H-montmorillonite, as well as a montmorillonitic soil clay, promoted the hydroysis of atrazine, the Ca- and Cu-clays enhanced that transformation to a far lesser extent. An allophanic soil clay did not catalyze atrazine hydrolysis with any of the exchangeable cations.

Theng (1974) summarized work on the degradation of aromatic amines when adsorbed on clays. A number of transformations were observed. Acid-base reactions were reported on clay surfaces which were capable of functioning as proton donors, such as surfaces of acid washed clays. Redox transformations of the amines were observed on clay surfaces in the presence of metal ions which are good reducing or oxidizing agents. In some cases, O_2 rather than metal ions, may oxidize organic molecules adsorbed on clay surfaces (e.g. 3,3',5,5'-tetramethyl benzidine on hectorite, McBride 1985).

Clays may catalyze (and in other cases inhibit) oligomerization reactions. The oligomerization of olefins, dienes, amino-acids, pyridine derivatives and other monomers was catalyzed by clay surfaces (e.g. Theng 1974; Lahav et al. 1978). Carbonium ion formation by surface protons was implicated in the polymerization of styrene and accordingly, H and Al-containing montmorillonites catalyzed the polymerization. Palygorskite and kaolinite catalyzed the polymerization of styrene more efficiently then montmorillonite (Solomon and Rosser 1965).

Al-montmorillonite acted as an acid catalyst in the formation of ethers from alcohols and alkenes (Adams et al. 1983). This reaction involved the formation of a carbonium ion immediately upon protonation. Practically any organic adsorbate, especially if it contains a carbon in an unsaturated bond, may interact with acidic sites at the surface to form a carbonium ion. The unstable carbonium ion may then act as an intermediate in numerous transformations. Two transformations which involve the formation of a carbonium ion were described above. The fungicide fenarimol, partially transformed at 100° C when adsorbed on montmorillonite into a carbonium ion, probably by interaction with the protons of the solvation water of the exchangeable cations (Fusi et al. 1983).

The catalysis of various types of rearrangement reactions by clays was also reported. Mingelgrin et al. (1978) and Mingelgrin and Saltzman (1979), described the rearrangement of parathion into 0,S-diethyl 0-paranitrophenyl phosphate at room temperature when adsorbed on clays (bentonite and kaolinite) in the absence of a liquid phase. The kaolinite catalyzed both the hydrolysis and rearrangement of parathion simultaneously and the rearrangement product underwent surface enhanced hydrolysis after its formation. The hydration shells of the exchangeable cations were suggested as the catalytically active sites. The rate of the rearrangement reaction increased with the polarity of the water of hydration.

Lopez Gonzales and Valenzuela-Calahorro (1970) reported that DDT transformed into DDE at clay mineral surfaces and that Na-bentonite catalyzed the transformation more efficiently then H-bentonite. This is an interesting example of a surface reaction for which increasing the surface acidity resulted in a retardation of the catalytic efficiency of the surface.

Adsorption on surfaces of minerals other then clays can also enhance transformations of organic chemicals. Metal (e.g. iron) oxides, are likely to enhance redox and other reactions (Fig. 1). Nash et al. (1973) demonstrated the capacity of MgO to enhance the degradation of DDT even at -5° C in dry soils. Such conditions suggest that the degradation is an abiotic, surface catalyzed reaction. Aluminum oxide catalyzed the hydrolysis and rearrangement of parathion (Mingelgrin and Saltzman 1979). Na_2CO_3 surfaces (Mingelgrin et al. 1977) as well as $CaSO_4$ and, to a lesser extent, $CaCO_3$ also enhanced the hydrolysis of parathion (Mingelgrin and Yaron 1974).

Heterogeneously catalyzed transformations, involve as a rule specific sites at the surface with which the reactant must be in contact for the transformation to take place. Surfaces may, however, enhance the transformations of molecules or ions which are found in the vicinity of the surface but are not adsorbed at specific sites. As discussed above, charged surfaces exert an electric field which decays with the distance from the surface. The effect of that field on the properties of that volume of the liquid phase which is under the field's influence was also discussed. Probably the most outstanding phenomenon in the diffuse electric double layer region is the rapid change in the concentration of charged solutes with distance from the surface. In the interfacial region near negatively charged surfaces there is a surplus of cations and a deficiency in anions as compared to the bulk solution. The concentration in the diffuse electric double layer region of potential inductors, catalysts (such as protons, hydroxyls or transition metal ions) and charged reactants and products may, therefore, be very different from their concentration in the bulk solution. A significant difference should thus exist between the kinetics of many transformations in the interfacial region and in the bulk liquid phase.

Adsorption measurements do not distinguish as a rule, between direct interactions with specific sites at the surface and accumulation in the interfacial region. Hence, some of the transformations which were observed in two phase (solid and liquid) systems and reported as surface catalyzed (e.g. Konrad and Chester 1969) may have actually been enhanced by the special conditions prevalent in the interfacial volume and not by direct interaction with specific sites at the solid surface. It is possible that in some cases, catalytic activity which was attributed to acid sites on the surface actually originated from the relatively high proton concentration in the diffuse electric double layer region at negatively charged solid surfaces.

One factor which may strongly affect the reactivity of organic pollutants in the diffuse double layer region is the influence of the electric field on the orientation, polarization and dissociation of chemical species. The fact that the orientation of polar molecules in the diffuse double layer region relative to each other is considerably more well defined than in the bulk solution may in some cases hinder and in other cases enhance their transformation. An electric field may polarize chemical species which thus become more likely to participate in, or enhance, various transformations. Polarization by the field may cause an increase in the dissociation constant of acidic and basic species. The increased tendency of polarizable species, and especially of water, to dissociate under the influence of an electric field should affect the rate of hydrolysis and of other reactions in the vicinity of charged surfaces. The well known increased tendency of water molecules in the first hydration shell of exchangeable cations to dissociate (e.g. Sposito 1984) is an example of the effect of an electric field on the dissociation of polarizable species. A similar effect exists throughout the diffuse double layer region, but it becomes less pronounced as the electric field weakens. The steep increase in the electric field as the surface (or a counter ion) is approached, implies that even catalytic processes which do not require an interaction with a specific adsorption site may often be more efficient in the first molecular layer near the surface than further away from it in the diffuse double layer region. Similarly, the reactivity of water molecules in the first hydration shell of an ion should be higher then that of water molecules one molecular layer away. Indeed, it was demonstrated that a number of surface catalyzed hydrolysis reactions were enhanced much more efficiently by water molecules in the first hydration shell of exchangeable cations than by other water molecules (e.g. Mingelgrin et al. 1977).

The activity of microorganisms, enzymes and other catalysts may be modified by their interaction with surfaces. As discussed below, there are enzymatic reactions which are enhanced by adsorption, while many other enzymatic reactions are retarded by it. Skujins (1967), has pointed out the importance of the excess of protons in the diffuse electric double layer region near negatively charged surfaces to enzymatic reactions in the vicinity of such surfaces. The significant effect surface interactions of both microorganisms and substrates have on the biotic transformations of organic pollutants is not, however, within the scope of this chapter.

4.7 Factors Affecting Surface Interactions

Environmental factors, such as temperature and moisture content, influence the kinetics of surface interactions very differently from the way they influence the kinetics of homogeneous reactions.

4.7.1 Temperature

Attempts were made to assay the importance of abiotic reactions in the transformations of biocides in soils by extrapolating rates of reaction at elevated temperatures to ambient temperatures (Hance 1967). These attempts may have in some cases underestimated the contribution of the abiotic reactions because of the dissimilar dependence of the rates of homogeneous and of surface reactions on the temperature. A continuous increase in the rate of degradation of amitrole in soils with temperature was observed by Ercegovich and Frear (1964) between 8° and 100° C. Such a temperature dependence indicates the presence of an abiotic transformation. Yet, taking into consideration the complexity of the temperature dependence of the kinetics of surface catalyzed reactions, the absence of a continuous increase in the rate of reaction with temperature would not have proven that the reaction is biologically controlled.

A suitable form of adsorption is a prerequisite for the occurrence of heterogeneous catalysis. The adsorbed molecule together with the site at the surface to which it is adsorbed form the intermediate complex through which the surface catalyzed reaction must pass. The temperature dependence of the kinetics of surface catalyzed transformations may thus be some composite of the effect of the temperature on the complex formation (adsorption) and breakdown steps. Frequently, the adsorption is exothermic and the adsorption coefficient is thus negatively correlated with the temperature (e.g. Bansal 1983), while the degradative step is endothermic and its rate is therefore positively correlated with the temperature.

Sufficiently large adsorbates (as many organic compounds are) may possess a number of functional groups which undergo various degradative and non-degradative interactions with the surface simultaneously. These interactions, which together define the observed adsorption, may each depend on the temperature in a different way. Non-degradative interactions may strongly affect the rate of the surface catalyzed reaction. Such interactions could modify the orientation of the adsorbate molecule relative to the surface or affect the electron charge distribution at the labile bond in the adsorbate molecule.

It is evident that the net temperature dependence of heterogeneously catalyzed transformations is hard to predict. It is also difficult to extract reliable thermodynamic parameters from the temperature dependence of the kinetics of such transformations.

The complexity of the effect of the temperature on the rate of surface reactions is exemplified by the degradation of organophosphate esters on clays. While the rate of hydrolysis of pirimiphos ethyl on Na-kaolinite was practically temperature independent between 23° and 47° C, the rate of that reaction on Na-bentonite increased strongly with the temperature (Mingelgrin et al. 1975). The rate of hydrolysis of parathion on some oven-dried homoionic kaolinites increased with temperature less than would have been expected for homogeneous hydrolysis and in the case of Ca-kaolinite, the rate was practically unaffected by the temperature (Saltzman et al. 1974).

4.7.2 Moisture Content

The moisture content has a drastic effect on the rate of many surface catalyzed transformations. Under sufficiently dry conditions, the moisture content will determine which sites on the surface will be anhydrous and which will be hydrated. The hydration status of a site will strongly influence the interaction of an adsorbate molecule with that site. For example, adding water to a benzidine-montmorillonite system inhibited the formation of the protonated benzidine-montmorillonite complex which was formed when benzidine came in contact with the dry clay (Theng 1974). Many surface transformations are enhanced by exchangeable cations. The hydration level of the cation will determine whether an adsorbate interacts through a water bridge or directly with the exchangeable cation thereby affecting the catalytic capacity of the cation. Some types of ester hydrolysis, among other reactions, were shown to be enhanced by the hydration water of cations. Their rates were, accordingly, controlled by the cation's hydration status. Oven-dried Na-kaolinite, which is almost anhydrous, catalyzed parathion hydrolysis much less efficiently than the same clay at a higher hydration level (Saltzman et al. 1976).

The hydration status of the exchangeable cation will determine the polarization and hence the dissociability of the cation's hydration water. The fewer water molecules in the hydration shell, the more polarized are these molecules and the higher their ability to act as Bronsted acids (e.g. Sposito 1984). Thus, the capacity of hydrated cations to catalyze certain transformations may diminish when the moisture content increases above some level. A sharp decrease in the rate of hydrolysis of parathion on a number of homoionic kaolinites (Saltzman et al. 1976) and of 1-(4-metoxyphenyl)- 2,3-epoxypropane on some homoionic montmorillonites and on Na-kaolinite (El-Amamy and Mill 1984) was observed when the moisture content went over a certain critical level. The above reactions were probably catalyzed by hydrated exchangeable cations.

When the surface is in contact with a free liquid phase, its catalytic properties may differ strongly from what they were in the absence of such a phase. The catalytic behavior of the surface in the presence of a liquid phase is important not only under saturated conditions, since even at moisture contents below saturation, a three dimensional liquid phase may be present, for example in capillary pores.

From the above discussion, it is evident that the ability of hydration water molecules to act as Bronsted acids or bases should be at a minimum in the presence of a free aqueous phase. The hydrolysis of organophosphate esters on homoionic kaolinites (e.g. Saltzman et al. 1976) required the presence of hydration water. Yet, at moisture contents above the saturation point of the surface, this hydrolysis proceeded considerably slower than on the partially hydrated clays.

The presence of an aqueous phase may affect the catalytic behavior of the surface by a number of other mechanisms. The above discussed diffuse electric double layer region, for example, can exist only in the presence of a liquid phase. Under saturated conditions, swelling clays may disperse or at least swell. Bulky solute molecules may then reach interlayer surface sites which, under drier conditions, were inaccessible to them. Dissociable groups in the solid fraction may exist in their dissociated form in the presence of an aqueous phase but not in its absence, making the presence of a liquid phase necessary for the occurrence of surface transformations which are enhanced by dissociated groups. Konrad and Chesters (1969), suggested that the surface degradation of ciodrin in water saturated soils was catalyzed by dissociated acidic groups in the soil's organic matter. Yet, dissociated acidic groups on mineral surfaces may similarly catalyze surface reactions. When a liquid phase is present, the transport of the reactants and the products of surface reactions is often considerably faster than in the absence of a liquid phase. A possible exception is the transport of volatile reactants or products. For non-volatile species, the very slow surface diffusion might dominate the transport in the absence of a liquid phase. The rate of transport will, in turn, affect the kinetics of surface transformations.

When a liquid phase is not present, those reactant molecules which are already adsorbed at the proper sites may transform promptly, but the transformation of additional reactant molecules is often hindered by the slow rate at which reactant molecules can reach the active sites and product molecules leave these sites. Accordingly, Saltzman et al. (1974), observed that, in the absence of a liquid phase, the kinetics of parathion degradation on Ca-kaolinite could be well approximated by two first order stages. The rate of degradation was much higher during the first stage than during the second one. In the presence of a liquid phase, a continuous, relatively fast equilibration between the bulk solution and the surface may maintain a significant rate of reaction even after the total number of molecules transformed surpassed the number of molecules initially adsorbed at the active sites.

The fixation (binding) of the reaction product to the active site may at times diminish the capacity of the reactant to replace the product at that site even when a liquid phase is present. Such a fixation was observed in the degradation of parathion on some kaolinites (e.g. Saltzman et al. 1976). Sanchez Camazano and Sanchez Martin (1983b), studied the hydrolysis of phosmet in suspensions of homoionic montmorillonites. At pH 6, the half life of the pesticide was 500 times shorter in the presence of Ca-montmorillonite than in the clay-free water. Two stages were again observed in the kinetics of this reaction and the rate of hydrolysis in the first stage was significantly higher than in the second. While the rate of the first stage might have been controlled by an actual surface reaction, the rate of the second stage was possibly controlled by the slow exchange at the active site between an adsorbed product and the dissolved reactant. This slow exchange could have resulted from a considerably stronger interaction between the active site and the product than between that site and the reactant. Soluble Cu and, to a lesser extent, Ca salts also catalyzed the hydrolysis of phosmet, but only a single, continuous stage was observed in the kinetics of these homogeneous transformations.

The moisture content may affect the distribution of organic pollutants between the different phases. The lower the moisture content, the higher is the fraction of the solute which is adsorbed. The drier the porous medium, the more of its volume is occupied by the gas phase. The fraction of the total pore volume which the gas phase occupies may strongly affect the distribution of volatile substances between the phases and hence the interactions of such substances with the surface of the solid phase.

Many of the abiotic transformations that organic pollutants may undergo in the unsaturated zone are acido-basic catalyzed reactions. The number, dissociability and extent of dissociation of the acido-basic groups at the surfaces of the solid phase may have therefore, a strong effect on the fate of many organic pollutants. Surface acidity is the parameter which is most frequently used to define the efficiency of a surface as an acid catalyst. When a liquid phase is present, its pH may become the most important parameter in defining the tendency of acido-basic catalysis to occur. The liquid phase may either supply acido-basic groups (e.g. protons) to the surface or serve as a sink for such groups, depending on the pH. Thus, the pH of the liquid phase influences not only acido-basic catalyzed reactions in the bulk solution or in the diffuse electric double layer region but also the properties of the catalytic sites at the surface itself. The possible influence of the pH in a soil suspension on the rate of adsorption catalyzed hydrolysis was suggested by Konrad and Chesters (1969) in their study on the degradation of ciodrin in soils.

The moisture content may, finally, determine the relative importance of biotic and abiotic processes. In sufficiently dry soils, microbial activity is inhibited, and the relative contribution of abiotic processes to the dissipation of many organic pollutants is likely to increase.

4.7.3. Intrinsic Properties of the Solid Phase

Adsorbed molecules may interact either with a component of the solid (e.g. oxidation by iron containing adsorbents) or with another adsorbed species (e.g. water of hydration, or adsorbed enzymes). In either case, the intrinsic properties of the solid, such as its surface charge density or cation exchange capacity will affect both adsorption and surface transformations. For the many transformations which require an interaction with an exchangeable cation or with its hydration water, the CEC will dictate the catalytic capacity (as distinguished from catalytic efficiency) of the solid. The surface charge density may affect surface interactions in a number of ways. For example, a higher surface charge density means a higher surface concentration of exchangeable counter ions which implies, in turn, a lower probability of interaction of nonionic species directly with the charged surface.

Mortland and Raman (1967), suggested that the efficiency of exchangeable Cu(II) as a catalyst in the surface enhanced hydrolysis of organophosphate esters, is determined by the strength of the electric field exerted by the charged solid. The stronger the field, the more it distorts the electron charge distribution in the exchangeable cation and its ligands. A strong field will reduce, for example, the polarization by the exchangeable cation (and hence the dissociability) of those ligands which are complexed with the cation in such a position that they project away from the surface. Accordingly, in the case of the above Cu(II) catalyzed hydrolysis, an increase in the strength of the electric field at the surface reduced the catalytic activity.

A surface property which is of great importance for many adsorption enhanced transformations, is the above defined (see section on 'Interrelations between the phases') surface acidity. The surface acidity of clay minerals was implicated in the surface catalyzed degradation of several pesticides (e.g. Brown and White 1969). The capacity of clays to catalyze both the hydrolysis of ethyl acetate and the inversion of sucrose was enhanced by the presence of exchangeable protons (McAuliffe and Coleman 1955). Mortland (1970), suggested that the dissimilarity between the capacity of H-montmorillonite and of allophane to catalyze the hydroxylation of atrazine resulted from the difference in surface acidity between the two minerals. Intrinsic properties of the solid, such as the presence of exposed structural acidic groups at its surface and their acidic strength, as well as environmental factors such as the nature of the exchangeable cation, the moisture content and the temperature, may all strongly influence the surface acidity.

In some cases, the acidity of the surface retarded its efficiency as a catalyst. The aforementioned transformation of DDT to DDE which was catalyzed more efficiently by Na- than by H-bentonite, may be such a case. It is not surprising, that a dehydrochlorination reaction, such as the conversion of DDT to DDE, is slowed down by an acid medium.

Both Bronsted and Lewis (electron-accepting) acidity are involved in the adsorption and surface transformations of organic substances at clay surfaces (e.g. Theng 1974). Fripiat (1968), discussed the increase in the acidity of water brought about by the high electric field at clay mineral surfaces and the related enhancement of surface reactions which involve protons. In their work on the clay catalyzed polymerization of styrene, Bittles et al. (1964), suggested that surface reactions which are initiated by the formation of a carbonium ion may involve a proton transfer from sites associated with tetrahedrally coordinated aluminum.

Ponec et al. (1974), discussed the role of surface acidity in a number of adsorption catalyzed reactions such as the polymerization of dienes and the dehydration of alcohols. Two pathways were proposed by these authors for reactions which involve acido-basic surface catalysts: the formation of a carbonium ion intermediate and the formation of a covalent alkoxy-metal bond with the surface. The second pathway was suggested, for example, for the dehydration of alcohols.

Solomon (1968), suggested that oxidized transition metals located in the silicate layers of clay minerals, may act as electron acceptors in surface catalyzed polymerization of organic monomers. While ferric ions may thus serve as electron acceptors, ferrous ions (and other ions of transition metals in their reduced state) located in the lattice of clay minerals may serve as electron donors (e.g. Theng 1974).

4.7.4 The Composition of the Exchange Complex

The adsorption of numerous organic molecules, especially polar and anionic ones, is through the exchangeable cations and their water of hydration (e.g. Praffit and Mortland 1968; Bowman and Sans 1977; Ristori et al. 1981; Sanchez Camazano and Sanchez Martin 1983a; Yariv et al. 1969; Mingelgrin and Tsvetkov 1985a). Due to their localized electric charge, the exchangeable cations strongly polarize their hydration water. Consequently, the dissociability of the water molecules in the first hydration shell of the exchangeable cations is likely to be considerably larger than that of water molecules adsorbed at practically any other site on charged mineral surfaces (e.g. Sposito 1984; Theng 1974). The higher the valence and the smaller the radius of the exchangeable cation, the stronger the effect it has on the behavior of its ligand water molecules. The enhanced dissociability of the hydration water of exchangeable cations plays an important role in the interaction of many organic species with surfaces of charged solids. Amino groups, for example, may take up a proton released by the strongly polarized hydration water and the resulting protonated species may then interact with the surface as an exchangeable cation (e.g. Mingelgrin and Tsvetkov 1985a).

Exchangeable cations and their hydration shells are apparently the sites of many surface transformations. The strong effect the type of the exchangeable cation dominating the exchange complex has on the kinetics of various surface reactions was reported by a number of authors (e.g. Mingelgrin et al. 1977; Sanchez Camazano and Sanchez Martin 1983b). Due to their high dissociability, water molecules in the first hydration shell of exchangeable cations may function as efficient Bronsted acids in surface enhanced transformations. When some transition metal cations, such as iron, are found in the exchange complex, they may catalyze redox reactions, just as they were reported to do when found in the silicate layers of clay minerals (e.g. Solomon 1968).

Among the factors which determine the catalytic efficiency of an exchangeable cation in transforming organic adsorbates are the charge of the cation, its radius, the arrangement of ligand molecules (in particular water) about it and the availability of competing modes of adsorption for the adsorbate. When the radius of the cation is sufficiently small, other groups at the surface and non-interacting groups in the adsorbate molecule may sterically hinder the adsorbed molecule from assuming the configuration which is most favorable for the transformation. The importance of steric factors in charge transfer interactions on clay surfaces was pointed out by Theng (1974). Al^{3+} with one hydration shell has a smaller radius than the similarly hydrated Ca^{2+}. This was suggested as the cause for the much lower efficiency of Al-kaolinite as compared to that of Ca-kaolinite in catalyzing the hydrolysis of parathion in the absence of a liquid phase (Mingelgrin et al. 1977).

The cationic composition of the exchange complex in the unsaturated zone is a function of the parent material and of the environmental conditions. This composition may change with time due, for example, to agricultural practices or disposal of industrial effluents. By altering the composition of the exchange complex, practices such as fertilization or liming may influence the interactions organic pollutants may undergo with the solid phase of the unsaturated zone.

4.7.5 Other Factors

The analogy between heterogeneous catalysis and enzymatic reactions was already mentioned above. It is demonstrated, for example, by the effect that small changes in the properties of the surface (other than the presence of the catalytically active site) have on the efficiency of the catalysis. Thus, water molecules located in the first hydration shell of the exchangeable cations are most likely responsible for the hydrolysis of organophosphate esters at the surface of negatively charged solids (Mingelgrin et al. 1977; Mingelgrin and Saltzman 1979). Yet, large differences between the kinetics of the hydrolysis at surfaces of different negatively charged minerals were observed even when the exchangeable cations and their hydration status were the same. The rates of the hydrolysis of parathion by two types of kaolinite were rather dissimilar and the rate of hydrolysis by a montmorillonite was considerably lower than the rates of hydrolysis by either kaolinite (Mingelgrin et al. 1977). The comparison between the rates was made for clays with the same cationic composition of their exchange complex and at the same hydration status. A rearrangement reaction seemed to dominate the transformation of parathion on montmorillonite (Mingelgrin and Saltzman 1979).

The accessibility of active sites may strongly affect the efficiency of heterogeneous catalysis. Groups other than those participating in the reaction, may, if bulky enough, sterically hinder the adsorbate molecule from reaching the active site at the surface or from assuming the orientation required for the surface reaction to take place. Restrictions on the orientation of adsorbate molecules relative to the surface in the interlayer spaces of smectites may play an important role in determining the catalytic activity of these clays.

A number of interactions between an adsorbate molecule and the surface may occur simultaneously. Only one (or some) of these interactions may directly catalyze the transformation of the adsorbed molecules. It was suggested that the interaction with hydrated exchangeable cations is responsible for the surface enhanced degradation of parathion (e.g. Fig. 2). Even if this interaction is similar on various surfaces, the fate of the adsorbed organophosphate ester may be very different at each surface. The aromatic moiety of parathion will interact with the hydroxyl dominated surface of the octahedral layer of kaolinite (e.g. Sahay and Low 1974) differently then with the oxygen dominated surface of the tetrahedral layer of that clay. In montmorillonite, only the latter type of surface is exposed, other than at broken edges. The strength and nature of the interaction between the surface and the aromatic moiety will affect the electron charge distribution about the phosphate ester bond and hence the lability of this bond to hydrolysis. Correspondingly, the rate of surface hydrolysis of the phosphate ester bond is

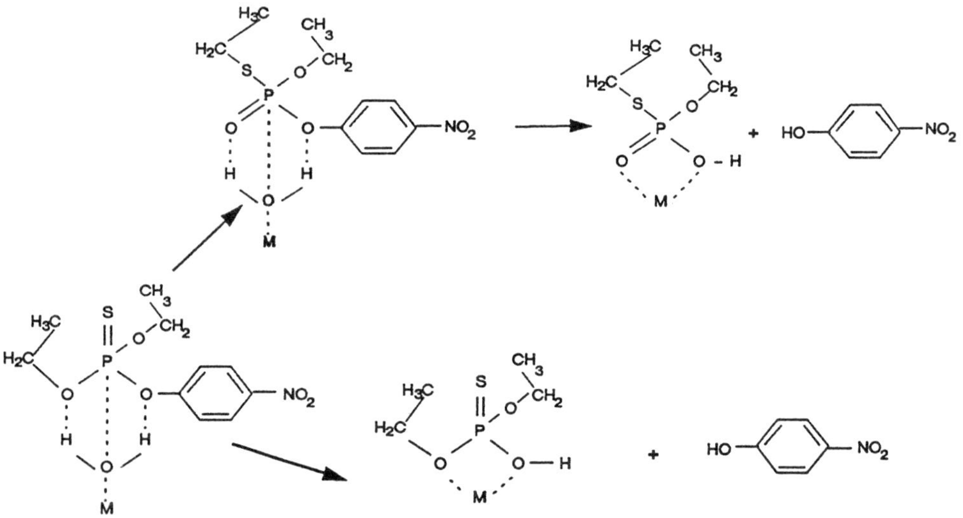

Fig. 2. A schematic representation of the surface catalyzed degradation of phosphorothioate esters exemplified by parathion. M represents an exchangeable cation (adapted from Mingelgrin and Saltzman 1979)

strongly influenced by the structure of the molecule as a whole (e.g. Mingelgrin et al. 1975; Saltzman et al. 1976). This influence may arise not only from the intrinsic difference in the strength of the hydrolyzed bond between various organophosphate esters, but also from the difference between the overall interactions of the various phosphate esters with the surface. In some cases, the adsorbate molecule possesses groups the interaction of which with the surface prevents the catalytic interaction of interest from occurring, but when that catalytic interaction does take place, the rate of hydrolysis will be strongly affected by simultaneous non degradative interactions with the surface.

One or more of the products of the surface catalyzed transformation may remain chemically sorbed (i.e. connected by a covalent bond) or otherwise strongly attached to the reaction site. Such a product is defined as fixed, or bound, to the surface. The binding of a reaction product to a catalytically active site will prevent this site from enhancing the transformation of additional adsorbate molecules. Fixation was reported, for example, for the phosphate moiety produced in the degradation of parathion on kaolinite (Saltzman et al. 1976). At times, the fixed product may retain some toxic activity.

Bound residues are xenobiotics or their degradation products which become bound to surfaces of the solid phase. Frequently it is the degradation products rather than the original molecules introduced to the environment which become bound. One reason for this may be the fact that the partial degradation of many organic molecules leads to the formation of species with a high chemical reactivity which may react with different sites on mineral or organic surfaces (Stevenson 1982).

4.8 Mechanisms and Kinetic Considerations

It may be instructive to present some mechanisms proposed for surface interactions of organic molecules. Russell et al. (1968a,b) suggested that when 3-aminotriazole or s-triazines are adsorbed on montmorillonite they become protonated. The protonation takes place at N atoms in the ring rather than at N atoms in the side chains. This is in agreement with the mechanism proposed by Armstrong and co-workers (Armstrong and Chesters 1968; Armstrong and Konrad 1974) for the hydrolysis of chloro-s-triazines on the surface of soil organic matter. H-bonding between N atoms in the atrazine ring and protonated surface carboxyl groups was implicated as the catalytic mode of interaction. The H-bonding (or protonation) of the ring N adjacent to the chlorinated carbon depletes the electron density at that C atom, making it more susceptible to an attack by nucleophilic agents and thus to hydrolysis. The highly polarized (and hence dissociable) water molecules present at a mineral surface (e.g. the hydration water of exchangeable cations) could interact with an N atom in the triazine ring through an H-bond just as the undissociated carboxyl groups at the surface of the organic matter do. Accordingly, Russell et al. (1968a) suggested that the more highly dissociable water on clay surfaces serves as the source of protons for the protonation of adsorbed chloro s-triazines when, as is often the case, exchangeable H^+ is not present and that this protonation is the pathway through which montmorillonite catalyzed the hydrolysis of the triazines.

Konrad and co-workers, in a series of articles, discussed the surface enhanced hydrolysis of organophosphate esters in soils. Their results are summarized by Armstrong and Konrad (1974). Diazinon underwent direct hydrolysis of the phosphate ester bond. Malathion hydrolyzed first at the phosphate ester bond and then proceeded to hydrolyze at the carboxyl ester bond which this pesticide contains. Ciodrin, which also contains both phosphate ester and carboxyl ester bonds, hydrolyzed first at the carboxyl ester bond and only then at the phosphate ester bond. Although detailed mechanisms for these surface enhanced transformations were not offered by the authors, a possible role of soil bound metal ions in the catalysis was suggested. Mingelgrin and co-workers (1977) and Mingelgrin and Saltzman (1979) proposed a mechanism for the surface enhanced hydrolysis of organophosphate esters which involves the interaction of the phosphate moiety with the highly polarized water molecules in the first hydration shell of exchangeable cations. It was suggested that in the case of phosphorothioate esters, this interaction may catalyze both the rearrangement and the hydrolysis of the molecule (Fig. 2). The relative importance of the two pathways depicted in Fig. 2 varies with the nature of the surface and the environmental conditions. Parathion degradation on bentonite powder, for example, proceeded predominantly through a rearrangement followed by hydrolysis. The proposed mode of interaction (Fig. 2), accounts for the above mentioned fixation of the phosphate (or thiophosphate) hydrolysis product.

The rate of a reaction depends, as a rule, on the concentrations of the participating species. The concentration relevant to adsorption catalyzed reactions is the surface concentration which will tend, in the presence of a liquid phase, to equilibrate with the bulk concentration. The dependence of the rate of surface reactions on the concentration of the reactants may be very different from that expected for homogeneous reactions. In the absence of a liquid phase, for example, the mobility of adsorbed molecules is controlled by the slow process of surface diffusion or by transport through the gaseous phase, which for non volatile organic contaminants is rather slow. Therefore, for non volatile adsorbates, only those molecules which were initially adsorbed at the active sites in the proper configuration may degrade at a high rate, while the molecules which are held at the surface through other modes of adsorption degrade subsequently at a much slower rate. The kinetics of hydrolysis of parathion on the surface of kaolinite in the absence of a liquid phase displays such a two stages kinetic pattern (Fig. 3). As long as the initial surface concentration was sufficiently low, the hydrolysis of parathion displayed to a good approximation first order kinetics in both the fast and the slow stages. When the initial surface concentration became

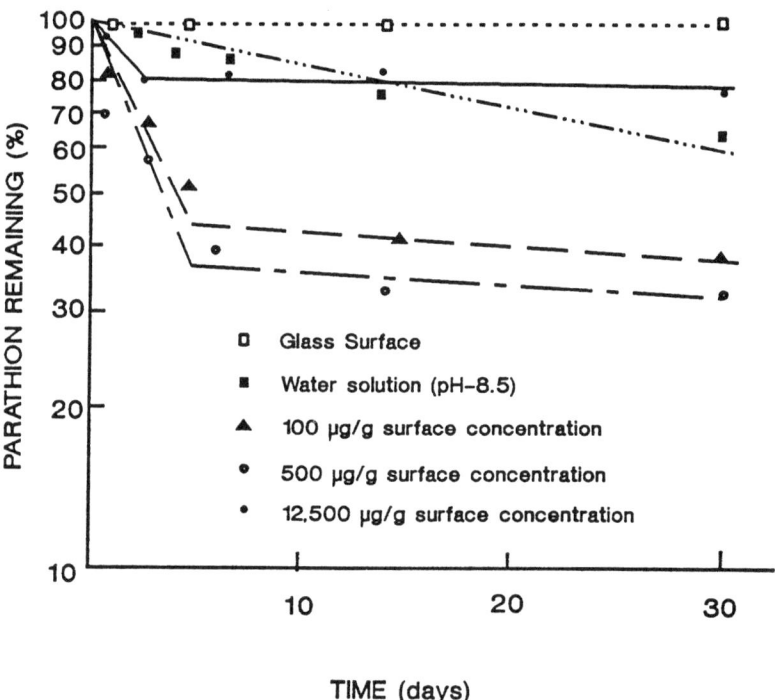

Fig. 3. Log parathion remaining vs. time. Surface hydrolysis of parathion on Ca-kaolinite at various initial surface concentrations as compared to the rates of degradation on glass surface and in water at pH 8.5. The reaction took place at 22° C; the clay was oven dried (105° C, 24h) prior to loading with parathion

high enough (i.e. when adsorption at the active sites neared saturation), the kinetics of the first, fast stage approached a pseudo zero order (see discussion following Eqs. 13–18) and the fraction of the adsorbate which degraded during the fast stage began to decrease as the initial surface concentration increased (Saltzman et al. 1974; Mingelgrin et al. 1977). It is thus apparent, that in dry systems the number of adsorbate molecules transformed by the surface at a significant rate will, in many cases, not exceed the number of the active sites present. When the catalyzing surfaces are in contact with a liquid phase, the number of reactant molecules which degrade at a substantial rate may be considerably larger than the number of active sites, due to the relatively high mobility of both reactants and products in the presence of a bulk liquid phase and the resulting faster exchange between them at the active sites.

Restrictions on the transport to or from the active sites may have a dominant effect on the kinetics of surface transformations even if a liquid phase is present. This can occur, for example, when fixation of the reaction products takes place, or when the active sites are located in the interlayer spaces of expandable clays and the reactant molecule is sufficiently bulky.

The kinetics of surface catalyzed reactions are considerably more difficult to describe with the aid of rate equations than the kinetics of homogeneous reactions. Limitations on the free access of potential reactants to the active surface sites and the slow exchange between non volatile reactants and products at the active sites when a liquid phase is not present are among the factors which contribute to this difficulty. Ponec et al. (1974) discussed the formal treatment of the kinetics of surface reactions at solid-gas interfaces. A similar treatment of surface reactions in the unsaturated zone will only rarely be useful because of the ideal conditions assumed by those authors (e.g. a mass flow rapid enough to avoid interference with the reaction kinetics).

Rate equations which successfully describe the kinetics of surface transformations under certain well defined conditions can be developed. Some equations which are applicable to surface reactions under various conditions common to the unsaturated zone are derived below. In most studies on the kinetics of reactions of organic pollutants in the unsaturated zone an empirical approach was adapted to define the rate equation for the investigated reaction. The expressions used included first order (e.g. Konrad and Chesters 1969), higher order and hyperbolic (e.g. Hamaker 1972) rate equations. An empirical approach is suitable for fitting experimental data to a desired mathematical expression but it is less appropriate for extrapolation purposes. Graham-Bryce (1981), pointed out the risk involved in applying empirically derived first order rate equations to pesticide concentrations which are outside the concentration range at which the empirical fit was performed. Half life values calculated by empirically fitting experimental data to first order rate equations are often valid only at concentrations which are sufficiently close to the concentration range at which the reaction kinetics were studied. A concentration independent half life exists only for truly first order reactions.

Hamaker (1972), suggested that surface catalyzed reactions may in some cases be well described by a hyperbolic rate model

$$-dc/dt = k_1 c/(k_2 + c), \tag{1}$$

where k_1 is the maximum rate achieved at sufficiently high concentrations and k_2 is the pseudo-equilibrium constant for the formation of a transition complex between the active site at the surface and the transformed molecule. The definition of k_2 arises from the formal similarity between the hyperbolic model and the Michaelis-Menten rate equation. The Michaelis-Menten equation describes enzymatic reactions and assumes the formation of a transition complex between the substrate and the enzyme. Thus, it may seem possible that the rate of a surface enhanced transformation which involves the formation of a complex between the active site and the adsorbed molecule may in some cases be approximated by a rate equation similar in form to the Michaelis-Menten equation. The Michaelis-Menten formulation assumes, however, that the concentration of the catalyst is very small as compared to the concentration of the substrate and that the concentration of the transition complex is maintained at a steady level. Another assumption inherent in the definition of enzymatic reactions is that the enzyme is regenerated as the reaction products are formed. The analogous assumption for surface enhanced reactions would be that the reaction products are removed from the active sites immediately after the transformation occurs, so that the adsorption of the products at the active sites is negligible. It is evident that the above assumptions do not hold for many surface reactions, especially in the absence of a liquid phase. Under some limiting conditions in the presence of a liquid phase (e.g. when only a small fraction of the organic pollutant is adsorbed, or when the catalytically active mode of adsorption is practically saturated), the hyperbolic rate model may be applicable to some surface reactions provided that the adsorption of the reaction products at the active sites is negligible. Farmer and Aochi (1987) stated that very few surface catalyzed reactions were shown to follow hyperbolic kinetics.

It is possible to construct a set of rate equations for surface reactions each equation being applicable to a limited and well defined range of conditions. Together, these equations could cover most situations which are likely to occur in the unsaturated zone. A few such equations are presented below.

4.8.1 A Two Phase (Solid-Liquid) System in Which the Adsorption of the Reactant Follows a Linear Adsorption Isotherm

The assumption of a linear adsorption isotherm is valid, for example, for many pesticides in wet soils or subsoils due to the low loads in which pesticides are usually found in the unsaturated zone.

For the following development to be valid, the rate of reaching an equilibrium between the adsorbed and dissolved states of the reactant should be considerably faster than the rate of the surface enhanced degradation step. In the presence of a liquid phase, this is in most cases a very reasonable approximation. Cases in which the rate of reaching an equilibrium between the adsorbed and dissolved states of the reactant is not much faster than the rate of the degradation step can be easily treated, but are beyond the scope of this illustrative presentation. The following development will hold, even if the rate of desorption of the degradation products is slow or if these products become bound to the surface provided that the total quantity of the reactant in the system is much smaller than the maximum amount of reactant which can be adsorbed in the catalytic mode of adsorption. Such a situation is not uncommon for organic contaminants in the unsaturated zone.

Let us assume, for the time being, that the catalytically active mode is the only mode of adsorption. If the adsorption of the reactant is well described by a linear isotherm, then

$$S' = k'C', \tag{2}$$

where S' is the quantity of reactant adsorbed per unit weight of solid, k' is a constant and C' is the equilibrium concentration in the liquid phase.

$$\text{Let } S = WS' \tag{3}$$
$$\text{and } C = \phi C', \tag{3a}$$

where S is the amount adsorbed per unit volume, W is the bulk density of the dry porous medium, C is the amount of unadsorbed (dissolved) reactant per unit volume and ϕ is the volume fraction occupied by the liquid phase.

$$\text{Let } k = k'W/\phi, \tag{3b}$$
$$\text{Then } S = kC \tag{4}$$
$$\text{and } S + C = R, \tag{5}$$

where R is the total quantity of reactant per unit volume.

$$\text{Hence, } S = Rk/(k+1). \tag{6}$$

If adsorption in modes which do not catalyze the degradation is significant, but all the modes of adsorption can be described by linear isotherms (a reasonable assumption at sufficiently low reactant concentrations), the constant k will simply be replaced in Eq. (6) by another constant.

$$\text{Thus, } S_i = k_i C \tag{4a}$$
$$\text{and } S = \Sigma S_i = (\Sigma k_i)C, \tag{4b}$$

where the subscript i denotes the ith mode of adsorption.

$$\text{Let } K = \Sigma k_i, \tag{3c}$$
$$\text{then } S = KC = RK/(K+1), \tag{6a}$$
$$\text{or } S_1 = k_1 C = Rk_1/(K+1) \tag{6b}$$

where the subscript 1 denotes the catalytic mode of adsorption.

The rate of disappearance of the reactant in the surface reaction is

$$dR/dt = -QS_1, \qquad (7)$$

where Q is a constant. That the rate is first order in S_1 is obvious, since the rate of degradation must be directly proportional to the amount adsorbed in the catalytic mode at any time (t).

Upon substituting R for S_1 in Eq. (7) from Eq. (6b), the solution of Eq. (7) becomes

$$R = R(0)\exp[-Qk_1 t/(K+1)], \qquad (8)$$

where (0) denotes the initial time.

In those cases in which the catalytically active mode dominates the adsorption, $S_1 \approx S$ and from Eq. (6), the solution of Eq. (7) is reduced to

$$R = R(0)\exp[-Qkt/(k+1)]. \qquad (8a)$$

When the catalytically active mode dominates the adsorption, k_1 is large enough so that

$$k_1 \gg k_i \text{ for all } i \neq 1, \qquad (9)$$

and $\qquad K \approx k_1 \approx k. \qquad (9a)$

Equations (8) and (8a) become then equivalent. In order to enhance a surface reaction at a significant rate, the interaction between the reactant molecule and the surface should be relatively strong, suggesting a large k_1. Thus, in situations in which linear adsorption isotherms are a reasonable approximation (as is usually the case when the adsorption sites are far from saturation), Eqs. (9) and (8a) may often be valid.

The rate of surface reactions in systems for which Eqs. (8) or (8a) hold is first order in the total concentration of the reactant in the system (R). The rate constant (Q') is

$$Qk/(k+1)$$

when only a single mode of adsorption exists and

$$Qk_1/(K+1)$$

when modes of adsorption other than the catalytically active one are important. The constant Q' and hence the half life of the reactant are functions of the adsorption coefficients (k or K and k_1) as well as of the intrinsic reaction rate constant (Q). The functional dependence of Q' on the adsorption coefficients offers some insight into the relationship between the extent of adsorption and the kinetics of the surface enhanced degradation.

For example, in systems in which Eq. (9) is valid, as k_1 (or k) becomes much larger than unity, the degradation rate becomes independent of the adsorption coefficients (Eqs. (9a), (8) and (8a)). This is so, because once k_1 is large enough to have practically all the reactant present adsorbed, a further increase in the adsorption coefficient cannot have a significant effect on the rate of the surface reaction.

When k (or K, if more than one adsorption mode contribute significantly to the adsorption) is sufficiently smaller than 1, the rate constant for the disappearance (Q′) becomes equal to the product Qk_1 or Qk (Eq. (8) or (8a) respectively). Thus, in systems in which only a small fraction of the reactant is adsorbed, the weight of the distribution of the reactant between the catalytically active mode of adsorption and the liquid phase is equal to that of the intrinsic rate constant of the surface degradation step in determining the rate of the surface enhanced reaction.

The half life of the transformed species is

$$-\ln(0.5)(K+1)/Qk_1. \tag{10}$$

If Eq. (9) holds, Eq. (10) is reduced to

$$-\ln(0.5)(k+1)/Qk. \tag{10a}$$

As k_1 increases, the half life decreases, but when Eq. (9) holds and k_1 becomes much larger than unity (i.e. practically all the reactant is adsorbed in the catalytically active mode), the half life becomes independent of the adsorption coefficient. Increasing K while k_1 is kept constant (namely, increasing the fraction of the reactant which is adsorbed on competing, non catalytic modes of adsorption), will, as expected, increase the half life.

4.8.2 A Two Phase (Solid-Liquid) System in Which the Catalytically Active Surface Sites are Saturated with the Reactant

The following development may be relevant to concentrated toxic waste disposal sites or to sites of accidental spills of toxic organic substances.

When the catalytic mode of adsorption is saturated with the reactant, S_1 becomes independent of C (or R). In the many cases in which the adsorption equilibrium is attained at a considerably faster rate than the rate of the surface enhanced degradation step, S_1 can be approximated by S_1(max) as long as C is large enough to maintain the catalytic adsorption mode saturated and there is no significant competing adsorption on the active sites. This requires, for example, that

$$Sp_1 \ll S_1, \tag{11}$$

where Sp_1 is the amount of the reaction products per unit volume which is adsorbed at the catalytically active sites. Under conditions in which $S_1 \approx S_1$(max), the solution of the rate equation (Eq. 7) becomes

$$R = R(0) - QS_1(\max)t, \tag{12}$$

where S_1(max) is the maximum amount of reactant which can be adsorbed in the catalytic mode. Equation (12) is the integrated form of a zero order rate equation.

Since $R \geq 0$, Equation (12) cannot apply at $t > R(0)/QS_1(max)$. As a matter of fact, the kinetics of the reaction can only be zero order for a considerably shorter time. Only as long as C is large enough and the concentration of the products is small enough (Eq. 11) to maintain $S_1 \approx S_1(max)$, will Equation (12) hold. As the reaction transforms enough of the reactant to have at equilibrium $S_1 < S_1(max)$, Equation (12) will cease to apply since S_1 can no longer be treated as a constant. Thus, in most cases in which $R(0)$ is large enough to have the active mode of adsorption saturated with the reactant, it is the initial rate of reaction, rather than the rate throughout the degradation process, which will be zero order (namely, independent of R).

4.8.3 A Porous Medium not in Contact with a Free Liquid Phase

This approximation is applicable to a considerable portion of the unsaturated zone, since by definition this zone contains solid surfaces which are not in contact with a free liquid phase. Surfaces in the unsaturated zone which are not in contact with a free liquid phase, are often loaded with organic pollutants which were introduced during a prior contact with a liquid phase. Such surfaces are in contact with the gas phase. The derivations below apply only to organic contaminants with a sufficiently low vapor pressure in the adsorbed state to make the content of the reactant in the gas phase insignificant and the rate of transfer of the reactant between the surface and the gas phase negligible as compared to the rate of the surface enhanced reaction.

Since practically all of the reactant exists in the adsorbed state,

$$R = S. \tag{13}$$
$$\text{Let } S = S_1 + S_2 + \ldots, \tag{13a}$$

assuming only sites of type 1 enhance the degradation of the organic pollutant. Then,

$$dR/dt = dS/dt = -QS_1. \tag{14}$$

Since no significant transport between a liquid or gas phase and the surface or between the different modes of adsorption takes place,

$$dS = dS_1 \tag{14a}$$

and solving Eq. (14) for S gives

$$S(t) = S_1(0)\exp(-Qt) + S_2(0) + S_3(0) + \ldots. \tag{15}$$

Equation (15) indicates that the reaction rate is first order in S_1. As long as S_1 is the dominant adsorption mode ($S_1 \gg \Sigma S_i$ where the sum is over all $i > 1$), the reaction rate is approximately first order in S. A dominance of the adsorption by S_1 (and hence a reaction rate which is approximately first order in S) is more likely to occur at low surface loads ($S \ll S_1(max)$) and it cannot occur if S is significantly larger than $S_1(max)$. The kinetics of parathion hydrolysis catalyzed by dry kaolinite (Fig. 3) is well approximated by Eq. (15) and the dependence of the kinetics of that reaction on the initial surface concentration of parathion suggests that as S increases, the relative contribution of the non degradative adsorption modes to the total adsorption increases as well.

In the absence of a free liquid phase, the Fermi-Dirac distribution law (e.g. Hill 1960) may often be successfully employed to describe the distribution of an adsorbate between the different modes of adsorption. Occupying a site in a particular mode of adsorption is analogous to the occupation of a state with a maximum occupancy of 1. The number of sites available for every mode of adsorption is the degeneracy of that mode. The application of the Fermi-Dirac distribution law gives

$$S_i = S_i(max)/(p_i + 1), \tag{16}$$
$$\text{where } p_i = e^{(E_i-\mu)/kT}, \tag{17}$$

E_i is the energy of adsorption in mode i and μ is the chemical potential of the adsorbate.

When S is considerably smaller than $\Sigma S_i(max)$, where the sum is over all significantly occupied modes of adsorption, namely when the adsorption is considerably below the adsorption capacity,

$$p_i = e^{(E_i/kT)} \Sigma S_j(max) e^{(-E_j/kT)}/S. \tag{17a}$$

The sum in Eq. (17a) is over all the available adsorption modes. As expected, when a single mode of adsorption dominates (e.g. when $E_1 \ll E_j$ and $S_1(max) \geq S_j(max)$, for all $j \neq 1$) and Eq. (17a) is used to define p_i, Eq. (16) will approach the correct limit, $S_1 = S$, only if $S \ll S_1(max)$.

Some of the limitations on the applicability of the above distribution should be pointed out. Eq. (16) will hold for a reactant only as long as the reaction did not proceed sufficiently to have a significant fraction of the reactant transformed into the reaction product. The above treatment also neglects adsorbate-adsorbate interactions and accordingly, multilayer adsorption. As a result, the approximation of the distribution of an adsorbate between the various modes of adsorption by the Fermi-Dirac distribution law improves as the surface concentration decreases. Finally, Eqs. (16), (17) and (17a) do not take into account the possibility of few modes of adsorption competing for the same sites. Correcting the above equations for competition between modes of adsorption for the same sites is straight forward but beyond the scope of this presentation.

When S is sufficiently small, $p_i \gg 1$ (Eq. 17a). The distribution of the adsorbate between the different modes of adsorption (Eq. 16) approaches then the classical Boltzmann distribution law. For large S, however, the Boltzmann distribution law must collapse, because, unlike Eq. (16), the Boltzmann distribution does not obey the restriction $S_i \leq S_i(max)$.

When $p_i \ll 1$, S_i becomes $S_i(max)$ (Eq. 16). The factor p_i will be considerably smaller than unity, if

$$\mu \gg E_i. \tag{18}$$

Since the chemical potential of the adsorbate increases with the number of adsorbed molecules, as expected, S_i will approach $S_i(max)$ at a sufficiently large S. The lower the E_i, the lower the μ (or S) at which condition (18) holds and S_i approaches $S_i(max)$. As pointed out above, it is reasonable to assume that a catalytically active mode of adsorption will display a relatively strong interaction with the surface (a low E_i). It may, therefore, be possible to approach the saturation of a catalytic adsorption mode even though the total adsorption is far from saturation. If the catalytically active mode is saturated with the reactant, the rate of disappearance of the reactant will not increase when more of it is adsorbed to the surface. In other words, when $S_1(0) \approx S_1(max)$, $S_1(0)$ will not change when $S(0)$ is increased and from Eqs. (14) and (15) it is apparent that adding adsorbate molecules to adsorption modes other than the active mode 1 will not affect the rate of the surface transformation. The kinetics of the surface reaction when $S_1(0) = S_1(max)$ may thus be defined as pseudo "zero order" since the reaction rate is then independent of $S(0)$. In systems subject to wetting and drying cycles, such as most soils, an increase in the surface concentration of an organic pollutant can occur during the drying period due to the gradual evaporation or uptake of the solvent. When the surface load becomes sufficiently high (even if it is still considerably below saturation) the catalytically active mode may become nearly saturated and hence the above defined pseudo "zero order" kinetics may be approached. In the surface hydrolysis of parathion on dry kaolinite depicted in Fig. 3, the kinetics of the reaction at an initial surface concentration of 12,500 μg/g seems to approach pseudo "zero order" (e.g. $S(t)/S(0)$ is an increasing function of $S(0)$ for all t), although a load of 12,500 μg/g is below the adsorption capacity.

4.9 A Unified Description of Surface Interactions

Adsorption measurements simply quantify the accumulation of a species at the interface between a solid and a non-solid phase in excess of the amount required to maintain in the interface the same concentration as in the bulk phase. Accordingly, negative adsorption means a lower concentration at the interface than in the bulk phase. The accumulation of a chemical species at a surface, does not necessarily imply an attraction between the surface and the adsorbate. Likewise, there may be cases in which a significant attraction between the adsorbate and the surface does exist and yet no measurable adsorption is observed. It is the free energy (or enthalpy and entropy) balance of a number of processes which will determine the distribution of a species between the bulk phase and the interface. For example, transferring a dissolved molecule from the bulk liquid phase to a solid surface, allows the approach of solvent molecules which surrounded the dissolved molecule towards other solvent molecules with the resultant change in the thermodynamic parameters (e.g. energy) of the system determined by the nature of the solvent-solvent interaction. Similarly, attaching a solute molecule to a site on a surface often necessitates the removal of a solvent molecule from that site, thereby reducing the net adsorption energy by the solvent-adsorbent interaction energy. The change in the thermodynamic parameters of the system resulting from the transfer of a molecule from the dissolved state in the bulk solution to the adsorbed state at a surface, determines how a substance will partition between the interface and the bulk phase and this change in the thermodynamic parameters is, in turn, a function of the solvent-solvent, solvent-solute, solvent-adsorbent as well as adsorbent-adsorbate interactions. Numerous cases of adsorption attributed to 'hydrophobic interactions' may in fact be examples of adsorption which takes place despite of the fact that the interaction between the adsorbate and the surface is negligible. If the solvent-solvent interaction is sufficiently strong and both the solvent-solute and solvent-adsorbent interactions are weak, a dissolved species will tend to accumulate at the surface even if the adsorbate-adsorbent interaction is very weak. It is thus not surprising that slightly soluble organic molecules often display considerable adsorption on nonpolar surfaces (namely, surfaces which do not interact strongly with water) from aqueous solutions even though the slightly soluble organic molecules display as a rule, rather weak surface-adsorbate interactions. In principle, 'hydrophobic adsorption' from aqueous solutions may be displayed by weakly polar molecules not only on nonpolar surfaces, but even on 'hydrophilic', namely considerably polar or charged, surface. This may happen, for example, if the solvent-solvent interaction is sufficiently stronger than the (rather strong) solvent-surface interaction, or if the surface contains both strongly polar (or charged) and nonpolar sites. Surface interactions do control the extent of adsorption of organic pollutants in the absence of a polar liquid phase and in some cases, also in the presence of such a phase. The thermodynamic treatment of the distribution of a species between the interface and the non-solid bulk phase at equilibrium is similar in many respects to the treatment of the solubility of a species and will not be further reviewed here.

As the polarity or polarizability of a molecule increases, so does the strength of its potential interaction with the surface. Similarly, as the charge or polarity of a site on the surface increases, so does the strength of its potential interaction with the adsorbate. Ions will, of course be repelled by sites on the surface possessing a similar charge. A large number of charged and strongly polar sites exists at the surfaces of both the organic and the mineral fractions in the unsaturated zone. Clay minerals often contribute most of the charged surface area of the mineral fraction. That the organic fraction is also rich in polar and charged sites, is attested by the high cation exchange capacity of the organic matter at the pH range which is encountered in the environment. The dominant role the organic fraction was observed to play in the uptake of non-polar or weakly polar organic molecules arises from the presence of weakly polar (or 'hydrophobic') sites on the surface of the organic matter in addition to the strongly polar and charged sites present there.

Two factors will determine the strength of an electrostatic interaction between an adsorbate and a site on the surface: the site's accessibility and the electric field at the site. An organic molecule may possess a number of groups with different polarities or charges. In order to maximize a potentially strong interaction between a site on the surface and a group in the molecule, neither the molecule nor the surface should contain groups which block the approach of the interacting moieties towards each other. Sites at the surface which are likely to display the strongest electrostatic interaction with adsorbates are thus those with both the strongest electric field about them and the highest accessibility. For non-ionic adsorbates which possess either a permanent or an induced dipole moment, it is the magnitude of the component of the electric field about the site of interaction perpendicular to the surface rather than its sign which is often the important factor in determining the strength of the interaction.

Most charged components of the solid phase of the unsaturated zone are negatively charged and exchangeable cations are therefore far more abundant in that zone than exchangeable anions. It is submitted that the exchangeable ions (which, as stated above, are predominantly cations), are the sites of the strongest surface interactions of neutral species in the unsaturated zone. This assertion will be discussed in detail below. The assumed dominant role of the exchangeable cations in surface interactions, enables the treatment of the interactions of organic pollutants with the surfaces of both the mineral and organic fractions of the unsaturated zone in a unified way since both fractions are negatively charged under the conditions prevailing in the environment and hence both possess exchangeable cations at their surface. For the sake of clarity, the important distinction between surface interaction and adsorption (namely, accumulation at the surface) which was discussed above, is reiterated. While a strong surface interaction may not always result in considerable adsorption from an aqueous solution, the strength of the surface interaction will determine the extent of adsorption from a gas or a non-polar liquid phase as well as the susceptibility of the interacting molecule to various reactions.

Exchangeable cations possess the two properties necessary to promote a strong electrostatic interaction: a high charge density (which creates a strong electric field near the cation) and a relatively easy accessibility. These cations possess by definition a permanent charge and are located at the surface. The strength of interaction between charged surfaces and organic adsorbates was often shown to increases with the valence (or more generally, with the electronegativity) of the exchangeable cation (e.g. Van Bladel and Moreale 1974). The ease of accessibility is enhanced by the fact that exchangeable cations protrude, as a rule, out of the plane of the charged solid's surface. In the absence of a bulk aqueous phase, the protrusion is more pronounced when the cation is hydrated. Thus, there is less steric hindrance by non-interacting groups to interactions with the water of hydration of the cation than to interactions with the cation itself due to the hydrated cation's larger radius. The strong polarization of the hydration water by the cation counteracts the long distance separating the interacting moiety in the adsorbate from the charged cation when the interaction is with the cation's hydration water as compared to the closer approach which may be possible if the interaction were with the anhydrous cation. The importance of the radius of the exchangeable cation in determining the efficiency of the cation in enhancing the surface hydrolysis of organophosphate esters was discussed above.

Both adsorption and abiotic degradation of many organic substances in the unsaturated zone displayed a strong dependence on the nature of the exchangeable cation dominating the exchange complex and on the cation's hydration status. Thus, Mingelgrin et al. (1977), demonstrated that the hydration status of the exchangeable cation has a strong effect on the rate of hydrolysis of organophosphate esters on homoionic kaolinites. Odom et al. (1979), reported a strong influence of the nature of the exchangeable cation on the binding of nucleotides on bentonite and Lawless and Levi (1979), demonstrated that the nature of the exchangeable cation affects the rate of oligomerization of glycine in the presence of bentonite.

A good example of a polar adsorbate is water. If as suggested, exchangeable cations are the site of the strongest electrostatic interaction with neutral adsorbates, the hydration energy of the exchangeable cations (at least of the more electronegative, e.g. di- or polyvalent, ones) should be higher than that of any other site on the charged surfaces in the unsaturated zone. Accordingly, most exchangeable cations on clay surfaces are still hydrated at low relative humidities in which other sites on the surface are practically anhydrous (e.g. Grim 1968). When a bulk aqueous phase is present, the accumulation of an adsorbate about the exchangeable cations will be hindered by water molecules competing for interaction with the cations. Some organic contaminants, even those possessing a substantial dipole moment, may therefore, not display a significant adsorption while others may adsorb at sites in which the electric field is weaker. Water molecules in the first hydration shell of exchangeable cations are probably the sites at which the strongest electrostatic interaction next to direct interaction with the cations takes place. The role of the exchangeable cations in surface interactions is further complicated in the presence of a bulk aqueous phase by the distribution of the cations throughout the diffuse electric double layer region. Interaction with a cation, either directly or through a water bridge, does not require, however, that the cation will be contiguous to the surface. At any rate, the very high concentration of cations in the plane adjacent to a negatively charged surface and the frequent existence of a Stern layer, minimize the importance of this complication.

By definition, the unsaturated zone contains surfaces which are not in contact with a bulk liquid phase. In the absence of a bulk liquid phase, exchangeable cations (whether hydrated or not) are unambiguously the preferred sites for surface interactions of polar, polarizable or anionic organic species. The interaction of most organic cations with negatively charged surfaces will not be limited to simple cation exchange. Organic cations often contain a number of functional groups which may interact with the surface simultaneously with the positively charged group. These simultaneous interactions of the non cationic groups will be governed by the same physical principles which govern the surface interactions of other similar organic moieties.

Solids in the unsaturated zone may possess charged groups other than exchangeable ions at their surface. Such charged groups may in some cases be just as accessible to adsorbates as exchangeable cations are. Yet, the charge density of di- or polyvalent exchangeable cations is higher than that of most other charged groups which exist at the surfaces in the unsaturated zone. Furthermore, exchangeable counter ions are likely to protrude further away from the surface than other charged surface groups and therefore be more accessible.

The exposure of a molecule to an electric field will perturb its electron charge distribution. The stronger the field, the larger the perturbation and the more unstable is the molecule likely to become. The sites of the strongest surface interactions, (which, as suggested, above are often exchangeable cations or their hydration shell) are also therefore likely to be sites of surface enhanced transformations. The role of the exchangeable cations and their water of hydration in surface catalysis was already pointed out. The high reactivity of the hydration water as compared to the reactivity of bulk water which results from the strong polarization of the hydration water by the field emanating from the cation, was also discussed in this chapter.

Many substances which are defined as hydrophobic do have a dipole moment, in some cases a considerable one. A sufficiently large molecule may possess a high dipole moment and still have a large hydrophobic part (the hydrophobic surface area, HSA, of the molecule is large). Such a molecule may display a 'hydrophobic' behavior, namely, low aqueous solubility and a considerable adsorption at non polar sites. The fact that the solvent-solute interactions are weaker than the sum total of the solvent-solvent interactions enabled by the removal of the 'hydrophobic' molecules from the bulk solution to the interface, will drive the 'hydrophobic' solute to the surface and in particular to the non polar sites on it. The large HSA of the molecule may prevent it from successfully competing with polar solvent (e.g. water) molecules for adsorption on charged or polar sites at the surface. When the surface is not in contact with a free aqueous phase, competition with water for interaction with charged or polar sites may be less significant and adsorption on polar or charged surfaces should increase. The less volatile components of crude oil are a good example of compounds to which the present discussion applies. Most of these components are considered hydrophobic and yet they display considerable adsorption from the crude oil mixture onto the highly charged ('hydrophilic') clays in the absence of a bulk aqueous phase.

The expected strong interaction between exchangeable cations and many organic adsorbates should not obscure the role of other possible surface interactions. In the absence of competition with a solvent or with other adsorbates for adsorption sites, the distribution of an adsorbate between the various types of interaction sites is a function of the relative number of available sites of each type and not only of the strength of interaction with the site and its accessibility. In addition, when the load on the surface is sufficiently high, the strongly interacting sites may be saturated and other sites may begin to dominate the surface-adsorbate interactions. Yet, even if in some cases interactions with other sites predominate, the strongest surface interactions of non-cationic organic pollutants in the unsaturated zone are, by and large, with exchangeable cations and their hydration water.

Many studies demonstrated the strong interaction between the soil's organic matter and di- or multivalent cations. Lewis acidity was invoked in these interactions and inner and outer shell complexes were described (e.g. Bloom and McBride 1979). Cation bridges were suggested as a mechanism for the formation of the clay-organic matter complexes which are so prevalent in the environment (e.g. Greenland 1971). The interaction between exchangeable cations and organic contaminants is just a mirror image of the above mentioned complexation of cations by the soil's organic matter. It is the cation's accessibility rather than their exchangeability that enhances the interaction of adsorbates with the exchangeable cations. Hence, interactions with ions which are located at the surface in an accessible position but are not exchangeable may be just as strong as the interactions with exchangeable cations. Aluminum and iron ions at broken edges of clay minerals and dissociated groups at organic surfaces are examples of sites which may in some cases display strong interactions with organic adsorbates. At any rate, exchangeable cations are probably the most important highly charged and accessible sites at surfaces in the unsaturated zone as far as the surface interactions of organic pollutants are concerned.

An adsorbate may interact with a cation directly or through a water bridge. Thus, when the cation is hydrated, both ligand exchange and outer shell complexation may take place and water can act either as a competing ligand or as an adsorption site. If the interaction with the cation involves a replacement of a water molecule, the heat of hydration of the cation will contribute negatively to the magnitude of the enthalpy (or net heat) of interaction. It is likely, therefore, that the moisture content together with the types of exchangeable cations present in the system may often be the dominant factors in determining which surface interactions (including surface catalyzed transformations) an organic pollutant will undergo in the unsaturated zone. Since adsorption and interaction with the surface are not synonymous, the effect of the moisture content and of the composition of the exchange complex on the extent of adsorption should be distinguished from their effect on the nature and strength of the surface interactions which may take place.

The hydration energy of any given cation increases as the number of water molecules attached to it decreases. The increase in the strength of interaction between the cation and its hydration water upon partial drying results in a similar increase in the polarization and hence in the dissociability of the water molecules (e.g. Mortland 1970; Sposito 1984). A higher polarization of the hydration water molecules means, in turn, a potentially stronger electrostatic interaction between an adsorbate and the water of hydration, namely, a stronger interaction through a water bridge. The importance of water bridges in surface interactions was suggested by a number of authors (e.g. Yariv et al. 1969; Mortland 1970). Figure 4, for example, represents two possible modes of interaction between amino groups and exchangeable cations through a water bridge. An increase in the

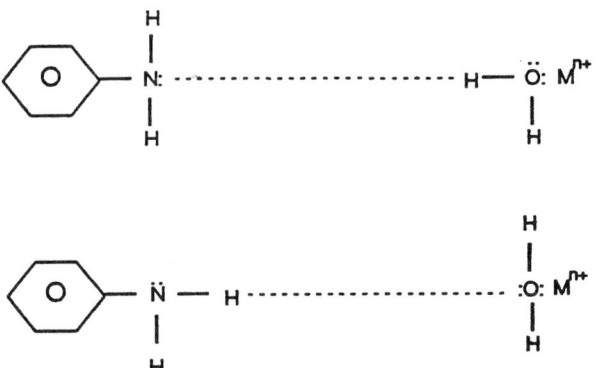

Fig. 4. Two possible water bridge interactions between amino groups and hydrated exchangeable cations

dissociability of water means that both the acidity and basicity of the system increases, since the dissociation of a water molecule creates both a proton and a hydroxyl ion. The rates of many acido-basic surface enhanced reactions should, therefore, be sensitive to the hydration status of the surface. This was shown to be true in a number of cases (e.g. Saltzman et al. 1976).

The strong electric field about a cation may affect the interactions between the adsorbate and sites at the surface other than the cation or its hydration shell. For example, basic (e.g. amino) groups may be protonated by the strongly polarized hydration water and the resultant cationic protonation product will interact with the surface as an exchangeable cation (e.g. Mingelgrin and Tsvetkov 1985a). Adsorbate molecules interacting with the cation or its hydration water may enhance some specific adsorbate-adsorbate interactions. They may, for example, serve as nuclei for surface condensation (e.g. Mingelgrin and Tsvetkov 1985b). Surface condensation is a process in which the molecules of a substance are stacked at the surface in a 3-dimensional array (a 'phase') with bulk properties (e.g. phase transition temperatures or solubility) different from those of the free substance. This condensation is not synonymous with multi layer adsorption, a form of adsorption in which the adsorbate is arranged in consecutive layers but the properties of the layers which are not contiguous to the surface quickly approach those of the free substance. Surface condensation is more likely to occur with adsorbates which are solids or viscous liquids at the temperature of interest. A species undergoing surface condensation will follow a BET-like adsorption isotherm. Under certain conditions, the adsorption of some organophosphate esters was shown to follow a BET-like isotherm while the solubility of the organophosphate esters in the presence of the adsorbing solid was different from that of the free esters and it displayed a significant dependence on the nature of the adsorbing solid and of the dominant exchangeable cation (Mingelgrin and Tsvetkov 1985b).

4.10 Enzymatic Reactions

The proteinaceous nature of enzymes suggests that at the pH range prevalent in the unsaturated zone, most of the extracellular enzyme content is adsorbed on clay and organic surfaces (e.g. Skujins 1967). The extraction of enzymes from soils or subsoils is often difficult, probably because of the aforementioned tendency of enzymes to adsorb on the solid surfaces.

Enzymes are frequently more stable in the environment then in vitro. Their role in the transformations of organic pollutants in the unsaturated zone may, therefore, be more important than might have been concluded from the lability of enzymes to microbial and other degradation processes. The added stability of the enzymes in the unsaturated zone results, most likely, from their strong association with solid surfaces. This association may, however, affect not only the stability, but also the activity of the enzymes. While in many cases adsorption reduces the enzymatic activity, in other cases adsorption may enhance this activity, or at least leave it unchanged. The activity of urease, for example, was hardly affected by adsorption on bentonite (Durand 1965) and the addition of montmorillonite to soil did not affect the activity of phosphatase (Kroll and Kramer 1955). Hoffmann (1959) demonstrated that urease released from lysed cells adsorbed on the clay fraction of the soil and remained active. Boyd and Mortland (1985), demonstrated that urease did not lose its activity upon adsorption on a smectite-alkyl ammonium complex and suggested that the enzyme may likewise retain its activity when adsorbed on the natural clay-organic matter complexes which are so common in the unsaturated zone. Ross and McNeilly (1972) found that glucose oxidase adsorbed very strongly on illite, montmorillonite and allophane. A comparison between the enzymatic activity in suspensions of the three clays and the enzymatic activity in a clay-free solution, indicated that when the systems were buffered with an acetate-phosphate buffer, each of these clays inhibited the enzyme's activity to some extent. In the unbuffered systems, the enzyme was more active when adsorbed on either the allophane or illite than in the clay-free aqueous solution. Many biological reactions occur at liquid-solid interfaces (e.g. Skujins 1967). It is, therefore, not surprising that adsorbed enzymes often retain their activity.

Not only the adsorption of the enzymes, but also the adsorption of their substrates may have a considerable influence on the catalytic efficiency of extracellular enzymes. The rate of the enzymatic degradation of adsorbed substrates was reported to be in some cases higher, but more frequently lower, than the rate of the enzymatic degradation of non-adsorbed substrates (e.g. Skujins 1967; Estermann et al. 1959; Estermann and McLaren 1959). Heuer et al. (1976) suggested that the degradation of some organophosphate insecticides by phosphatases was inhibited in the soil by the adsorption of both the enzymes and the substrates.

The enzymatic activity near the surface of charged solids may also be influenced by factors other than the direct interactions between the substrate or the enzyme and the surface. An example of such a factor is the concentration of ions near charged surfaces. In the presence of a liquid phase, the concentration of protons increases and that of hydroxyls decreases as a negatively charged surface is approached. Accordingly, the pH at the negatively charged surface is considerably lower than the pH of the bulk solution. Since the activity of enzymes is, as a rule, sensitive to the pH, this activity should be strongly affected by the distance of the enzyme from the charged surface. The pH values measured in suspensions are mean values for the whole system and if the solid component of the suspension possesses a negatively charged surface, the measured pH is higher than the actual pH near the negatively charged surfaces. Thus, as a result of both the strong interaction of enzymes and many substrates with charged surfaces and the discrepancy between the measured pH and the actual pH near the charged surfaces, the measured pH at which an enzymatic reaction displays a maximum rate in the unsaturated zone may not be the same as the optimum pH for the reaction in a solid-free aqueous solution. The availability of ions other than H^+ and OH^- (for example ions of transition elements) may play an important role in enzymatic reactions (e.g. Juma and Tabatabai 1977; Press et al. 1985). The change in the concentration of these ions with distance from charged surfaces further complicates the effect of proximity to surfaces on the activity of enzymes.

The considerable adsorption displayed by both enzymes and many substrates in the unsaturated zone suggests the possibility of enzyme catalyzed formation of bound residues. Liu et al. (1985) for example, reported that the binding of amino acid esters to phenolic humus constituents was catalyzed by an extracellular enzyme. Similarly, enzyme catalyzed, covalent bonding between organic species and some sites on mineral surfaces is also conceivable.

4.11 General Discussion and Conclusions

Adsorption at surfaces is a major mechanism by which transport downwards of toxic residues is checked, allowing time for degradative processes to eliminate the residues. Surface interactions play an important role not only in the retardation of the transport but also in the deactivation and breakdown of contaminants in the unsaturated zone. This role is particularly important under conditions in which biotic transformations are not likely to take place. Whenever the organic matter content is sufficiently low (e.g. in arid areas), mineral surfaces are likely to dominate the surface interactions.

Toxic residues may be leached below the root zone and at times reach the ground water. The concentration of nutrients, the organic matter content and the rate of gas exchange usually decrease with depth. As a result, the microbial population and, with it, the prevalence of biotic transformations also decline with depth, while interactions with mineral surfaces become more important in determining the fate of organic toxicants.

There is an ever increasing number of cases of pollution resulting from the leakage of toxic organic substances from concentrated wastes disposal sites. The high concentration of toxicants in these wastes often inhibits microbial degradation and abiotic transformations may then be the major pathway for the breakdown of the toxic organic substances.

Below some critical moisture content, microbial activity is strongly reduced. Under dry condition, physical and chemical phenomena such as surface interactions may, therefore, determine the rate of disappearance of organic pollutants. In arid zones, damage to crops from herbicide residues is often controlled by the adsorption or surface transformation of the herbicide.

Detoxification of contaminated areas may be achieved by manipulating the conditions in these areas so as to enhance the abiotic transformation of the contaminant. Systems containing high concentrations of organophosphate pesticides may, for example, be detoxified by adding a Cu^{+2} salt at a low concentration and by adjusting the pH. Abiotic transformations do not, however, always reduce the biocidal activity in the contaminated systems, since products of abiotic transformations of toxic substances may still be toxic (e.g. Joiner et al. 1973).

Bound residues are often much more persistent then the unbound parent molecule and they may retain a biocidal activity. Fixation to surfaces may, therefore, bring about an accumulation of toxic residues to harmful levels, for example, as a result of repeated application of pesticides. A change in the prevailing conditions, such as may be expected following a modification in agricultural practices may cause the eventual release of the bound residues.

Compatibility is an important factor in the design of pesticide formulations. Formulation chemists are well aware, for example, of the fact that pesticide degradation may be catalyzed by unsuitable clay carriers. The persistence of the active ingredient may, on the other hand, increase by a protective association with the carrier. Thus, the Cu catalyzed hydrolysis of some pesticides was inhibited by their formulants (Chapman and Harris 1984). Associations with the carrier may hinder the adsorption of the active ingredient on the solid surfaces in the unsaturated zone, possibly facilitating thereby the transport of the active ingredient.

Considerable research effort is devoted to the development of mathematical models which are capable of predicting the fate of organic pollutants in the environment. Such a model, if successful, will help make the application of pesticides or the disposal of toxic organic wastes more efficient and less environmentally hazardous. A recognition of the important role the interactions of organic pollutants with the mineral surfaces in the unsaturated zone play in the transport and degradation of the organic pollutants is essential for the development of a successful predictive model.

References

Adams JM, Clement DE, Graham SH (1983) Reactions of alcohols with alkenes over an aluminum-exchanged montmorillonite. Clays Clay Minerals 31:129-136

Ahlrichs JL (1972) The soil environment. In: Goring AI, Hamaker JW (eds) Organic chemicals in the soil environment, vol 1. Dekker, New York, pp 3-43

Amberger A, Vilsmeier K (1978) Anorganisch-katalytische Umsetzungen von Cyanamid und dessen Metaboliten in Quarzsand. 1: Mechanismus des Cyanamidabbaues unter dem Einfluß von Eisenoxiden und feuchtigkeit. Z Pflanzenernähr Bodenkdd 141:665-676

Armstrong DE, Chesters G (1968) Adsorption catalyzed chemical hydrolysis of atrazine. Environ Sci Technol 2:683-689

Armstrong DE, Konrad JG (1974) Nonbiological degradation of pesticides. In: Guenzi GW (ed) Pesticides in soil and water. Soil Sci Soc Am Madison, Wisc, pp 123-130

Bailey GW, White JL, Rothberg T (1968) Adsorption of organic herbicides by montmorillonite. Role of pH and chemical character of adsorbate. Soil Sci Soc Am Proc 32:222-234

Bansal OP (1983) Adsorption of oxamyl and dimecron in montmorillonite suspensions. Soil Sci Soc Am J 47:877-883

Barnsley GE, Gabbott PA (1966) A new herbicide 2-azido-4-ethylamino-6-t-butylamino-1,3,5-triazine. Proc 8th Br Weed Control Conf, pp 372-376

Bittles JA, Chaudhuri AK, Benson SW (1964) Clay-catalyzed reactions of olefins. II: Catalyst acidity and mechanisms. J Polymer Sci A2:1847-1862

Bloom PR, McBride MB (1979) Metal ion binding and exchange with hydrogen ions in acids washed peat. Soil Sci Soc Am J 43:687-692

Bowman BT, Sans WW (1977) Adsorption of parathion, fenitrothion, methyl-parathion, aminoparathion and paraoxon by Na^+, Ca^{2+} and Fe^{3+} montmorillonite suspensions. Soil Sci Soc Am J 41:514-519

Boyd SA, Mortland MM (1985) Urease activity on a clay-organic complex. Soil Sci Soc Am J 49:619-622

Brown CB, White JL (1969) Reactions of 12 s-triazines with soil clays. Soil Sci Soc Am Proc 33:863-867

Chapman RA, Harris C (1984) The chemical stability of formulations of some hydrolyzable insecticides in aqueous mixtures with hydrolysis catalysts. J Environ Sci Health B19:397-407

Crosby DG (1970) The nonbiological degradation of pesticides in soils. In: Pesticides in the soil: ecology, degradation and movement. Proc Int Symp Pesticides in soil. Michigan State Univ, East Lansing, pp 86-94

Dixon JB, Weed SB (eds) (1977) Minerals in soil environments. Soil Sci Soc Am Madison, Wisc

Durand G (1965) Enzymatic splitting of urea in the presence of bentonite. Ann Inst Pasteur 109: Suppl 3, pp 121-132

El-Amamy MM, Mill T (1984) Hydrolysis kinetics of organic chemicals on montmorillonite and kaolinite surfaces as related to moisture content. Clays Clay Minerals 32:67-73

Ercegovich CD, Frear DEH (1964) The fate of 3-amino-1,2,4 triazole in soils. J Agric Food Chem 12:26-31

Estermann EF, McLaren AD (1959) Simulation of bacterial proteolysis by adsorbents. J Soil Sci 10:64-78

Estermann EF, Peterson GH, McLaren AD (1959) Dilgestion of clay-protein, lignin-protein, and silica-protein complexes by enzymes and bacteria. Soil Sci Soc Am Proc 23:31-36

Farmer WJ, Aochi Y (1987) Chemical conversion of pesticides in the soil water environment. In: Biggar JW, Siber JM (eds) Fate of pesticides in the environment. Agric Exp St, Univ Cal Spec Publ 3320:69-74

Fripiat JJ (1968) Surface fields and transformation of adsorbed molecules in soil colloids. Trans 9th Int Congr Soil Sci 1:679-689

Fusi P, Ristori GG, Cecconi S, Franci M (1983) Adsorption and degradation of fenarimol on montmorillonite. Clays Clay Montmorillonite 31:312-314

Gorbunov NI, Dzyadevich GS, Tunik BM (1961) Methods of determining non-silicate amorphous and crystalline sesquioxides. Sov Soil Sci (English Transl) 11:1251-1259

Graham-Bryce IJ (1981) The behaviour of pesticides in soil. In: Greenland DJ, Hayes MHB (eds) The chemistry of soil processes. John Wiley & Sons, Chichester New York, pp 621–671

Greenland DJ (1965) Interaction between clays and organic compounds in soils. Part 1: Mechanisms of interaction between clays and defined organic compounds. Soils Fertil 28:415–420

Greenland DJ (1971) Interactions between humic and fulvic acids and clays. Soil Sci 111:34–39

Grim RE (1968) Clay mineralogy, 2nd edn. McGraw-Hill, New York 596 pp

Hamaker JW (1972) Decomposition: quantitative aspects. In: Goring AI, Hamaker JW (eds) Organic chemicals in the soil environment, vol 1. Dekker, New York, pp 253–341

Hance RJ (1967) Decomposition of herbicides in the soil by non-biological chemical processes. J Sci Food Agric 18:544

Hance RJ (1970) Influence of adsorption on the decomposition of pesticides. In: Sorption and transport processes in soils. SCI Monogr 37, Soc Chem Ind London, pp 92–104

Helling CS, Kearney PC, Alexander M (1971) Behavior of pesticides in soil. Adv Agron 23:147–240

Heuer B, Birk Y, Yaron B (1976) Effect of phosphatase on the persistence of organophosphorus insecticides in soil and water. J Agric Food Chem 24:611–614

Hill TL (1960) An introduction to statistical thermodynamics. Wesley, Reading, Mass, 508 pp

Hoffman G (1959) Verteilung und Herkunft einiger Enzyme im Boden. Z Pflanzenernähr Dung Bodenkd 85:97–104

Hurle K, Walker A (1980) Persistence and its prediction. In: Hance RJ (ed) Interactions between herbicides and the soil. Academic Press, New York London, pp 83–122

Joiner RL, Chambers HW, Baetcke KP (1973) Comparative inhibition of Boll Weevil, Golden Shiner and White Rat cholinesterases by selected photoalteration products of parathion. Pesticide Biochem Physiol 2:371–376

Juma NJ, Tabatabai KA (1977) Effect of trace elements on phosphatase activity in soils. Soil Sci Soc Am J 41:343–346

Konrad JG, Chesters G (1969) Degradation in soils of ciodrin, an organophosphate insecticide. J Agric Food Chem 17:226–230

Konrad JG, Armstrong DE, Chesters G (1967) Soil degradation of diazinon, a phosphorothioate insecticide. Agron J 59:591–594

Kroll L, Kramer M (1955) Der Einfluß der Tonmineralien auf die Enzym-Aktivität der Bodenphosphatase. Naturwissenschaften 42:157–158

Lahav N, White D, Chang S (1978) Peptide formation in the prebiotic era: Thermal condensation of glycine in fluctuating clay environments. Science 201:67–69

Lawless JG, Levi N (1979) The role of metal ions in chemical evolution: polymerization of alanine and glycine in a cation exchanged clay environment. J Mol Evol 13:281–286

Liu Shu-Yen, Freyer AJ, Minard RD, Bollag J-M (1985) Enzyme-catalyzed complex-formation of amino acid esters and phenolic humus constituents. Soil Sci Soc Am J 49:337–342

Lopez-Gonzales, J De D, Valenzuela-Calahorro C (1970) Associated decomposition of DDT to DDE in the diffusion of DDT on homionic clays. J Agric Food Chem 18:520–523

Macalady DL, Wolfe NL (1985) Effects of sediments sorption and abiotic hydrolyses. 1: Organophosphorothioate esters. J Agric Food Chem 33:167–173

McAuliffe C, Coleman NT (1955) H-ion catalysis by acid clays and exchange resins. Soil Sci Soc Am Proc 19:156–160

McBride MB (1985) Surface reactions of 3,3', 5,5'-tetramethyl benzidine on hectorite. Clays Clay Minerals 33:510–516

Mingelgrin U, Saltzman S (1979) Surface reactions of parathion on clays. Clays Clay Minerals 27:72–78

Mingelgrin U, Tsvetkov F (1985a) Adsorption of dimethylanilines on montmorillonite in high-pressure liquid chromatography. Clays Clay Minerals 33:285–294

Mingelgrin U, Tsvetkov F (1985b) Surface condensation of organophosphate esters smectites. Clays Clay Minerals 33:62–70

Mingelgrin U, Yaron B (1974) The effect of calcium salts on the degradation of parathion in sand and soil. Soil Sci Soc Am Proc 38:914–917

Mingelgrin U, Gerstl Z, Yaron B (1975) Pirimiphos ethyl-clay surface interactions. Soil Sci Soc Am Proc 39:834–837

Mingelgrin U, Saltzman S, Yaron B (1977) A possible model for the surface-induced hydrolysis of organophosphorus pesticides on kaolinite clays. Soil Sci Soc Am J 41:519–523

Mingelgrin U, Yariv S, Saltzman S (1978) Differential infrared spectroscopy in the study of parathion-bentonite complexes. Soil Sci Soc Am J 42:664–665

Morrill LG, Mahilum BC, Mohiuddin SH (1982) Organic compounds in soils. Ann Arbor Science, Ann Arbor, Mich

Mortland MM (1966) Urea complexes with montmorillonite: an infrared absorption study. Clay Miner 6:143–156

Mortland MM (1970) Clay-organic complexes and interactions. Adv Agron 22:75–117

Mortland MM, Raman KV (1967) Catalytic hydrolysis of some organic phosphate pesticides by copper (II). J Agric Food Chem 15:163–167

Nash RG, Harris WG, Lewis CC (1973) Soil pH and metalic amendment effects of DDT conversions to DDE. J Environ Qual 2:390–394

Odom DG, Rao M, Lawless JG, Oro J (1979) Association of nucleotides with homoionic clays. J Mol Evol 12:365–367

Ponec V, Knor Z, Cerny S (1974) Adsorption on solids. Butterworths, London

Praffit RL, Mortland MM (1968) Ketone adsorption on montmorillonite. Soil Sci Soc Am Proc 32:355–363

Press MC, Henderson J, Lee JA (1985) Arylsulphatase activity in peat in relation to acidic deposition. Soil Biol Biochem 17:99–103

Probst GW, Tepe JB (1969) Trifluralin and related compounds. In: Kearney PC, Kaufman DD (eds) Degradation of herbicides. Dekker, New York, pp 255–282

Ristori GG, Fusi P, Franci M (1981) Montmorillonite-asulam interactions. II: Catalytic decomposition of asulam adsorbed on Mg, Ba, Ca, Li, Na, K and Cs-clay. Clay Minerals 16:125–137

Rosenfield C, Van Valkenburg W (1965) Decomposition of (O,O-dimethyl-0-2,4,5-trichlorophenyl) phosphorothioate (ronnel) adsorbed on bentonite and other clays. J Agric Food Chem 13:68–72

Ross DJ, McNeilly BA (1972) Some influences of different soils and clay minerals on the activity of glucose oxidase. Soil Biol Biochem 4:9–18

Russell JD, Cruz M, White JL (1968a) Mode of chemical degradation of s-triazines by montmorillonite. Science 160:1340–1342

Russell JD, Cruz M, White JL (1968b) The adsorption of 3-aminotriazole by montmorillonite. J Agric Food Chem 16:21–24

Sahay BK, Low MJD (1974) Interactions between surface hydroxyl groups and adsorbed molecules. V: Fluorobenzene adsorption on Germania. J Coll Interface Sci 48:20–31

Saltzman S, Yaron B, Mingelgrin U (1974) The surface catalyzed hydrolysis of parathion on kaolinite. Soil Sci Soc Am Proc 38:231–234

Saltzman S, Mingelgrin U, Yaron B (1976) Role of water in the hydrolysis of parathion and methylparathion on kaolinite. J Agric Food Chem 24:739–743

Sanchez Camazano M, Sanchez Martin MJ (1983a) Factors influencing interactions of organophosphorous pesticides with montmorillonite. Geoderma 29:107–118

Sanchez Camazano M, Sanchez Martin MJ (1983b) Montmorillonite catalyzed hydrolysis of phosmet. Soil Sci 136:89–93

Schwertman U, Taylor RM (1977) Iron oxides. In: Dixon JB, Weed SB (eds) Minerals in soil environments. Soil Sci Soc Am Madison, Wisc, pp 145–180

Skipper HD, Volk VV, Mortland MM, Raman KV (1978) Hydrolysis of atrazine on soil colloids. Weed Sci 26:46–51

Skujins JJ (1967) Enzymes in soil. In: McLaren AD, Peterson GH (eds) Soil biochemistry. Arnold, London; Dekker, New York, pp 371–416

Soil Conservation Service (ed) (1975) Soil taxonomy. Agric Handb 436, USDA

Solomon DH (1968) Clay minerals as electron acceptors and/or electron donors in organic reactions. Clays Clay Minerals 16:31–39

Solomon DH, Rosser MJ (1965) Reactions catalyzed by minerals. I: polymerization of styrene. J Appl Polymer Sci 9:1261–1271

Spencer WF, Shoup TD, Spear RC (1980) Conversion of parathion to paraoxon on soil dusts as related to atmospheric oxidants at three California locations. J Agric Food Chem 28:1295–1300

Sposito G (1984) The surface chemistry of soils. Oxford Univ Press, New York

Stevenson FJ (1982) Humus chemistry. Wiley Interscience, New York

Theng BKG (1974) The chemistry of clay-organic reactions. Hilger, London

Van Bladel R, Moreale A (1974) Adsorption of fenuron and monuron by two montmorillonite clays. Soil Sci Soc Am Proc 38:244–249

Van Olphen H (1966) An introduction to clay colloid chemistry. Wiley Interscience, New York

Yariv S, Heller L, Kaufherr N (1969) Effect of acidity in montmorillonite interlayers on the sorption of aniline derivatives. Clays Clay Minerals 17:301–306

Zelazny LW, Calhoun FG (1977) Palygorskite (attapulgite), sepiolite, talc, pyrophyllite, and zeolites. In: Dixon JB, Weed SB (eds) Minerals in soil environments. Soil Sci Soc Am Madison, Wisc, pp 435–470

5 Abiotic Transformations of Toxic Organic Chemicals in the Liquid Phase and Sediments

N. LEE WOLFE[1]

5.1 Introduction

Increasing concern over the disposal of hazardous chemicals in commercial waste sites and municipal landfills in the United States has served to focus additional emphasis on the transport and transformation of organic compounds in ground waters, sediments and soils. In particular, mechanisms of transformations of the parent compounds as well as daughter products are receiving increased attention. Because of the large number of disposal sites ($>$ 20,000) and an expanding list of hazardous organic chemicals to be assessed, models are being increasingly used to predict the impact of these chemicals on the environment as well as their potential for human exposure (Brown et al. 1986).

An integral part of exposure analysis models is a quantitative description of transformation processes (Baughman and Burns 1980). Not accounting for transformation reactions can result in an unrealistic over estimation in the exposure concentration. Although this over estimation is conservative on the side of the environment, it has the disadvantage of under estimating or even ignoring the concentrations of transformation products. For example, the U.S. Environmental Protection Agency was recently charged with regulation of the land disposal of some 454 compounds (Appendix Eight List). Of these 362 chemicals listed, 298 were organic compounds. Of these 298 chemicals, it was reported that, under reaction conditions common to most natural waters, 85 would have hydrolysis half-lives of less than one year, 72 would have half-lives greater than a year but less than 10 years, 79 would have half-lives greater than 10 years, and 49 would have no hydrolyzable groups (Wolfe 1985). This wide range of reactivity emphasizes the importance of developing accurate descriptions of transformation processes in estimating exposure concentrations.

This paper addresses transformation processes in the water column, sediments, and groundwater systems. Transformations in soils have been studied and the results reported by others (Mingelgrin et al. 1977; Saltzman et al. 1976; Gerstl and Yaron 1983). The focus is on hydrolysis, biolysis, redox reactions, and to a limited extent, photolysis in the water column.

---

[1]Environmental Research Laboratory, U.S. Environmental Protection Agency, Athens, GA. 30613, USA

5.2 Transformations in the Water Column

Process level models have evolved that describe specific acid-base hydrolysis, direct photolysis, and biolysis in the water column. Generally these are first- or second-order kinetic expressions that include the rate constant, the inherent reactivity of the compound, a term for the concentration of the compound, and a term for the concentration or activity of the system reactant (or at least a measurable quantity that is proportional to the activity) (Zepp and Cline 1977; Wolfe et al. 1977). Kinetic expressions that are applicable to the water column are discussed below.

Hydrolysis. Hydrolysis of organic compounds in the water column is described by a kinetic expression that embodies acid, neutral, and alkaline pathways. Of these three pathways, neutral and alkaline routes generally are the most facile. Such a reaction scheme for the hydrolysis of chlorpyrifos, an organophosporothioate pesticide, is given in Fig. 1 (Macalady and Wolfe 1983). For chlorpyrifos, neutral hydrolysis is favored below pH 9 whereas alkaline hydrolysis dominates above pH 9. Thus, not only is the disappearance half-life dependent on pH but so are the products.

The general kinetic expression that describes acid, base, and neutral hydrolysis reactions is given by Eq. (1).

$$-\text{rate} = (k_H[H^+] + k_{H_2O} + k_{OH}[OH^-])\,[C] \tag{1}$$

where: k_H and k_{OH} are second-order rate constants for acid and base hydrolysis, respectively; k_{H_2O} is the pseudo-first-order rate constant for neutral hydrolysis; C is the concentration of reacting compound; and H^+ and OH^- are proton and hydroxide activities, respectively.

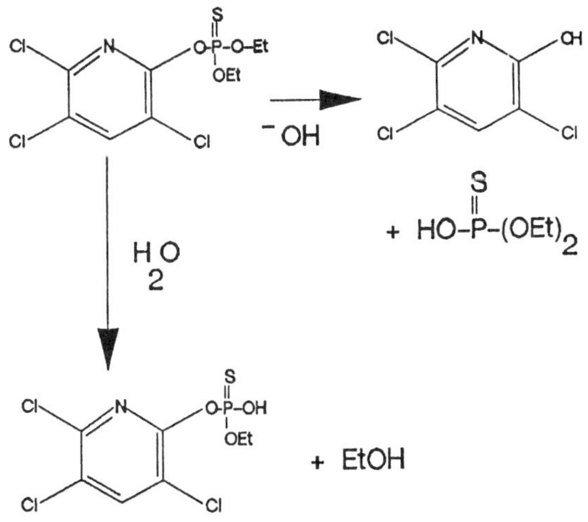

Fig. 1. Alkaline and neutral hydrolysis pathways for chlorpyrifos in natural water samples

Several organic and inorganic species common to natural waters have been shown in laboratory studies to enhance or catalyze hydrolysis reactions. These species include humic and fulvic acids (Khan 1978; Perdue and Wolfe 1982) along with general acids and bases such as bicarbonate and phosphate (Perdue and Wolfe 1983). Metal catalysis has been demonstrated by iron, copper, and manganese species, among others (Blanchet and St. George 1982). Extrapolation of these findings to the ambient concentrations of these catalytic species common to most natural water columns, however, suggests that generally they do not contribute significantly to the overall hydrolytic process.

Photolysis. Photolysis has been studied extensively under reaction conditions common to natural waters (Zepp and Wolfe 1987; Zepp et al. 1975). Photochemists generally divide light-induced transformations into direct and indirect processes. Direct photolysis occurs as the result of sorption of light by the molecule followed by bond cleavage and formation. Direct photolysis occurs for a large number of compounds under laboratory and field conditions.

Direct photolysis is modeled using a first-order kinetic expression given below (Zepp and Cline 1977):

$$P \xrightarrow{h\nu} P^* \longrightarrow \text{products} \tag{2}$$

where: P is the ground state molecule, P* the electronically excited species, and $h\nu$ is a photon of light.

The rate of disappearance of P is given by

$$-\text{rate} = k_p[P] \tag{3}$$

where k_p is a first-order rate constant given by the product of system-specific light adsorption, Ia, multiplied by the quantum yield, Φ, the efficiency of the reaction.

Indirect photolysis can occur by two pathways (Draper and Wolfe 1987). In the first case, a photosensitizer adsorbs the sunlight and transfers the energy to the molecule of interest, which then undergoes bond cleavage and formation. In the second case, light absorption results in homolytic cleavage of a bond with formation of a radical and subsequent reaction of the radical with the molecule. No general overall equations have been developed to describe these processes.

Biolysis. Because biolysis competes with hydrolysis in many environmental situations, it deserves a brief discussion. Biolysis is loosely defined as an enzyme- or bacteria-mediated hydrolysis. In many cases, the biolysis products are the same as those of alkaline hydrolysis. For example, both biolysis and hydrolysis of 2,4-D esters (Fig. 2) result in the formation of 2,4-D acid.

A second-order kinetic expression, which is a limiting case of the Monod equation, has been proposed to describe biolysis (Paris and Rogers 1986). This equation, Eq. (4), has been used successfully to model the disappearance of compounds at low concentrations in the water column (Schnoor et al. 1987).

$$-\text{rate} = k_b[C][B] \tag{4}$$

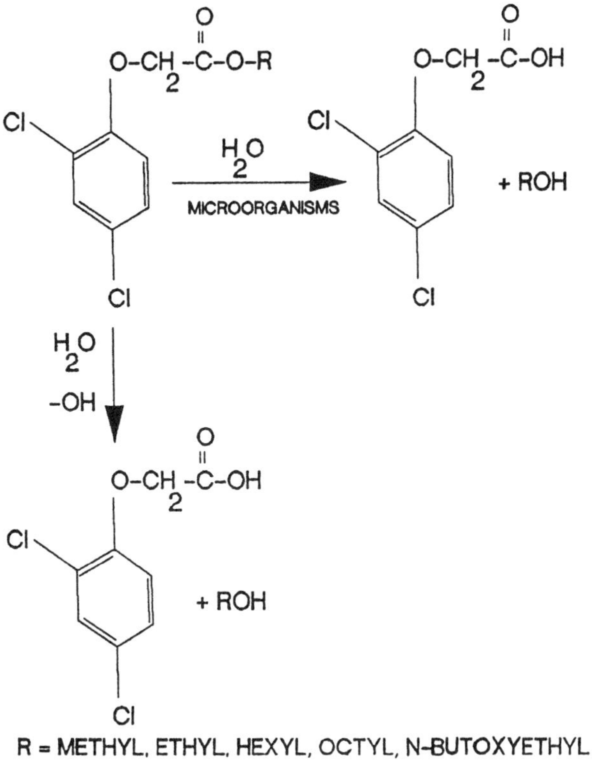

Abiotic Transformations of Toxic Organic Chemicals

Fig. 2. Biolysis and alkaline hydrolysis pathways of 2, 4 esters in natural water samples

where: k_b (L/org hr) is the second-order rate constant, C is the concentration of the reacting compound, and B is the active bacterial concentration.

Support for this model is given by the correlation of second-order biolysis rate constants with second-order alkaline hydrolysis rate constants (Wolfe et al. 1980). The rationale for these correlations is that, in the water column of unpolluted water bodies with low nutrient sources, enzyme systems of microorganisms must be relatively non-stereospecific to survive. Thus, the hydroxide ion is a good global descriptor of the bond breaking and making processes in biolysis.

Redox Reactions. Redox processes, for the most part, have not been quantified at the process level in the water column. Of the two general redox pathways, chemical and biological, the latter has received the most attention. Although results from field and monitoring studies indicate that organic compounds undergo abiotic oxidation in the water column, for the most part they do not permit one to separate the chemical and biological contributions.

The photochemical generation of reactive chemical species that can oxidize organics has been shown to occur in the water column. Estimates of steady-state concentrations of singlet oxygen, solvated electrons, hydrogen peroxide, peroxy radicals, and hydroxy radicals along with known rate constants for reaction with organics suggest that oxidation by these species can be an important fate process in natural waters (Draper and Wolfe 1987).

5.3 Transformations in Sediments

Although significant advances have been made in the kinetic description of transformation processes in the water column, progress has been much slower for sediment systems. The most obvious reason is the introduction of a second phase. The solids introduce the potential for heterogenous reaction kinetics and thus greater complexity. The most obvious role of the solids is to sorb the compound, reducing the concentration of the compound in the aqueous phase that normally could react by homogenous pathways, and introducing the potential for reaction to be mediated by the second phase either at the interface or in the sorbed phase.

5.3.1 Hydrolysis

Hydrolysis of selected organic compounds has been investigated in sediment-water samples under laboratory conditions. A model, consistent with the limited kinetic data, has been proposed to account for the effect of sorption on hydrolysis in the two-phase system (Macalady and Wolfe 1983; Burkhard and Guth 1981). Equation (5) outlines the proposed mechanism for hydrolysis.

$$P + S \underset{k^{1}}{\overset{k^{-1}}{\rightleftarrows}} P{:}S \quad (5)$$
$$k_w \downarrow \qquad k_s \downarrow$$
$$\text{pdts} \qquad \text{pdts}$$

where: P is the organic compound, S is sediment, P:S is the compound in the sorbed phase, k_1 and k_{-1} are the sorption and desorption rate constants, k_w and k_s are the hydrolysis rate constants.

In this model, sorption of the compound to the sediment organic carbon is by the hydrophobic mechanism, which is described by the partition coefficient K_p. The organic matrix can be a reactive or nonreactive sink depending on the hydrolytic process.

Detailed kinetic studies for selected organic compounds indicate that, in the solid associated phase:

— alkaline hydrolysis is retarded
— neutral hydrolysis is unaffected
— acid hydrolysis is accelerated

Although it is possible to rationalize this phenomenon qualitatively, it is not yet possible with the present state of knowledge to predict the magnitude of acceleration or retardation.

5.3.2 Redox Reactions

An example of the different types of redox reactions that are known to occur in sediments and water is shown in Fig. 3 for aldicarb, a carbamate pesticide (Bromilow et al. 1986; Freeman and McCarty 1984; Miles and Delfino 1985; Ou et al. 1985; Smelt et al. 1983). Laboratory and field studies demonstrate that aldicarb undergoes oxidation at the sulfur bond to give the sulfone and sulfoxide in surface waters. Under anoxic conditions existing in sediments, aldicarb has been shown to undergo reduction at the carbon-nitrogen bond to give the nitrile and aldehyde.

Only recently have quantitative expressions been proposed to describe the reduction of organic compounds in the presence of sediment solids in anaerobic environments. The obvious reaction parameters to be elucidated are Eh, pH, solid-water partitioning, microbial activities, and the activities of the reducing agents.

Although a few investigators have looked for quantitative relationships between the redox potential of sediments and rates of reaction of organics, no

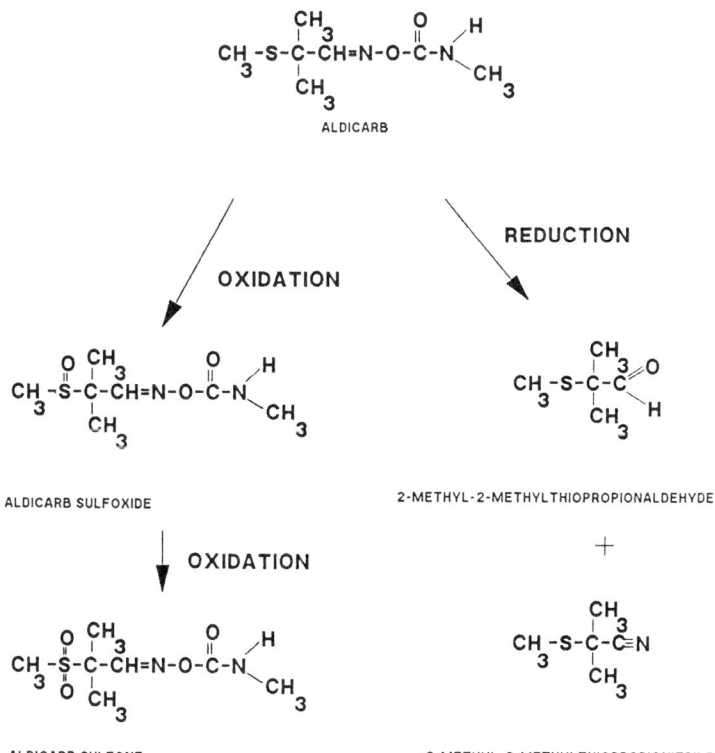

Fig. 3. Redox transformation pathways of aldicarb in natural waters and sediment systems

basis exists to suggest that any such relationship should be any more than fortuitous (Macalady et al. 1986). One recent study of the reduction of nitroaromatic compounds in sediments of varying Eh failed to show a significant correlation of disappearance rate constants with the Eh of the systems. Also there was no major effect of pH on the rate of reduction (Wolfe et al. 1986).

A process model to describe abiotic reduction in sediments that accounts for sorption to the solid phase is described below. In this model, Eq. (6), there are two sorptive sites (Jafvert and Wolfe 1987; Weber and Wolfe 1987). One is a nonreactive sorptive sink that is consistent with partitioning of the compound to the organic carbon matrix of the solids. The other is a reactive sorptive site that is independent of the nonreactive sites.

$$P:S' \underset{k_{-2}}{\overset{k_2}{\rightleftarrows}} P + S \underset{k_{-1}}{\overset{k_1}{\rightleftarrows}} P:S \qquad (6)$$

$$k_c \downarrow \qquad k_w \downarrow \qquad k_s \downarrow$$
$$\text{pdts} \qquad \text{pdts} \qquad \text{pdts}$$

where: P:S' is the compound at the reactive sorbed site; P is the compound in the aqueous phase; P:S is the compound in the nonreactive sink; k_2, k_{-2}, k_1, and k_{-1} are the sorption-desorption rate constants, k_c, k_w and k_s are the respective reaction rate constants.

Available kinetic data indicate that reaction constants k_w and k_s are very small and can be neglected relative to k_c. For the compounds that have been studied to date, it appears that the kinetics of sorption-desorption are not rate limiting. Two limiting situations then arise – one is that transport to the reactive sites is rate limiting and the other is that reduction at the reactive site is rate limiting. At this time based on the available kinetic data, it is not possible to distinguish between the two mechanisms.

5.4 Transformations in Groundwaters

It is possible to put boundary conditions on system reactants based on our extensive knowledge of the chemistry and biology of the water column; unfortunately, this is not the case for ground waters. For example in subsurface ecosystems, little is known about the menu of oxidants and reductants likely to be present, let alone their concentrations. This puts severe constraints on our ability to assess fate processes.

Brown and coworkers (Brown et al. 1986) have provided some insight on the probable chemistry of these systems through their efforts in modeling metals in groundwater environments. They analyzed Eh and pH data contained in the EPA's Storet Data Base. The results of their analysis of Eh and pH data are shown in Fig. 4. This figure shows the distribution of the Eh and pH values for several thousand measurements. The average pH is 6.8 with a standard deviation of 1.3 pH units. The average Eh is –50 mv with a standard deviation of 156 mv.

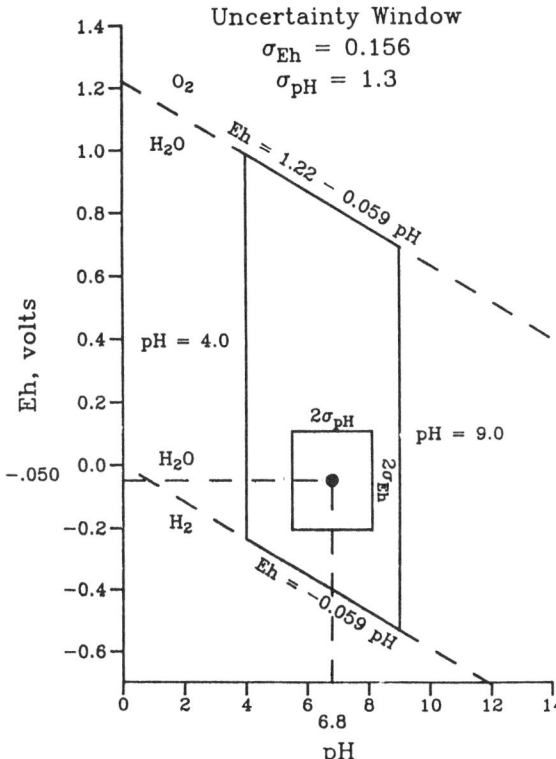

Fig. 4. Eh vs pH diagram showing the range of Eh and pH in groundwaters based on measured data

5.4.1 Hydrolysis

Based on the distribution of pH in groundwaters, it would appear that neutral hydrolysis reaction processes are likely to be more important than acid or base mediated pathways. It should be noted, however, that acid-catalyzed pathways even at pHs 5 and 6 could be important for selected compounds because of the potential acceleration of hydrolysis in the sorbed phase. At the present time, it is not possible to predict the magnitude of the acceleration.

At the present time, laboratory studies dealing with transformation of organics in aquifer samples suggest that for hyrolysis that occurs in the aqueous phase, the solid materials do not enhance or catalyze the aqueous phase reaction. This observation is supported by studies of epoxides, aziridines, and halogenated hydrocarbons. Similarly detailed studies of the hydrolysis of four organic compounds in clay suspensions showed no catalysis or enhanced degradation of the compounds when compared to hydrolysis in distilled water (El-Amamy and Mill 1984). These finding are consistent with the model proposed in Eq. (5).

5.4.2 Redox Reactions

Based on the average redox state of groundwaters as shown in Figure 4, the reduction of organics would be a likely process in many groundwater systems. Data accumulated on halogenated hydrocarbons appear to support this statement (Lage et al. 1986, 1987; Bouwer and Wright 1987; Jafvert and Wolfe 1987). Most of the studies have focused on the methanes and ethanes; a partial listing of some of these compounds is given in Table 1.

Two general types of redox reactions have been studied in the laboratory and are believed to occur under field conditions (Fig. 5).

It is difficult to separate the chemical and biological contributions to the degradation of these compounds. It is very likely that the biology of the system dictates the activity of the chemical reductants. Degradation studies of polyhalogenated methanes under laboratory conditions suggested that degradation occurred by a chemical pathway. Similar studies with halogenated ethanes showed that they reacted by an abiotic process.

The oxidation of organics and consequent reduction of manganese and iron by pure mineral surfaces has been studied under controlled laboratory conditions. (Stone 1986; Stone and Morgan 1984). Detailed studies on selected organic compounds have shown that the rate determining step is the electron transfer at the surface and not diffusion to the reaction site.

Table 1. Partial List of Organohalides that have been investigated under reducing conditions

ETHANES
- 1,1-dichloroethane
- 1,1,1-trichloroethane
- 1,1,1,2-tetrachloroethane
- 1,2-dichloroethane
- 1,1,2,2-tetrachloroethane
- 1,1,1,2,2,2-hexachloroethane
- 1,2-dibromoethane
- 1,2-diiodoethane

ETHENES
- trichloroethene
- 1,2-dichloroethene
- 1,1-dichloroethene

METHANES
- bromoform
- dibromomethane
- Bromomethane
- carbon tetrachloride
- chloroform
- dichloromethane
- chloromethane

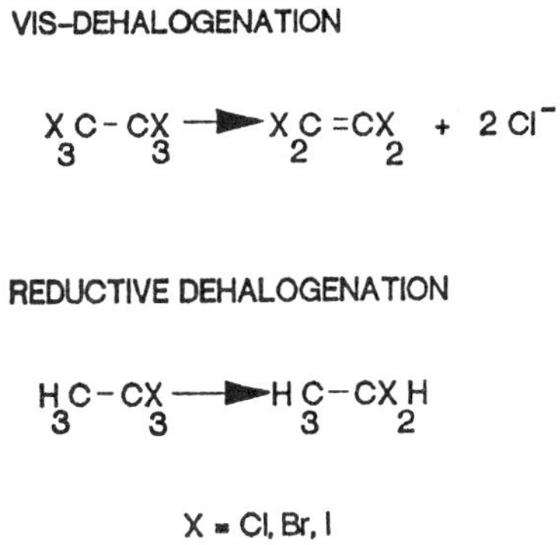

VIS-DEHALOGENATION

$$X_3C-CX_3 \longrightarrow X_2C=CX_2 + 2\,Cl^-$$

REDUCTIVE DEHALOGENATION

$$H_3C-CX_3 \longrightarrow H_3C-CX_2H$$

$$X = Cl, Br, I$$

Fig. 5. Reductive transformation pathways of halogenated hydrocarbons under anaerobic conditions

Disappearance kinetics of halogenated ethanes have been studied under a variety of laboratory conditions. Transformations have been shown to occur under controlled conditions of denitrification, sulfate respiration, and methanogenous. These reactions have been shown to occur in aquifer materials, sediment samples, and soils and have been observed under field conditions, in microcosms, and in biofilm reactors. Based on the information presented in these reports, it is not possible to definitively state whether the reductions occurred by chemical or biological mechanisms or by a combination of both.

5.5 Future Direction

It is not enough to react to pollutants. It is best to forecast their behavior and consequently their impact on the environment in order to avoid potential problems. Thus process models will continue to be developed and refined. Chemical and biological processes will have to be decoupled to better model hydrolytic and redox processes in sediments and groundwaters. Improved experimental techniques are needed to control biological parameters while the chemical redox transformations are studied. Measurement techniques are needed to identify and quantify oxidizing and reducing species in sediments and groundwaters. This is particularly true for reducing species in ground waters where the systems can be poorly poised. Further studies are needed to delineate the kinetics of transformation processes in heterogeneous media. Estimation techniques and measurement methods need to be developed to provide processes coefficients at acceptable cost with short turn-around times.

References

Baughman GL, Burns LA (1980) Transport and transformation of chemicals a perspective. In: Hutsinger (ed) The Handbook of Environmental Chemistry, Vol 2/Part A. Springer, Berlin Heidelberg New York

Blanchet PF, St-George A (1982) Kinetics of chemical degradation of organo-phosphorus pesticides; Hydrolysis of chlorpyrifos and chlorpyrifos-methyl in the presence of copper(II). J Pestic Sci 13:85–91

Bouwer EJ, Wright JP (1987) Transformations of trace halogenated aliphatics in subsurface microcosms with anoxic biofilms. J Contam Hydrol 1–21

Bromilow RH, Briggs GG, Williams MR, Smelt JH, Thuinstra L, Traag WA (1986) The role of ferrous ions in the rapid degradation of oxamyl, methomyl and aldicarb in anaerobic soils. J Pestic Sci 17:535–547

Brown DS, Carlton RE, Mulkey LA (1986) Development of land disposal decisions for metals using minteq sensitivity analyses. US Environmental Protection Agency EPA/600/3-86/030, pp 1–8

Burkhard N, Guth JA (1981) Chemical hydrolysis of 2-chloro-4,6 bis (alkylamino)-1,3,5-triazine herbicides and their breakdown in soil under the influence of adsorption. J Pestic Sci 12:45–52

Burns LA, Cline DM (1985) Exposure Analysis Modeling System: Reference Manual For EXAMS II. US Environmental Protection Agency, EPA-600/3-85/038, pp 1–83

Draper WM, Wolfe NL (1988) Abiotic processes in the degradation of pesticides in natural waters. Proceedings of the Sixth International Congress of Pesticide Chemistry, pp 1–8

El-Amamy MM, Mill T (1984) Hydrolysis kinetics of organic chemicals on montmorillonite and kaolinite surfaces as related to moisture content. Clays Clay Miner 32:67–73

Freeman PR, McCarthy KD (1984) Photochemistry of oxime carbamates. 1. Photo-transformations of aldicarb. J Agric Food Chem 32:873–877

Gerstl Z, Yaron B (1983) Behavior of bromacil and napropamide in soils: I. Adsorption and degradation. J Soil Sci Am 47:474–478

Jafvert CT, Wolfe NL (1987) Degradation of selected halogenated ethanes in anoxic sediment-water systems. Environ Toxicol Chem 6:827–37

Khan SU (1978) Kinetics of hydrolysis of atrazine in aqueous fulvic acid solution. J Pestic Sci 9:39–43

Lage GB, Parsons FZ, Nassar RS, Lorenzo PA (1986) Sequential dehalogenation of chlorinated ethenes. Environ Sci Technol 20:96–99

Lage GB, Parsons FZ, Nassar RS (1987) Kinetics of the depletion of trichloroethane. Environ Sci Technol 21:366–370

Macalady DL, Wolfe NL (1983) New perspectives on the hydrolytic degradation of the organo-phosphorothioate insecticide chlorpyrifos. J Agric Food Chem 31:1139–1147

Macalady DL, Wolfe NL (1984) Abiotic hydrolysis of sorbed pesticides. In: Krueger RF, Seiber JN (eds) Treatment and Disposal of Pesticide Wastes. ACS Symposium Series 259, chapt 14, pp 221–244

Macalady DL, Tratnyek PG, Grundl TJ (1986) Abiotic reduction reactions of anthropogenic organic chemicals in anaerobic systems. A critical review. J Contam Hydrol 1:1–28

Miles CJ, Delfino JJ (1985) Fate of aldicarb, aldicarb sulfoxide, and aldicarb sulfone in Florida groundwater. J Agric Food Chem 33:455–460

Mingelgrin U, Saltzman S, Yaron B (1977) A possible model for the surface-induced hydrolysis of organophosphorous pesticides on koalinite clays. J Am Soil Sci Soc 41:519–523

Ou L, Edvardsson KSV, Rao PS (1985) Aerobic and anaerobic degradation of aldicarb sulfone in soils. J Agric Food Chem 33:72–78

Paris DF, Rogers JE (1986) Kinetic concepts for measuring microbial rate constants: Effects of nutrients on rate constants. Appl Environ Microbiol 51:221–225

Perdue EM, Wolfe NL (1982) Modification of pollutant hydrolysis kinetics in the presence of humic substances. Environ Sci Technol 16:847–852

Perdue EM, Wolfe NL (1983) Prediction of buffer catalysis in field and laboratory studies of pollutant hydrolysis reactions. Environ Sci Technol 17:635–642

Saltzman S, Mingelrin U, Yaron B (1976) The role of water in the hydrolysis of parathion and methyl parathion on kaolinite. J Agric Food Chem 24:739–743

Schnoor JL, Sato C, McKechnie D, Sahoo D (1987) Processes, coefficients and models for simulating toxic organics and heavy metals in surface waters. US Environmental Protection Agency, EPA/600/3-87/015

Smelt JH, Dekker A, Leistra M, Houx NWH (1983) Conversion of four carbamoyloximes in soil samples from above and below soil water table. J Pestic Sci 14:173–181

Stone AT (1986) Adsorption of organic reductants and subsequent electron transfer on metal oxide surfaces. ACS Symp Ser 323:446–461

Stone AT, Morgan JJ (1984) Reduction and dissolution of manganese (III) and manganese (IV) oxides by organics: 2. Survey of the reactivity of organics. J Environ Sci Technol 18:617–624

Weber EJ, Wolfe NL (1987) Kinetic studies of the reduction of aromatic azo compounds in anaerobic sediment/water systems. Environ Toxicol Chem 6:911–19

Wolfe NL, Zepp RG, Gordon JA, Baughman GL, Cline DM (1977) Kinetics of chemical degradation of malathion in water. Environ Sci Technol 11:88–93

Wolfe NL, Paris DF, Steen WC, Baughman GL (1980) Correlation of microbial degradation rates with chemical structure. Environ Sci Technol 14:1143–1144

Wolfe NL, Kitchens BE, Macalady DL, Grundl TJ (1986) Physical and chemical factors that influence the anaerobic degradation of methyl parathion in sediment systems. Environ Sci Technol 5:1019–1026

Zepp RG, Cline DM (1977) Rates of direct photolysis in aquatic environments. Environ Sci Technol 11:359–366

Zepp RG, Wolfe NL (1987) Abiotic transformation of organic chemicals at the particle-water interface. In: Stumm W (ed) Aquatic Surface Chemistry: Chemical Processes at the Particle-Water Interface. Wiley, New York, pp 423–455

Zepp RG, Wolfe NL, Gordon JA, Baughman GL (1975) Dynamics of 2,4-D esters in surface waters, hydrolysis, photolysis, and vaporization. Environ Sci Technol 9:1144–1150

Part III. Pesticides in Porous Media

Introductory Comments

Pesticides are a fact of life. Without them the yields of nearly all commercially grown crops would be severely diminished, domesticated livestock would suffer and large tracts of land would become uninhabitable. Attempts at replacing pesticides by natural or biological methods, as well as attempts at doing away with them entirely, have yet to prove themselves viable alternatives. We must, therefore, learn to cope with the threat that accidents or improper use of pesticides pose to the quality of the environment in general and to pollution of soils and groundwater in particular.

When pesticides reach the soil, either by direct application, runoff from foliage applications or by accidental spills, their ultimate fate will be determined by three major processes; volatilization, degradation and adsorption. The degree to which one of these processes will dominate depends on the properties of both the soil and the pesticide itself. The properties of most pesticides and toxic organic compounds are well known, however the nature and properties of soils vary both spatially and with depth. This requires that the effect of soil properties on these three processes be well understood and in fact a tremendous amount of effort has been devoted in recent years to elucidating these relationships.

The papers in Part III deal with selected aspects of the relationship between soil properties and organic compounds. The first paper in this section presents a mechanistic explanation for adsorption of pesticides and other organic compounds in porous media. Chiou (Chapter 7) presents a wide range of sorption data for nonionic organic compounds in both dry and wet soils. The author accounts for the observed behavior by treating the soil as a dual adsorbent where the mineral fraction is considered to behave as a conventional solid adsorber while the organic fraction functions as a partition medium. A second paper in this section deals with adsorption from an entirely different angle. Ignoring the actual mechanism of adsorption, Gerstl (Chapter 6) reviews attempts at predicting adsorption of organic compounds in soils, and thus indirectly their movement and availability. For most practical purposes the adsorption of a compound for a given situation as predicted by a compounds solubility or octanol-water partition coefficient is found to be adequate if one is aware of the limitations of these correlations. The author proposes using (QSAR) Quantitative Structure-Activity Relationships between adsorption data and the chemicals structure to improve these relationships and possibly even to help design new and more effective pesticides. One method, based on molecular connectivity, is dealt with in some detail.

The need to better understand complex mixtures of pollutants and solvents has arisen due to their presence at hazardous wastes disposal and spill sites. Rao and coauthors (Chapter 8) review recent experimental and theoretical developments in describing the adsorption and movement of complex mixtures of organic pollutants in soils. Unlike the previous two papers which deal primarily with nonionic compounds the present approach can deal with nonpolar hydrophobic compounds, hydrophobic ionizable compounds and ionic organic chemicals in either water or mixtures of water with a miscible or immiscible solvent. A solvophobic theory has been developed to describe the observed data and numerous cases and examples are presented.

Degradation of pesticides is usually looked upon as desirable. If all pesticides were as persistent as DDT our soils and groundwater would be seriously endangered. We count on the soil microbial population to slowly remove excess pesticides. Lately, cases of accelerated biodegradation resulting in loss of pesticidal efficacy have been reported. Chapter 9 by Katan and Aharonson presents a state of the art summary of this problem which occurs to herbicides, insecticides and fungicides and which has been found to carry over from one class of compounds to another.

The papers presented in this chapter by no means cover the entire topic of organic compounds in porous media. To do this would require several volumes and even then there would be much left unresolved. We have, however, been able to present recent work on topics which are of major importance today in our effort to understand and solve the problems our environment faces from our dependence on synthetic chemicals.

6 Predicting the Mobility and Availability of Toxic Organic Chemicals

Z. GERSTL[1]

6.1 Introduction

The behavior of organic chemicals in the unsaturated zone is governed by a variety of complex processes, the most important of which is adsorption. Adsorption of organic compounds can, and does, influence the mobility and biological activity of many chemicals (Leistra 1980). This is the direct result of the adsorption process which determines the distribution of the chemical in question between the aqueous phase and the soil solid phase (in the following discussion we shall disregard volatile compounds for the sake of simplicity). Adsorption can be defined as: the excess of solute concentration at the solid-liquid interface over the concentration in the bulk solution regardless of the nature of the interface region or of the interaction between the solute and the solid surface causing the excess (Mingelgrin and Gerstl 1983). Any process which proceeds at a more rapid rate in the soil solution than in the adsorbed state, such as transport, will be retarded as a result of adsorption. Conversely, reactions such as degradation can either be enhanced or impeded by adsorption depending on the exact nature of the degradation process. Surface catalyzed reactions, such as the degradation of parathion on kaolinite surfaces, will be enhanced by adsorption whereas solution phase reactions will be slowed down due to adsorption of one of the reactants. Accordingly, adsorption has received a tremendous amount of attention and any method which can reliably predict the adsorption of a solute will be of great importance to scientists and decision makers. An advanced method for predicting adsorption will be discussed below.

A large number of pesticides have been synthesized during the last 40 years. Together with the vast number of other organic compounds which are known to be potential environmental pollutants they can serve as a fairly large data base for defining quantitative structure-activity relationships (QSAR). QSAR are quantitative models in which the variation in some selected measured property (e.g. biological activity, adsorption, decomposition) is related to the variation in the chemical structure of a group of compounds.

The measurement of any particular property for a series of compounds, be it chemical or biological in nature, is usually a straightforward matter in which investigators of the appropriate discipline are well versed. However, the

---

[1]Institute of Soils and Water, ARO, Volcani Center, Bet Dagan 50250, Israel

quantification of a chemical's structure in such a way that will yield useful QSAR is a more complicated process and one with which most chemists or biologists are not familiar.

In the present paper we will deal primarily with one of the methods proposed for quantifying chemical structures, namely, molecular connectivity, although several other approaches to chemical structure quantification, such as STERIMOL (Verloop 1982; Anderson et al. 1982) and molecular fragments (Hansch and Leo 1979) have been studied.

6.2 Current Methods for Predicting Adsorption

Numerous studies have shown that the adsorption of nonionic organic compounds by soils and sediments is often directly related to the organic carbon (matter) content of the adsorbing phase (e.g. Chiou et al. 1979; Karickhoff 1981; Briggs 1981). Thus by expressing adsorption on an organic carbon basis we can define an adsorption constant, K_{oc}, which should be practically independent of the soil type. This was shown to be true in many cases although some exceptions have been reported. Gerstl and Mingelgrin (1984) found, for example, that sediments from the Sea of Galilee and from the Dead Sea adsorbed pesticides belonging to a number of chemical classes several times more than would be expected from their organic matter content and their K_{oc} as determined from their adsorption on soils. They postulated that the organic matter in the lake sediments studied was derived from different starting material than soil organic matter and might have undergone different transformations in the nearly anaerobic lake bottom environment.

The K_{oc} of pesticides and other nonionic chemicals is often well correlated with some molecular properties of the compounds, such as their aqueous solubility (S) and their octanol-water partition coefficient (K_{ow}). Consequently, all that is needed to determine the adsorption of any compound in any soil is their solubility or K_{ow} and the organic matter content of the adsorbing medium (soil, sediment or aquifer material). Despite the limitations inherent in the use of these parameters for predicting adsorption (Mingelgrin and Gerstl 1983; Banerjee et al. 1980) they can provide a first approximation to the adsorption of organic chemicals in the environment. Since frequently, it is not possible to obtain an accurate value for the sorption of a substance in a given environment, one is required to settle for a reasonable estimate.

A close look at the equations obtained for the relationship between K_{oc} and either S or K_{ow} (Mingelgrin and Gerstl 1983; Lyman 1982) reveals some interesting findings (Table 1). The various equations for Log K_{oc} vs Log S are all approximately parallel but displaced from one another by a certain finite factor. The nearly identical slope could indicate that the nature of the governing process is similar in all cases. The different intercepts indicate a systematic shift in the regression lines which may be due to differences in the nature of the organic

Table 1. Regression models for the relationship between K_{oc} and K_{ow} and between K_{oc} and solubility

K_{oc} vs K_{ow}:

Log K_{oc} = 1.00 Log K_{ow} − 0.194 r^2 = 1.00	Karickhoff et al. 1979
Log K_{oc} = 0.87 Log K_{ow} + 0.056 r^2 = 0.73	Mingelgrin and Gerstl 1983
Log K_{oc} = 0.38 Log K_{ow} + 0.119	Chiou et al. 1979
Log K_{oc} = 0.52 Log K_{ow} + 0.854 r^2 = 0.84	Briggs 1974
Log K_{oc} = 0.49 Log K_{ow} + 1.25 r^2 = 0.55	Gerstl (unpublished data)

K_{oc} vs S:

Log K_{oc} = −0.54 Log S + 4.70 r^2 = 0.94	Karickhoff et al. 1979
Log K_{oc} = −0.576 Log S + 4.24 r^2 = 0.64	Mingelgrin and Gerstl 1983
Log K_{oc} = −0.557 Log S + 4.28 r^2 = 0.99	Chiou et al. 1979
Log K_{oc} = −0.54 Log S + 3.72 r^2 = 0.95	Felsot and Dahm 1979
Log K_{oc} = −0.46 Log S + 4.13 r^2 = 0.63	Gerstl (unpublished data)

matter in the adsorbing media (see above) or to the nature of the different groups of compounds dealt with in the different studies. The two extreme intercepts in the Log K_{oc}-Log S equations in Table 1 are for fairly homogenous groups of compounds whereas the other equations describe collections of data from the literature and include a diverse assortment of chemicals. In the relationship between K_{ow} and K_{oc} the regression lines are not parallel and intersect each other at a Log K_{ow} of about 2 (with the exception of the data of Chiou et al. (1979)). A relationship of this kind suggests that the nature of the processes governing adsorption and octanol-water partitioning differ. Furthermore, use of the K_{ow}-K_{oc} relations might result in large errors especially when the value of K_{ow} for the compound in question is considerably different from 2.

Adsorption, solubility and partition data were collected from the literature for over 400 compounds including solvents, pesticides and other toxic organic compounds. Regressions for the K_{oc} − K_{ow} and K_{oc} − S relationships were calculated for this much larger and heterogeneous data base (Gerstl, unpublished data). The preliminary results (Figs. 1 and 2, Table 1) show the same general trend as the previously mentioned studies with only a slight difference in the slope of the K_{oc} − S equation. It should be emphasized that the data used for these calculations are the average K_{oc}, K_{ow} and S values (representing over 4000 data points for K_{oc} alone). A further breakdown of the data into individual chemical classes may result in several similar, but slightly different and more significant, relationships. In general, however, the results show that the average values for most compounds fall within two standard deviations of the predicted values (Figures 1b & 2b). The spread in published values of K_{oc}, K_{ow}, and S for many compounds is rather large making the use of the average values essential. The effect that this spread in data has on any predictive method is beyond the scope of the present report. A more thorough evaluation of these data for outliers, solubility corrections and multivariate analysis is presently under way.

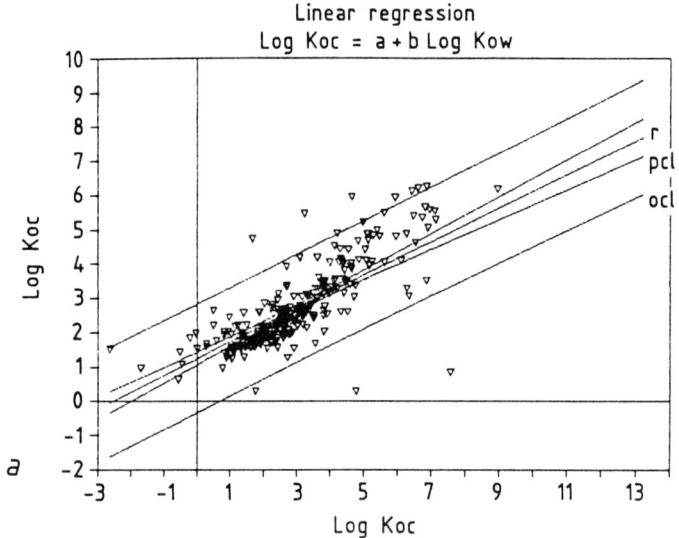
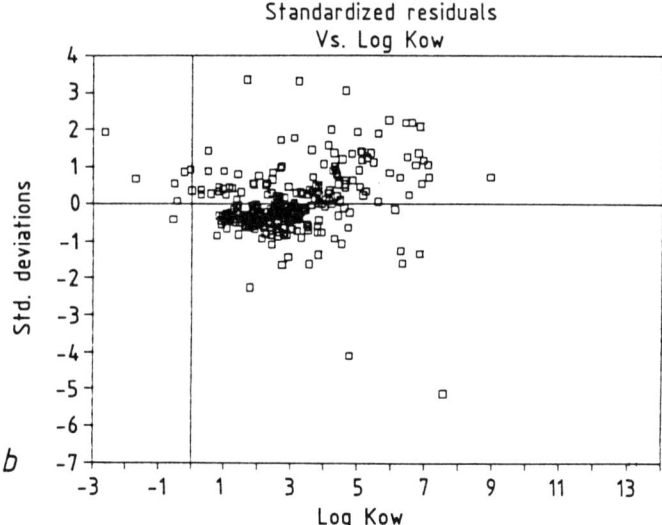

Fig. 1a. Relationship between the average Log K_{oc} and the average Log K_{ow} for 272 sets of data points for organic compounds. **r** – regression line; **pcl** – predicted confidence limits; **ocl** – observed confidence limits. Regression data can be found in Table 1. **b** Residuals for Log K_{oc} as a function of Log K_{ow}

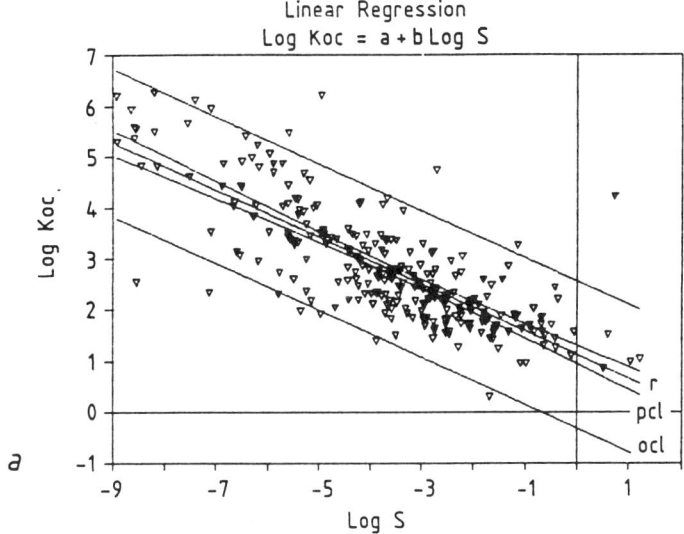
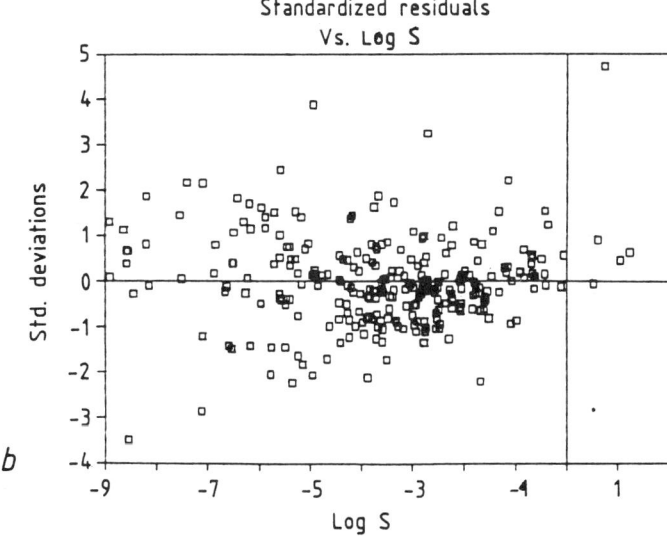

Fig. 2a. Relationship between the average Log K_{oc} and the average Log S for 264 sets of data points for organic compounds. **r** — regression line; **pcl** — predicted confidence limits; **ocl** — observed confidence limits. Regression data can be found in Table 1. **b** Residuals of Log K_{oc} as a function of Log S

As discussed above, the most commonly used estimators for the K_{oc} of a given compound are the compound's aqueous solubility and the K_{ow}. These properties have been shown to be highly correlated with molecular connectivity (MC) derived indices, (MC is a method of obtaining topological indices for a compound based on it's chemical structure, see below) of many groups of compounds (Hall et al. 1975; Murray et al. 1975; Kier and Hall 1976). It would seem, therefore, to be possible to predict adsorption for new compounds even before their synthesis if one could relate their K_{oc}, K_{ow}, or solubilities to MC indices which can in turn be calculated from a compound's structure.

Some workers have found that the relationships obtained between aqueous solubility and either K_{ow} or K_{oc} are improved if one takes into account the heat of fusion for solid compounds in expressing the solubility of a compound (Chiou and Schmedding 1981; Steen and Karickhoff 1981). They proposed that the following relationships describe the aqueous solubility (S) of solid compounds:

for liquid compounds: $S = 1/(\gamma_w V_w)$

and

for solid compounds: $S' = [1/(\gamma_w V_w)] (f_s/f_1)$

and

$$\ln(f_s/f_1) = - \frac{(\Delta H_f)(T_m - T)}{(R)(T*T_m)}$$

where: γ_w is the solute activity coefficient in water, V_w is the molar volume of water, f_s/f_1 is the ratio of the fugacities of the solid and its super-cooled liquid, ΔH_f is the solute's heat of fusion, R is the gas constant and T_m is the solutes melting point. The term f_s/f_1 is a function of the melting point and heat of fusion and is less than 1 at system temperatures. Thus, the solubility of a solid is less than that of its super-cooled liquid by a factor of f_s/f_1.

This implies that the state of the compound, either solid or liquid, which is determined by the nature and strength of the bonds in a molecule determines, in turn, the behavior (adsorption) of the compound. As will be shown in the next section the relationship between the nature of the molecule and the property of interest, such as adsorption, is exactly the type of information which is encoded in the MC indexes.

6.3 Molecular Connectivity

The term "molecular connectivity" has been adopted to describe the general method leading to the calculation of indices derived from the molecular structure of a compound. The molecular structure is defined by a set of atoms and the connections between them. MC is a topological method in which the interconnection between a molecule's atoms, which determine the ultimate three dimensional topography of the molecule, can be described. The various indices encode within themselves information on molecular size, branching, cyclization, unsaturation and heteroatom content (Kier and Hall 1976). Simple connectivity indices arise from assigning each atom (or "vertex"), other than hydrogens, a numerical value equivalent to the number of non-hydrogen atoms to which the vertex in question is bonded and is called the delta value (δ). In this approach no attempt is made to differentiate between different elements or the multiplicity of bonding. While this approach is sufficient to describe general properties (size and shape) of molecules, it does not differentiate between similar compounds with obvious chemical differences (e.g. acetone and acetic acid). Kier and Hall (1981) proposed a method which includes these structural differences by accounting for valence electrons in σ, π, and lone-pair orbitals of each atom (excluding bonds to hydrogen). The valence delta values are based on the number of valence electrons not involved in bonds to hydrogen. All other calculations, described later, are identical for both simple and valence connectivity indices. The indices which can be calculated are of the general form $^m\chi_t$, where m is the order of the molecular fragment and t is the type. Valence values are indicated by a superscript, v, on the right ($^m\chi^v_t$). The order, m, defines the number of bonds or edges used for calculations. Zero order defines individual atoms (vertices), first order deals with single bond length, and so on. For m = 3, two types of path lengths are possible, three consecutive bonds or a cluster of three bonds emanating from a single vertex. For order 4, both path (p) and path-clusters (pc) are possible; at order 6, a chain (c) term may be generated. Thus, four types exist, path, cluster, path-cluster, and chains.

$^0\chi$ is defined as the sum of the reciprocal of the square root of the delta values of all paths of length 0 (individual vertices,)

or $^0\chi = \Sigma \dfrac{1}{\sqrt{\delta_i}}$ $^0\chi^v = \Sigma \dfrac{1}{\sqrt{\delta^1_v}}$

where i is each atom and δ and δ_v are the delta and valence delta values.

Similarly, $^1\chi^v = \Sigma \dfrac{1}{\sqrt{\delta_i^v \delta_j^v}}$ where i and j are bonded. It follows that $^2\chi^v = \Sigma \dfrac{1}{\sqrt{\delta_i^v \delta_j^v \delta_k^v}}$

where i is bonded to j and j is bonded to k. In a similar manner χ values are calculated to whichever order is desired.

Molecular connectivity indexes are not just empirical values obtained by some algorithm but contain structural information related to the topology, geometry and spatial attributes of the molecules. For example, $^0\chi$ indices encode information about the number of atoms in a compound and the $^4\chi_{pc}$ index provides useful information to the structure analysis of substituted rings (Kier and Hall 1986).

6.4 MC-SAR Studies

The various MC indices have been found to be well correlated with physicochemical properties such as density, water solubility, heat of vaporization, boiling point and octanol-water partition coefficients (K_{ow}) of homogeneous classes of chemical compounds. They have also been found to correlate well with biological properties as varied as bioconcentration of a chemical in an organism (Sabljic and Protic 1982a), anesthetics (DiPaolo et al. 1977), and toxicity (Koch 1983; Schultz et al. 1982).

Sabljic and Protic (1982a) found that the use of MC indexes to model and predict the bioconcentration factor for halocarbons and organochlorinated pesticides in fish gave much better results than if aqueous solubility or K_{ow} had been used. The authors did not imply any mechanistic explanation of their results and treated it as a completely empirical finding. Schultz et al. (1982) noted that the $^1\chi^v$ index of 23 selected nitrogenous heterocyclic compounds produced a better correlation with the biological response data than did the K_{ow}. The use of the $^1\chi^v$ index enabled them to describe some of the effect of molecular structure on toxicity. A similar study on the toxicity of phenols to fish (Hall and Kier 1984) also presented direct structural interpretations for the correlation of toxicity to MC indexes.

Comparable work has been carried out for adsorption of nonionic compounds by soils and sediments. Dragun et al. (1980) presented results on the correlation of Log K_{ow}, Log S (solubility) and R_f (soil TLC mobility) with several molecular parameters, including simple MC indices, for a fairly large group of pesticides. Their results indicated that these indices did not predict R_f, S or K_{ow} very accurately and they concluded that MC and the other parameters studied did not describe the hydrophobic-hydrophilic balance of the compounds studied. On the other hand, Sabljic and Protic (1982b) found an excellent correlation between the Log K_{oc} of polyaromatic hydrocarbons (PAH) and the $^2\chi^v$ molecular connectivity index. The $^2\chi$ indexes encode information for three atom fragments which is the minimum number necessary to describe a plane (Kier and Hall 1986). Since most of the PAH compounds are planar this relationship seems to be logical if one assumes an adsorption mechanism based on the interaction of two parallel planes. In a study of 37 polyaromatic hydrocarbons and halogenated hydrocarbons, Sabljic (1984) reported that the first order molecular connectivity index, $^1\chi$, correlated best with Log K_{oc} values. $^1\chi$ encodes information on the molecular bulk or volume and the molecular surface area. It was also reported that the $^1\chi$ index was highly correlated with the calculated van der Waals surface area and the conclusion was reached, therefore, that the adsorption process may be viewed as an attractive interaction between two planes, the magnitude of which is proportional to the surface area of the solute molecule. A more recent study (Gerstl and Helling 1987) of over 40 pesticides from all classes of commonly used pesticides showed that the predictive ability of MC for such a heterogeneous group of chemicals was rather low; yet they were not inferior in relation to the predictive ability of S or K_{ow}. As in the previous study $^1\chi^v$ was found to be the best descriptor of the adsorption process indicating the non-empirical nature of MC indices.

In both of the latter studies (Sabljic 1984; Gerstl and Helling 1987) the relationships found between MC and K_{oc} were statistically better than those between MC and K_{ow}. Sabljic (1984) concluded that this suggests that the adsorption and partitioning processes reflect altogether different mechanisms. Gerstl and Helling (1987) proposed that phase partitioning is only one of several mechanisms responsible for adsorption so that adsorption as expressed by the K_{oc} may actually mask differences between different chemical classes by averaging the contributions of the various mechanisms to the adsorption process.

6.5 The Need for Future Work

The question that should be asked at this point is whether the use of S and/or K_{ow} is adequate for predicting adsorption or whether another predictive method for K_{oc} is necessary. The presently used methods are limited in their accuracy and even experimentally derived K_{oc} value are limited in their reliability to describe actual field behavior of pesticides.

Rao et al. (1986) have shown that the organic matter distribution in a citrus grove in Florida and in a field in Georgia had a coefficient of variation of 20% and that the spatial patterns of the measured adsorption coefficient (K_d) and organic matter corresponded closely to each other. Similar data for the K_d of ethroprop (*O*-ethyl *S,S*-dipropyl phosphorodithioate) on a light sandy soil in Israel (Gerstl, unpublished data) shown in Fig. 3 shows how adsorption can vary from one end of a field to the other and follows the soil organic matter distribution fairly closely. Rao et al. (1986) also discussed the effects of extrinsic variability, such as method of pesticide application and tillage operation, on pesticide concentrations and their effects on the subsequent leaching of the pesticides. Their results showed that the range of concentrations within a field of uniformly applied compounds (Br and aldicarb) were found to differ by an order of magnitude immediately following their incorporation to a depth of 10 cm and that these initial differences persisted during the leaching process. Even if OM was uniformly distributed in a field the value of batch derived adsorption coefficients is questionable in light of the findings of Jury et al. (1986). They found that K_d values measured by the batch method and those measured on undisturbed cores were only negligibly correlated. Furthermore, it was found (Jury et al. 1986; White et al. 1986) that approximately 20% of the napropamide applied was found leached below the depth of maximum penetration expected on the basis of measured K_d and OM values due to by pass flow through the soil profile.

The use of predicted K_{oc} values based on a compounds aqueous solubility or MC (or its K_{ow} if need be) is not adequate. If one bears in mind their limitations (Mingelgrin and Gerstl 1983) the use of the predicted values may be used as first approximations of a compounds behavior.

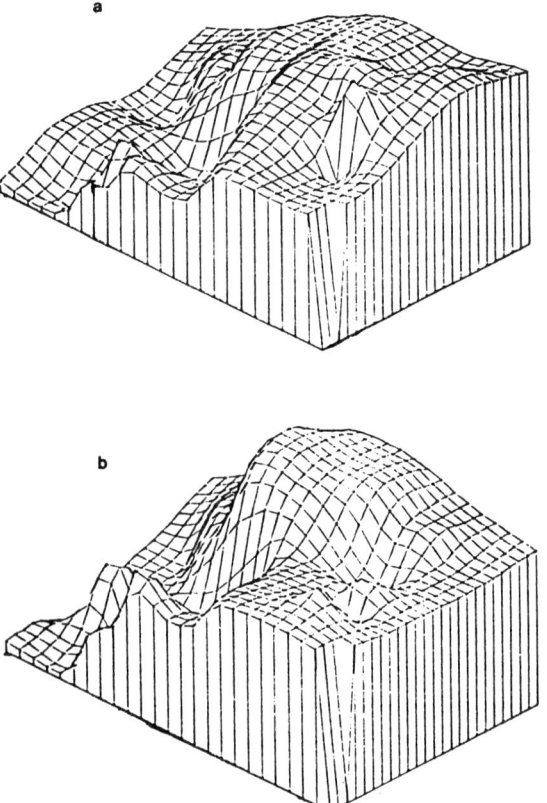

Fig. 3. Distribution of the K_d of ethroprop (**a**) and the percent organic matter (**b**) in a sandy soil (Gerstl, unpublished data). The field was approximately 30m × 30m

The use of MC for predicting adsorption values can still be of importance, however. There may be cases when the solubility or octanol-water partition coefficient of a compound are not known (or not accurately known) but the chemical structure of the compound in question is. In such cases the molecular connectivity can be used to predict the adsorption characteristics for the compound especially if the regression equation for the appropriate family of compounds is available. This is even more true for compounds which have not yet been synthesized but for which we would like some estimate of their behavior. In such a case the use of MC indexes and their correlations to S, K_{ow} and K_{oc} can provide information on the expected behavior of the compound. This approach is being widely used in the pharmaceutical industry in computer aided design of drugs and could be implemented in designing pesticides with desired mobility properties.

6.6 Summary

Molecular connectivity is a topological description of molecules which encodes in the various indices structural information describing the compound. These indexes have been found to be well correlated with numerous physicochemical and biological properties of molecules for homogenous groups of chemicals. For large, heterogeneous groups of chemicals, such as pesticides, the correlations with MC indices are presently no better than those with solubility or partitioning (K_{ow}). The main advantage of using parameters calculated from molecular structure for assessing adsorption and mobility is that it enables one to do so when measured values of S and K_{ow} are not available or for compounds that are as yet unsynthesized.

Acknowledgement. This work was supported in part by a grant from the National Council for Research and Development, Israel and the Commission of the European Community, General Directorate of Science, Research and Development, Bruxelles.

References

Anderson NH, Heritage J, Branch SK (1982) Co-formational studies and quantitative structure-activity relationships in 1-aryl-4, 4-dimethyl-1-(1,2,4-triazol-1-yl)-1-penten-3-one herbicides. In: Miyamoto J, Kearney PC (eds) Pesticide Chemistry: Human Welfare and the Environment vol 1. 5th International Congress of Pesticide Chemistry, Kyoto, Japan pp 345–350

Banerjee S, Yalkowsky SH, Valvani SC (1980) Water solubility and octanol/water partition coefficients of organics. Limitations of the solubility-partition coefficient correlation. Environ Sci Technol 14:1227–1229

Briggs GG (1974) A simple relationship between soil adsorption of organic chemicals and their octanol/water partition coefficients. Proc 7th Br Insectic Fungic Conf 1973, Brighton, pp 83–86. The Boots, Nottingham

Briggs GG (1981) Theoretical and experimental relationships between soil adsorption, octanol-water partition coefficients, water solubilities, bioconcentration factors, and the parachor. J Agric Food Chem 29:1050–1059

Chiou CT, Schmedding DW (1981) Measurement and interrelation of octanol-water partition coefficient and water solubility of organic chemicals. In: Test Protocols for Environmental Fate and Movement of Toxicants. Assoc Off Anal Chem, Arlington, Va

Chiou CT, Peters LJ, Freed VH (1979) A physical concept of soil-water equilibria for nonionic organic compounds. Science 206:831–832

DiPaolo T, Kier LB, Hall LH (1977) Molecular connectivity and structure-activity relationship of general anesthetics. Molec Pharmacol 13:31–37

Dragun J, Protezone R, Fowler CS Jr, Helling CS (1980) Evaluation of molecular modelling techniques to estimate soil chemical mobility. I: Molecular connectivity and charge related indices. Proc Res Sym 53rd Ann Mtg, Water Pollut Contr Fed, Las Vegas, Nevada

Felsot A, Dahm PA (1979) Sorption of organophosphorous and carbamate insecticides by soil. J Agric Food Chem 27:557–559

Gerstl Z, Helling CS (1987) Evaluation of molecular connectivity as a predictive method for the adsorption of pesticides by soils. J Environ Sci Health B22:55–69

Gerstl Z, Mingelgrin U (1984) Sorption of organic substances by soils and sediments. J Environ Sci Health B19:297–312

Hall LH, Kier LB (1984) Molecular connectivity of phenols and their toxicity to fish. Bull Environ Contam Toxicol 32:354–362

Hall LH, Kier LB, Murray WJ (1975) Molecular connectivity II: Relationship to water solubility and boiling point. J Pharmacol Sci 64:1974–1977

Hansch C, Leo AJ (1979) Substituent Constants for Correlation Analysis in Chemistry and Biology. Wiley, New York

Jury W, Elabd H, Clendening LD, Resketo M (1986) Evaluation of pesticide transport screening models under field conditions. In: Garner WY, Honeycutt RC, Nigg HN (eds) Evaluation of Pesticides in Groundwater. ACS Symposium Series 315, American Chemical Society, Washington DC

Karickhoff SW (1981) Semi-empirical estimation of sorption of hydrophobic pollutants on natural sediments and soils. Chemosphere 10:833–846

Karickhoff SW, Brown DS, Scott TA (1979) Sorption of hydrophobic pollutants on natural sediments. Water Res 13:241–248

Kier LB, Hall LH (1976) Molecular connectivity VII: Specific treatment of heteroatoms. J Pharmacol Sci 65:1806–1809

Kier LB, Hall LH (1981) Derivation and significance of valence molecular connectivity. J Pharmacol Sci 70:583–589

Kier LB, Hall LH (1986) Molecular Connectivity in Structure-Activity Analysis. Research Studies, Hertfordshire, England

Koch R (1983) Molecular connectivity index for assessing ecotoxicological behaviour of organic compounds. Toxicol Environ Chem 6:87–96

Leistra M (1980) Transport in solution. In: Hance RJ (ed) Interactions Between Herbicides and the Soil. Academic Press, London

Lyman WJ (1982) Adsorption coefficient for soil and sediment. In: Lyman WJ, Reehl WF, Rosenblatt DV (eds) Handbook of Chemical Property Estimation Methods, McGraw Hill, New York

Mingelgrin U, Gerstl Z (1983) Reevaluation of partitioning as a mechanism of nonionic chemical adsorption in soils. J Environ Qual 12:1–11

Murray WJ, Hall LH, Kier LB (1975) Molecular connectivity III: Relationship to partition coefficient. J Pharmacol Sci 64:1978–1981

Rao PSC, Edvardson SV, Ou LT, Jessup RE, Nkedi-Kizza P, Hornsby AG (1986) Spatial variability of pesticide sorption and degradation parameters. In: Garner WY, Honeycutt RC, Nigg HN (eds) Evaluation of Pesticides in Groundwater. ACS Symposium Series 315, American Chemical Society, Washington DC

Sabljic A (1984) Predictions of the nature and strength of soil sorption of organic pollutants by molecular topology. J Agric Food Chem 32:243–246

Sabljic A, Protic M (1982a) Molecular connectivity: a novel method for prediction of bioconcentration factor of hazardous chemicals. Chem Biol Interactions 42:301–310

Sabljic A, Protic M (1982b) Relationship between molecular connectivity indices and soil sorption coefficients of polycyclic aromatic hydrocarbons. Bull Environ Toxicol 28:162–165

Schultz TW, Kier LB, Hall LH (1982) Structure-toxicity relationships of selected nitrogenous heterocyclic compounds. III. Relations using molecular connectivity. Bull Environ Toxicol 28:373–378

Steen WC, Karickhoff SW (1981) Biosorption of hydrophobic organic pollutants by mixed microbial populations. Chemosphere 10:27–32

Verloop A (1982) The sterimol approach: further development of the applications. In: Miyamoto J, Kearney PC (eds) Pesticide Chemistry: Human Welfare and the Environment, vol 1. 5th International Congress of Pesticide Chemistry, Kyoto, Japan, pp 339–344

White RE, Dyson JS, Gerstl Z, Yaron B (1986) Leaching of herbicides through undisturbed cores of a structured clay soil. J Soil Sci Soc Am 50:277–283

7 Partition and Adsorption on Soil and Mobility of Organic Pollutants and Pesticides

C. T. CHIOU[1]

7.1 Introduction

The mechanism for sorption of organic pollutants and pesticides by soil has long been a subject of profound interest because of its direct impacts on the mobility and activity of the compounds in soil. Although a large volume of laboratory and field data on many aspects of soil behavior had been gathered between the 1950s and 1970s, during which period the use of organic pesticides was increased, no general agreement was reached regarding the sorptive mechanism involved. Since the 1970s, the outgrowth of public concern over environmental contamination further stimulated research in this subject. The development of this field of research has now reached a point that the diverse characteristics of soil sorption can be placed in a much better perspective. This enables researchers to reexamine old and new data for consistency and for assessing the activity of organic pollutants and pesticides in soil.

The complex and heterogeneous nature of soil composition has undoubtedly been a major retarding factor in the development of this field of science for some time. The crux of the problem lies in defining the roles of soil mineral matter and organic matter in sorption and the effect of moisture on the functions of organic and mineral matter. In the following text discussing the mechanistic functions of soil organic matter and minerals, some terminology is used frequently in referring to the mechanism involved. The term "sorption" is used to denote uptake of a solute or vapor by soil without reference to a specific mechanism. The term "adsorption" refers to condensation of vapor or solute (both being referred to as "adsorbate") on the surface or interior pores of a solid (adsorbent) by physical or chemical bonding forces. In contrast to adsorption, the term "partition" or "partitioning" is used to describe a model in which the sorbed material permeates (i.e. dissolves) into the network of an organic phase by forces common to solution (e.g. by van der Waals forces). This is analogous to the extraction of an organic compound (solute) from water into an organic phase. When the organic phase is a solid (e.g. soil organic matter), partition is distinguished from adsorption by the homogeneous distribution of the sorbed material throughout the entire volume of the solid phase. The partition heat of a solute between two solvent phases is relatively small and independent of concentration because of partial cancellation

---

[1] U.S. Geological Survey, Denver Federal Center, Box 25046, Denver, CO 80225-0046, USA

of heats of solution for the transfer of the solute from one phase (e.g. water) to another (e.g. an organic phase). In adsorption, the adsorbate occupies only the surface of the solid and this surface condensation is achieved through attractive forces, resulting in additional exothermic heat (over that resulting from condensation of the solute out of solution). Because of the constraint of adsorbent surfaces or active sites, the adsorption process is necessarily competitive between solutes (or vapors), whereas the partition process is largely noncompetitive. The partition isotherm of a solute between an organic phase and water is largely linear over a wide range of solute concentration because the heats of solution of the solute in individual phases are relatively independent of concentration. The adsorption isotherm is seldom linear because the adsorbent surface is rarely energetically homogeneous, in addition to constraints of available surfaces or active sites.

In earlier studies of the sorption of pesticides or pollutants by soil, the soil was often regarded as single adsorbent, or a mixed adsorbent, in analogy to some well-defined adsorbents. While this view largely accounts for the sorption data of (nonionic) pesticides with dry (or subsaturated) soils, it runs into difficulty in attempting to interpret the results with water-saturated soils (note that the saturation water content of a soil is not the same as the soil's field capacity, the former being substantially less than the latter). The adsorptive character of the soil was recognized unequivocally, for example, by the nonlinear isotherm, and suppression by water, in the vapor uptake of methyl bromide (Chisholm and Koblitsky 1943), chloropicrin (Stark 1948), ethylene dibromide (Hanson and Nex 1953; Wade 1954), and benzene and chlorobenzenes (Chiou and Shoup 1985) by dry soils and clays. Similarly, the uptake of parathion and lindane from hexane by soils was found to be depressed by soil moisture (Yaron and Saltzman 1972; Chiou et al. 1985). These effects are typical of surface adsorption. On the other hand, the sorption data in aqueous systems (or for water-saturated soils) display a set of unique features. Most notably, the extent of soil uptake for given compounds is closely related to the organic-matter content of the soil (Bailey and White 1964; Goring 1967; Lambert 1968; Hamaker and Thompson 1972; Saltzman et al. 1972; Browman and Chesters 1977; Chiou et al. 1979, 1983; Karickhoff et al. 1979; Kenaga and Goring 1980; Karickhoff 1981; Means et al. 1982). The uptake of organic vapors by wet soils shows similar effects (Wade 1954; Leistra 1970). The equilibrium sorption isotherms of compounds on hydrated soils are all essentially linear (Wade 1954; Swoboda and Thomas 1968; Leistra 1970; Yaron and Saltzman 1972; Chiou et al. 1979, 1983, 1985; Karickhoff et al. 1979), and are not strongly temperature dependent, giving only relatively small exothermic heats (Mills and Biggar 1969; Spencer and Cliath 1970; Yaron and Saltzman 1972; Hamaker and Thompson 1972; Pierce et al. 1974; Chiou et al. 1979, 1985). Moreover, soil uptake of binary organic solutes from water shows no apparent competition between the solutes (Chiou et al. 1983, 1985), in contrast to strong competitive effects recognized in soil uptake from the vapor phase, or from organic solvents. These features clearly indicate the controlling effect of soil organic matter on sorption by wet soils, and the inconsistency of the organic matter as an adsorbent.

7.2 Sorption by Soil in Aqueous Systems

In spite of the recognition that sorption of nonionic organic compounds by water-saturated soils is controlled predominately by the soil organic matter content, there had been a lack of agreement on the mechanism involved in the earlier studies. One prevalent hypothesis was that the organic matter functioned as a high-surface-area adsorbent (Bailey and White 1964; Haque 1975; Browman and Chesters 1977) capable of adsorbing nonionic organic compounds by hydrophobic interactions (Weber and Weed 1974; Carringer et al. 1975; Browman and Chesters 1977; Hassett et al. 1981; Stevenson 1982; Mingelgrin and Gerstl 1983). However, the hydrophobic adsorption concept is not in keeping with the soil sorption data and general adsorption criteria, as discussed in detail in a recent review article (Chiou 1989).

In attempts to rationalize the strong uptake of pesticides by muck soil, Hartley (1960) speculated that a solvent action of the "oily constituents" in soil organic matter might be responsible for the enhanced uptake (while making no effort to distinguish this effect from surface adsorption). This view was further emphasized by Goring (1967) in discussing the distribution of organic toxicants between soil and water. Lambert (1967, 1968) empirically assumed soil organic matter to be analogous to a solvent medium in partition chromatography and suggested a correspondence between sorption coefficient and solvent-water distribution coefficient. In recognition of a wide isotherm linearity in the uptake of parathion by soil from water, Swoboda and Thomas (1968) suggested that part of the parathion might be "adsorbed" as a liquid dissolved in the organic fraction of the soil, similar to the partition process in liquid-liquid extraction. However, strong evidence supporting the partition of organic compounds into soil organic matter was not available at that time, and consequently the validity and significance of soil uptake by partition in soil organic matter for dry and hydrated soils was not duly appreciated. Later, numerous correlations relating sorption coefficients of organic compounds between soil organic matter and water (K_{om}), or between soil organic carbon and water (K_{oc}), to compounds' octanol-water partition coefficients (K_{ow}) or water solubilities (S_w) have been reported for estimating the extent of sorption by soil (Hamaker and Thompson 1972; Briggs 1973, 1981; Chiou et al. 1979, 1983; Karickhoff et al. 1979; Kenaga and Goring 1980; Means et al. 1982; Karickhoff 1981; Hassett et al. 1981; McCall et al. 1983; Dzombak and Luthy 1984).

More rigorous analyses of the function of soil organic matter were given in a number of studies by Chiou et al. (1979, 1981, 1983, 1985). In addition to the recognized dependence of soil sorption on organic-matter content, these studies showed that the isotherms for sorption of nonionic organic compounds from water are virtually linear up to high relative concentrations (ratios of equilibrium concentrations to solute solubilities), and the soil uptake shows an absence of competition between solutes. An example is given in Fig. 1 for the sorption of parathion (S_w, about 12 mg/L) and lindane (S_w, about 7.8 mg/L) on Woodburn soil (1.9 percent organic matter, 9 percent sand, 68 percent silt, and 21 percent clay). Moreover, the equilibrium heat for solute sorption is generally less ex-

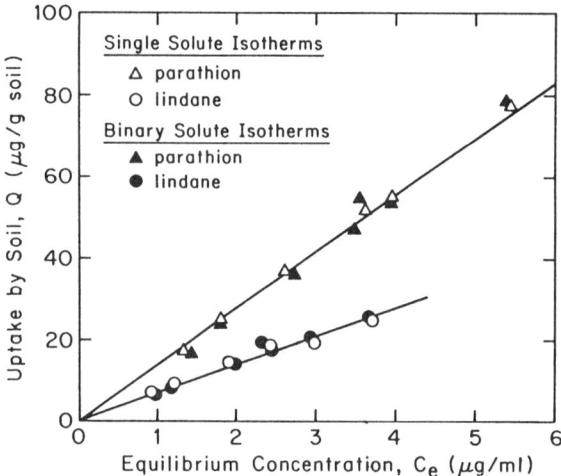

Fig. 1. Equilibrium sorption isotherms of parathion and lindane as single and binary solutes from water on Woodburn soil at 20° C

othermic than the heat of solute condensation from water (i.e. the reverse heat of solution in water). These observations led Chiou and coworkers to postulate that soil uptake of nonionic organic compounds from water consisted primarily of solute partition into soil organic matter. The inability of the soil mineral matter to adsorb these compounds from water was attributed to the strong dipole interaction of minerals with water, as reasoned by Goring (1967), that excludes the organic compounds from this portion of the soil. In keeping with this hypothesis, the sorption coefficients of organic compounds (K_{om} or K_{oc}) are strongly correlated with reciprocals of their water solubilities. By application of the Flory-Huggins theory for solute solubility in (amorphous) soil organic matter, Chiou et al. (1983) developed a partition equation to account for the magnitudes of the observed K_{om} values. This analysis indicates that the primary factor affecting K_{om} (or K_{oc}) is the water solubility of the solute (as liquid or supercooled liquid), as is for the partition coefficient of a solute between an organic solvent (e.g. octanol) and water (Chiou et al. 1982).

The diverse sorption characteristics with dry and wet soils can be reconciled by the postulate that the soil behaves as a dual sorbent, in which the mineral matter functions as a conventional solid adsorbent and the organic matter as a partition medium (Chiou and Shoup 1985; Chiou et al. 1985). As stated, in aqueous solution, adsorption of organic compounds by minerals is suppressed by water, and hence the soil uptake consists mainly of solute partitioning into soil organic matter. The strong vapor sorption by dry soil and subsequent suppression by moisture is attributed to mineral adsorption that predominates over the simultaneous uptake by partitioning into soil organic matter. The dual-sorbent character of the soil in sorption of organic compounds can be illustrated by a comparison of sorption data in aqueous systems with those in nonaqueous systems.

7.3 Sorption by Soil from Organic Solvents

A number of studies on the sorption of organic compounds and pesticides from organic solvents are useful for illustrating the sorptive behavior of soil. On the basis of the assumed roles of soil minerals and organic matter, one would immediately expect that uptake of nonionic compounds from organic solvents by partitioning into organic matter should be minimal because of the good solvency of the solution phase (Chiou et al. 1985). Thus, the extent of soil uptake would be controlled mainly by the ability of the compounds to compete with the solvent for adsorption on soil minerals. Because of the inherent polarity of minerals, the extent of adsorption of a compound (solute or solvent) is expected to be strongly governed by polar interactions with mineral surfaces.

Hance (1965) compared data on the sorption of diuron from aqueous and petroleum solutions by an oxidized soil (with about 3 percent organic matter) and a soil organic material (about 76 percent organic matter). The diuron uptake by the oxidized soil was markedly greater from petroleum solution than from water, whereas the sorption by the organic material showed the opposite effect. Hance thus concluded that there was a competition between diuron and water for adsorption sites in soil under aqueous slurry conditions and that diuron competed more effectively for soil organic matter than for soil mineral surfaces. The high uptake of diuron from petroleum by the (dry) oxidized soil is apparently the result of strong competitive adsorption of the polar solute (diuron) for soil minerals over weak competition of a relatively nonpolar solvent. On the other hand, the high uptake of diuron by soil organic material in aqueous solution is in keeping with solute partitioning into the organic matter, with concomitant suppression by water of adsorption on soil minerals.

Mills and Biggar (1969) reported similar characteristics in sorption of lindane by Venado clay (50 percent montmorillonite and 6 percent organic matter) and Staten muck (22 percent organic matter) from aqueous and hexane solutions. The lindane uptake in aqueous solution by the two soils was largely proportional to the organic-matter content, whereas in hexane solution the sorption by the dry Venado clay was markedly enhanced over that by the dry peat muck. Moreover, they noted that whereas the molar heat of lindane sorption in aqueous systems with both soils is less exothermic than the reverse heat of solution of lindane in water, the heats of sorption from hexane are much more exothermic than the reverse heat of solution in hexane.

Yaron and Saltzman (1972) studied the uptake of parathion by soils from water and from various organic solvents. Parathion shows a high uptake on dry soils from hexane, a lower uptake from benzene, and practically no uptake from such polar solvents as methanol, ethanol, acetone, chloroform, ethyl acetate, and dioxane. While the uptake from hexane by dry soils is markedly greater than from water, such uptake is strongly suppressed by humidity and approaches zero when the soils become fully water-saturated (in contrast to the finding that parathion shows definitive uptake from aqueous solution). These observations led the authors to conclude that different mechanisms are involved in parathion "adsorption" in aqueous systems and in hydrated soil-organic solvent systems.

In view of the results reported by Yaron and Saltzman (1972), there is little doubt that the mechanism of sorption of (nonionic) organic compounds by soil in organic-solvent systems is fundamentally different from that in aqueous systems. Results for the aqueous system are readily reconcilable with the assumed solute partitioning into soil organic matter as discussed. The high uptake from hexane by dry soils is attributable to competitive adsorption of the solute with the solvent for soil minerals (mainly clay components), while the bulk organic phase minimizes solute partition into the organic matter. Therefore, as expected, the solute uptake from hexane (or other nonpolar solvents) would be depressed by humidity (in this case, water is considered as a competing solute), and gives nearly zero uptake when the soil becomes fully saturated with water. The failure of the soil to sorb parathion or other organic compounds from polar organic solvents is ascribed to the fact that such solvents minimize solute (parathion) adsorption on minerals because of their polarity and reduce solute partitioning into organic matter by their good solvency.

Another significant observation from the study of Yaron and Saltzman (1972) is the increasing parathion uptake from hexane on partially hydrated soils with increasing temperature, as opposed to decreasing sorption with increasing temperature in aqueous solution. Such differences led Mingelgrin and Gerstl (1983) to postulate that the heat of "adsorption" of a solute can be either exothermic or endothermic and consequently that the associated entropy change for solute adsorption can be either negative or positive. The validity of this contention warrants careful consideration. In principle, a net adsorption of single vapors or solutes from solution must be accompanied by exothermic heat. While a solute may adsorb endothermically by competing with the solvent, adsorption of the solute must be very weak in this case because of the strong adsorption of the solvent. This is supposedly the case for the very weak adsorption of nonionic organic compounds on "pure minerals" in aqueous solution. In binary-solute systems, adsorption of the energetically weaker adsorbate may give rise to an anomalous temperature effect even though it nevertheless exhibits exothermic interactions with the adsorbent. Therefore, if the temperature dependence of the number of adsorption sites is not taken into account for individual solutes, one may mistakenly conclude that adsorption of the weaker adsorbate is endothermic in such systems.

In the case of parathion uptake from hexane by partially hydrated soils, parathion (a weaker adsorbate) competes with water (a more powerful adsorbate) for adsorption on soil minerals. An increase in temperature weakens the energetic interaction of minerals with water to a greater extent than with parathion, assisting the latter to compete more favorably for adsorption. The spurious temperature dependence of parathion adsorption may be more properly explained in terms of the higher exothermic heat of adsorption per unit surface area of minerals for water than for parathion. In all likelihood, adsorption of parathion per unit area is also decreasing with increasing temperature, but not as rapidly as the concomitant increase in the number of water-free sites (resulting from desorption of water) that become accessible to parathion. When excess water is added to fully saturate mineral sites, adsorption of parathion from hexane would therefore be nearly completely suppressed. Interpretations by Mingelgrin and Gerstl (1983) by assumption of an endothermic heat of adsorption for parathion (and consequently an increase in entropy for parathion adsorption) appears to be theoretically untenable.

Further results obtained by Chiou et al. (1985) on the sorption of parathion and lindane from aqueous and hexane solutions substantiate the roles of soil organic matter and minerals in soil sorption. The results with Woodburn soil (the same composition as stated) are presented here for discussion. Solubilities of parathion and lindane in hexane at 20° C are 5.7×10^4 mg/L and 1.3×10^4 mg/L, respectively. In aqueous systems, both parathion and lindane show linear isotherms over wide relative concentrations, and there is no apparent competition between the two solutes, as illustrated in Figure 1. In dry soil-hexane systems, the sorption of parathion is distinctly nonlinear with markedly greater sorption even at very low relative concentrations than uptake from water, as shown in Fig. 2. Such curvature is not evident in the study of Yaron and Saltzman (1972) as their study was limited to very low relative concentrations in hexane, which obviously fell within the Henry's law region. The isosteric heat of parathion sorption deter-

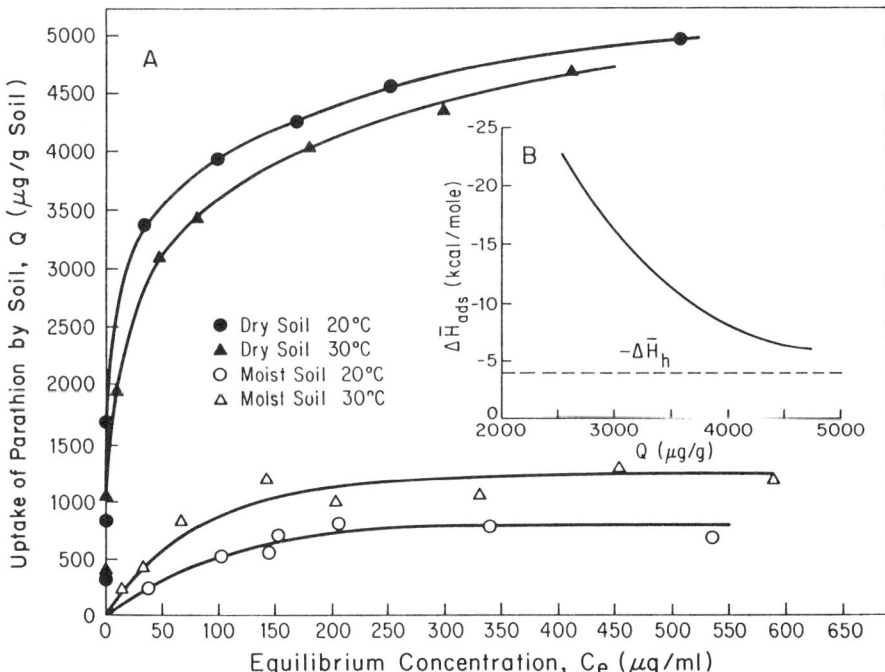

Fig. 2. (**A**) Equilibrium sorption isotherms of parathion from hexane on dry and partially hydrated (about 2.5% water) Woodburn soil at 20° C and 30° C. (**B**) The isosteric heat of sorption of parathion as a function of its loading on dry Woodburn soil

mined from the 20° C and 30° C isotherms with dry soil is highly exothermic (more than the reverse heat of solution in hexane, $-\Delta H_h$) and varies with parathion loading. In hexane solution, the parathion uptake is much lower with the moist soil (about 2.5 percent water) than with dry soil, and the system shows a similar anomalous temperature effect, as noted by Yaron and Saltzman (1972). When more water was added to the soil to reach the saturation point (about 5 percent water), the parathion uptake from hexane was virtually totally suppressed. Moreover, a competitive effect is also found between parathion and lindane in their simultaneous uptake from hexane on dry soil.

7.4 Sorption by Soil from the Vapor Phase

In light of the sorption data of organic compounds from organic solvents on dry and hydrated soils, it is evident that dry and subsaturated minerals in soil would contribute strongly to soil uptake. Stark (1948) studied the vapor sorption of chloropicrin on dry soils and found a close relation between the amount of uptake and clay content. Jurinak (1957) found that the vapor sorption of ethylene dibromide (EDB) on dehydrated clays (with subsaturated amounts of water) was related to the external clay surface areas of specific clays and the sorption data fit the BET adsorption model. Call (1957) studied the effect of relative humidity (R.H.) on the vapor uptake of EDB on several soils. On dry and partially hydrated soils, the EDB isotherms followed the Brunauer type-II shape, in which clay soils showed greater sorption capacity than soils low in clay content. An increase of R.H. from 0 to 50 percent progressively suppressed the EDB uptake with a concomitant change in isotherm shape toward linearity. The suppression of vapor uptake by moisture is essentially the same as is noted for solute uptake from nonpolar organic solvents, the only difference being that the organic solvent simultaneously minimizes partitioning in the organic matter (causing the solute uptake to approach zero at full water saturation) while such partitioning is largely unaffected by moisture in the vapor system.

The observed dependence of the equilibrium vapor concentrations of dieldrin and lindane in soil on soil water content, as reported by Spencer et al. (1969) and Spencer and Cliath (1970), also supports the assumed effects of soil minerals and organic matter. At soil water contents less than 2.2 percent on Gila silt loam (0.6 percent organic matter), the equilibrium vapor densities of lindane (at about 50 μg/g soil) and dieldrin (at 100 μg/g soil) were very small relative to saturation vapor densities of the pure compounds, indicating that amounts of the applied pesticides were much below the saturation limits on the soil. However, the vapor densities rose to saturation values with an increase of the soil water content to more than 3.9 percent. In addition, the vapor-sorption isotherm of lindane on Gila soil containing 3.9 percent water was practically linear and exhibited the same capacity as with the water-saturated soil. The low vapor densities associated with dry soil can be attributed to strong mineral adsorption, which overrides the effect of partitioning with organic matter. Upon wetting, water displaces the pesticides from soil minerals by adsorptive competition, and, as a result, the amount of pesticides in soil becomes more than is required to saturate the organic matter, giving rise to saturation vapor densities.

To substantiate the effect of humidity on the mechanism and magnitude of sorption of organic compounds by soil, Chiou and Shoup (1985) determined the vapor sorption of benzene, chlorobenzene, m-dichlorobenzene, p-dichlorobenzene, 1,2,4-trichlorobenzene, and water on Woodburn soil. Results with the dry soil closely fit the BET adsorption model. An increase of humidity sharply lowered the sorption capacity and made the isotherm more linear, as shown for m-dichlorobenzene in Fig. 3. The sorption capacity for the compound at 90 percent R.H. is about the same as found in aqueous systems (Chiou et al. 1983), being about two orders of magnitude lower than on dry soil. The data are in close agreement with the findings by Spencer et al. (1969), Spencer and Cliath (1970), and others and are in keeping with the assumed dual-sorbent character of the soil.

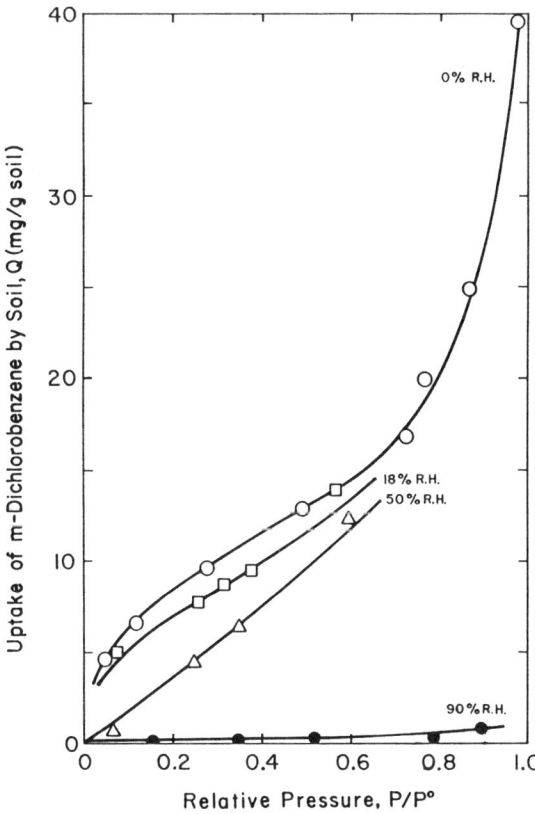

Fig. 3. Vapor sorption isotherms of m-dichlorobenzene on Woodburn soil as a function of relative humidity (R.H.) at 20° C

7.5 Influence of Soil Sorption on Pollutant and Pesticide Activity

The proposed roles of soil organic matter and soil minerals in sorption and the influence of soil water on their functions provide a theoretical basis for assessing the activity of organic pollutants and pesticides in soil. In aqueous systems, the important properties for soil sorption are the organic matter content of the soil and the water solubility of the compound. Given the equilibrium concentration of a compound in water, one can make a reasonable estimate of the amount of uptake by soil, based on soil organic content and its K_{om} value. Reasonable estimates of K_{om} values for various compounds can be obtained from known correlations with S_w or K_{ow}. Moreover, since the uptake by water-saturated soils is controlled mainly by partitioning in (macromolecular) organic matter (in which the solute solubility is small relative to that in an organic solvent), the amount of the organic contaminant sorbed by soil should be much less than the soil organic-matter content.

The activity of a pesticide (or contaminant) may be expected to be most sensitively affected by a changing soil-water content from above to below the saturation capacity in the field, because this drying-wetting cycle would cause a sharp change in mineral adsorptivity, especially for soils low in organic matter. The very top layer of surface soils usually undergoes a drying-wetting cycle to an extent depending on the change in ambient humidity as well as on agricultural practice. Under dry conditions, strong adsorption by minerals in addition to uptake by partitioning into organic matter lowers the activity of the pesticide in soil (as measured by its relative pressure, P/P^o). Upon wetting, the chemical activity would sharply rise due to displacement by water of the compound previously adsorbed on soil minerals, as shown earlier. Volatile pesticides originally sorbed by dry soils and clays are thus readily released by addition of water, a phenomenon widely documented (Goring 1967). While many pesticides may not show significant vapor losses from soil, their chemical activities are nonetheless strongly influenced by soil water content.

A change in chemical activity invariably leads to a change in apparent toxicity of a pesticide. Upchurch (1957) found that diuron was more toxic to cotton when applied to moist rather than dry soils. Barlow and Hadaway (1955) observed that toxicities of chlorinated hydrocarbon pesticides (lindane, DDT, and dieldrin) were greatly inactivated by dry clays but were reactivated under high humidity for control of mosquitoes. Harris (1964) reported correlations of insecticide toxicities of heptachlor, DDT, diazinon, V-C 13 (dichlofenthion), and parathion on a sand (0.52 percent organic matter) and a muck (65 percent organic matter) with soil moisture content. The results are shown in Table 1. Heptachlor was 7.8, DDT was 9.9, parathion was 24.4, diazinon was 132, and V-C 13 was 189 times more toxic to the cricket when the sand was moist (5.5 percent water) than when it was dry. By contrast, moisture in muck soil (162 percent water) had only a marginal effect on the insecticide toxicity. At field moisture capacity DDT was 38.5, heptachlor was 61.6, diazinon was 65.7, V-C 13 was 73.3, and parathion was 92.0 times less toxic in the muck than in the sand. Harris thus concluded that

Table 1. Influence of soil type and soil moisture on the toxicity of insecticides in soils to first instar nymphs of the common field cricket[1]

Insecticide	Soil Type	LD_{50}(ppm) Moist	Dry
Heptachlor	Plainfield Sand	0.068	0.53
	Muck	4.19	5.39
DDT	Plainfield Sand	1.75	17.3
	Muck	67.2	99.8
Diazinon	Plainfield Sand	0.26	34.1
	Muck	17.0	11.5
V-C 13	Plainfield Sand	3.80	717
	Muck	279	165
Parathion	Plainfield Sand	0.25	6.
	Muck	22.6	9.10

[1] Reproduced from Harris (1964).

inactivation of insecticides in moist soils was proportional to soil organic content, while in dry soils inactivation was related to the adsorption capacity of the mineral fraction. Harris and Bowman (1981) reported a similar moisture dependence of the toxicity of various organophosphates in a moist and air-dried sandy loam.

The observed magnitude of the effect of soil moisture on the apparent toxicity of soil-incorporated pesticides from these studies is closely related to the recognized effect of soil moisture on the functions of soil minerals and organic matter in soil uptake. The distinct effect of moisture on adsorption of organic compounds by minerals can best be appreciated by our common experience that the field gives a fragrant smell following a rain shower that succeeds a long period of drought. This phenomenon does not take place if the field has been wet prior to the rain shower. By these recognitions, we now have a better understanding of the influence of humidity on the activity of organic compounds in soil, as well as the distinct processes through which the organic matter and mineral matter of the soil function in sorption.

References

Bailey GW, White JL (1964) Review of adsorption and desorption of organic pesticides by soil colloids, with implications concerning pesticide bioactivity. J Agric Food Chem 12:324–332

Barlow F, Hadaway AB (1955) Studies on aqueous suspensions of insecticides. V. The adsorption of insecticides by soils. Bull Entomol Res 46:547–559

Briggs GG (1973) A simple relationship between soil adsorption of organic chemicals and their octanol-water partition coefficients. 7th British Insecticide and Fungicide Conf, pp 83–88

Briggs GG (1981) Theoretical and experimental relationships between soil adsorption, octanol-water partition coefficients, water solubilities, bioconcentration factors, and parachor. J Agric Food Chem 29:1050–1059

Browman MC, Chesters G (1977) The solid-water interfaces: Transfer of organic pollutants across the solid-water interface. In: Suffet IH (ed) Fate of Pollutants in the Air and Water Environments, Part 1. Wiley, New York, pp 49–105

Call F (1957) The mechanism of sorption of ethylene dibromide on moist soils. J Sci Food Agric 8:630–639
Carringer RD, Weber JB, Monaco TJ (1975) Adsorption-desorption of selected pesticides by organic matter and montmorillonite. J Agric Food Chem 23:568–572
Chiou CT (1981) Partition coefficient and water solubility in environmental chemistry. In: Saxena J, Fisher F (eds) Hazard Assessment of Chemicals: Current Developments. Vol 1. Academic Press, London, pp 117–153
Chiou CT (1989) Roles of organic matter, minerals and moisture in sorption of nonionic organic compounds and pesticides by soil. In: MacCarthy P, Malcolm RL, Clapp E, Bloom P (eds) Humic Substances in Soil and Crop Sciences. American Society of Agronomy, Madison, WI (in press)
Chiou CT, Shoup TD (1985) Soil sorption of organic vapors and effects of humidity on sorptive mechanism and capacity. Environ Sci Technol 19:1196–1200
Chiou CT, Peters LJ, Freed VH (1979) A physical concept of soil-water equilibria for nonionic organic compounds. Science 206:831–832
Chiou CT, Peters LJ, Freed VH (1981) Reply to comment on "Soil-Water Equilibria for Nonionic Organic Compounds". Science 213:684
Chiou CT, Schmedding DW, Manes M (1982) Partitioning of organic compounds in octanol-water systems. Environ Sci Technol 16:4–10
Chiou CT, Porter PE, Schmedding DW (1983) Partition equilibria of nonionic organic compounds between soil organic matter and water. Environ Sci Technol 17:227–231
Chiou CT, Shoup TD, Porter PE (1985) Mechanistic roles of soil humus and minerals in the sorption of nonionic organic compounds from aqueous and organic solutions. Org Geochem 8:9–14
Chisholm RC, Koblitsky L (1943) Sorption of methyl bromide by soil in a fumigation chamber. J Econ Entomol 36:549–551
Dzombak DA, Luthy RG (1984) Estimating adsorption of polycyclic aromatic hydrocarbons on soils. Soil Sci 137:292–308
Goring CAI (1967) Physical aspects of soil in relation to the action of soil fungicides. Annu Rev Phytopathol 5:285–318
Hamaker JW, Thompson JM (1972) Adsorption. In: Goring GAI, Hamaker JW (eds) Organic Chemicals in the Soil Environment, Vol 1. Dekker, New York, pp 49–143
Hance RJ (1965) Observations on the relationships between the adsorption of diuron and the nature of adsorbent. Weed Res 5:108–114
Hanson WJ, Nex RW (1953) Diffusion of ethylene dibromide in soil. Soil Sci 76:209–214
Haque R (1975) Role of adsorption in studying the dynamics of pesticides in a soil environment. In: Haque R, Freed VH (eds) Environmental Dynamics of Pesticides. Plenum, New York, pp 97–114
Harris CR (1964) Influence of soil type and soil moisture on the toxicity of insecticides in soils to insects. Nature 202:724
Harris CR, Bowman BT (1981) The relationship of insecticide solubility in water to toxicity in soil. J Econ Entomol 74:210–212
Hartley GS (1960) Physico-chemical aspects of the availability of herbicides in soil. In: Woodford EK, Sagar GR (eds) Herbicides and the Soil. Blackwell, Oxford, pp 63–78
Hassett JJ, Banwart WL, Wood SG, Means JC (1981) Sorption of α-naphthol: Implication concerning the limits of hydrophobic sorption. Soil Sci Soc Am J 45:38–42
Jurinak JJ (1957) The effect of clay minerals and exchangeable cations on the adsorption of ethylene dibromide vapor. Soil Sci Soc Am Proc 21:599–602
Karickhoff SW (1981) Semi-empirical estimation of sorption of hydrophobic pollutants on natural sediments and soils. Chemosphere 10:833–846
Karickhoff SW, Brown DS, Scott TA (1979) Sorption of hydrophobic pollutants on natural sediments. Water Res 13:241–248
Kenaga EE, Goring GAI (1980) Relationship between water solubility, soil sorption, octanol-water partitioning, and concentration of chemicals in biota. In: Eaton JC, Parrish PR, Hendricks AC (eds) Aquatic Toxicology. American Society for Testing and Materials, Philadelphia, PA, pp 78–115
Lambert SM (1967) Functional relationship between sorption in soil and chemical structure. J Agric Food Chem 15:572–576

Lambert SM (1968) Omega, a useful index of soil sorption equilibria. J Agric Food Chem 16:340-343
Leistra M (1970) Distribution of 1,3-dichloropropene over the phases in soil. J Agric Food Chem 18:1124-1126
McCall PJ, Laskowski DA, Swann RL, Dishburger HJ (1983) Estimation of environmental partitioning of organic chemicals in model ecosystems. Res Rev 85:231-244
Means JC, Wood SG, Hassett JJ, Banwart WL (1982) Sorption of carboxy-substituted polynuclear aromatic hydrocarbons by sediments and soils. Environ Sci Technol 16:93-98
Mills AC, Biggar JW (1969) Solubility-temperature effect on the adsorption of gamma- and beta-BHC from aqueous and hexane solutions by soil materials. Soil Sci Soc Am Proc 33:210-216
Mingelgrin U, Gerstl Z (1983) Reevaluation of partitioning as a mechanism of nonionic chemicals adsorption in soils. J Environ Qual 12:1-11
Pierce RH Jr, Olney CE, Felbeck GT Jr (1974) p,p'-DDT adsorption to suspended particulate matter in sea water. Geochim Cosmochim Acta 38:1061-1073
Saltzman S, Kliger L, Yaron B (1972) Adsorption-desorption of parathion as affected by soil organic matter. J Agric Food Chem 20:1224-1226
Spencer WF, Cliath MM (1970) Desorption of lindane from soil as related to vapor density. Soil Sci Soc Am Proc 34:574-578
Spencer WF, Cliath MM, Farmer WJ (1969) Vapor density of soil-applied dieldrin as related to soil-water content, temperature, and dieldrin concentration. Soil Sci Soc Am Proc 33:509-511
Stark FL Jr (1948) Investigation of chloropicrin as a soil fumigant. New York (Cornell) Agric Exp Stn Mem 178:1-61
Stevenson FJ (1982) Humus Chemistry, Chap 17. Wiley, New York
Swoboda AR, Thomas GW (1968) Movement of parathion in soil columns. J Agric Food Chem 16:923-927
Upchurch RP (1957) The influence of soil moisture content on response of cotton to diuron. Weeds 5:112-120
Wade PI (1954) The sorption of ethylene dibromide by soils. J Sci Food Agric 5:184-192
Weber JB, Weed SB (1974) Effects of soil on the biological activity of pesticides. In: Geunzi WD (ed) Pesticides in Soil and Water. American Society of Agronomy, Madison, WI, pp 223-256
Yaron B, Saltzman S (1972) Influence of water and temperature on adsorption of parathion by soils. Soil Sci Soc Am Proc 36:583-586

8 Sorption and Transport of Organic Pollutants at Waste Disposal Sites

P.S.C. RAO[1], L.S. LEE[1], P. NKEDI-KIZZA[1], and S.H. YALKOWSKY[2]

8.1 Introduction

Chemodynamic properties of organic pollutants, such as solubility, sorption, and transport, have usually been characterized with water and/or aqueous electrolyte solutions as the solvent and with single organic solute systems. We define here **complex mixtures** as those systems having multiple solutes and multiple solvents. The solute mixtures of interest may consist of various combinations of nonpolar hydrophobic organic chemicals (HOC), hydrophobic ionizable organic chemicals (HIOC), and ionizable organic chemicals (IOC). The solvent may be a mixture of water and one or more of completely-miscible organic solvents (CMOS) and partially-miscible organic solvents (PMOS). We denote solvents consisting only of water and CMOS in a single homogeneous liquid phase as **mixed solvents**. Those solvents with water plus PMOS and CMOS and form at least two distinct liquid phases will be referred to as **multi-phase solvents**. Numerous examples of complex mixtures, as defined above, are found at or near all hazardous waste disposal and spill sites. Thus, examining the chemodynamics of complex mixtures is essential for predicting the environmental impact from the disposal of hazardous wastes on land.

In recent years, the need to study the chemodynamics of **complex mixtures** has been recognized and coordinated efforts have been initiated in the U.S. over the past five years to develop theoretical bases and laboratory data for characterizing the chemodynamics of complex mixtures. In this paper, we will review recent advances in characterizing solubility, sorption, and transport of complex mixtures. Conceptual models proposed to predict these processes are examined, and the shortcomings in the data needed to evaluate these models are briefly discussed.

---

[1] Prof., Chemist-III, and Asst. Prof., respectively, Soil Science Department, University of Florida, Gainesville, FL 32611, USA
[2] Prof., Dept. of Pharm. Science, University of Arizona, Tucson, AZ, USA

8.2 Solubility in Complex Solvents

8.2.1 Solubility in Mixed Solvents

Many properties of mixed solvents can be approximated by treating the mixed solvent as a linear combination of the pure component solvents. This is true for the logarithm of the solubility of nonpolar hydrophobic organic chemicals (HOC). Yalkowsky and Roseman (1981) and Rubino and Yalkowsky (1987a,b) have derived the following equation relating the solubility of a nonpolar solute in a binary mixed solvent (S_m) to that in pure water (S_w):

$$\log S_m = \log S_w + \sigma_s f_s \tag{1}$$

where

$$\sigma_s = (A \log K_{ow} + B), \tag{2}$$

f_s is the volume fraction of the cosolvent in the binary mixed solvent; A and B are empirical constants dependent on the cosolvent properties; and K_{ow} is the octanol-water partition coefficient of the solute.

Equation (1) has been found to work quite well for systems in which the cosolvent has minimal solubility in the solute phase, i.e. the solute phase is unchanged by the addition of the cosolvent, and the only effect of adding the cosolvent is to decrease the polarity of the aqueous phase. Solubilization curves determined for a number of environmentally important organic chemicals have shown a reasonable log-linear increase in solubility with increasing cosolvent composition (Yalkowsky 1985; Fu and Luthy 1986a) (Fig. 1 A,B). Several indices of cosolvent polarity, including dielectric constant, surface tension, and interfacial tension, have also been found to be strongly correlated to σ_s (Rubino and Yalkowsky 1987a).

Deviations from the predicted log-linear solubility behavior [Eq. (1)] in binary mixed solvents have been observed by several investigators, and the nonlinearity in the curves may be sigmoidal or concave at higher cosolvent fractions (Fu and Luthy 1986a; Rubino and Yalkowsky 1987b). Excess solubility is conceptually analogous to excess free energy. Excess free energy may be defined as the free energy not predicted by ideal treatment of a system. Comparison and correlation of excess solubilities with excess properties may provide a mechanism for predicting solute solubility and elucidating the interactions causing the deviation. Rubino and Yalkowsky (1987b) found that the excess density maximum corresponded most closely to the excess solubility maximum for several drug-cosolvent systems. They also inferred that the nonideality is largely due to cosolvent-water interactions and is generally independent of the nature of the solute.

In a manner similar to the treatment of the binary mixed solvents, the solubility in a mixed solvent with multiple CMOS can be determined by

$$\log S_m = \log S_w + \Sigma \sigma_i f_i \tag{3}$$

where the subscript i indicates the i-th cosolvent. Studies by Yalkowsky (1986) verified Eq. (3) for a range of pollutants in several mixed solvents (Fig. 1C). This

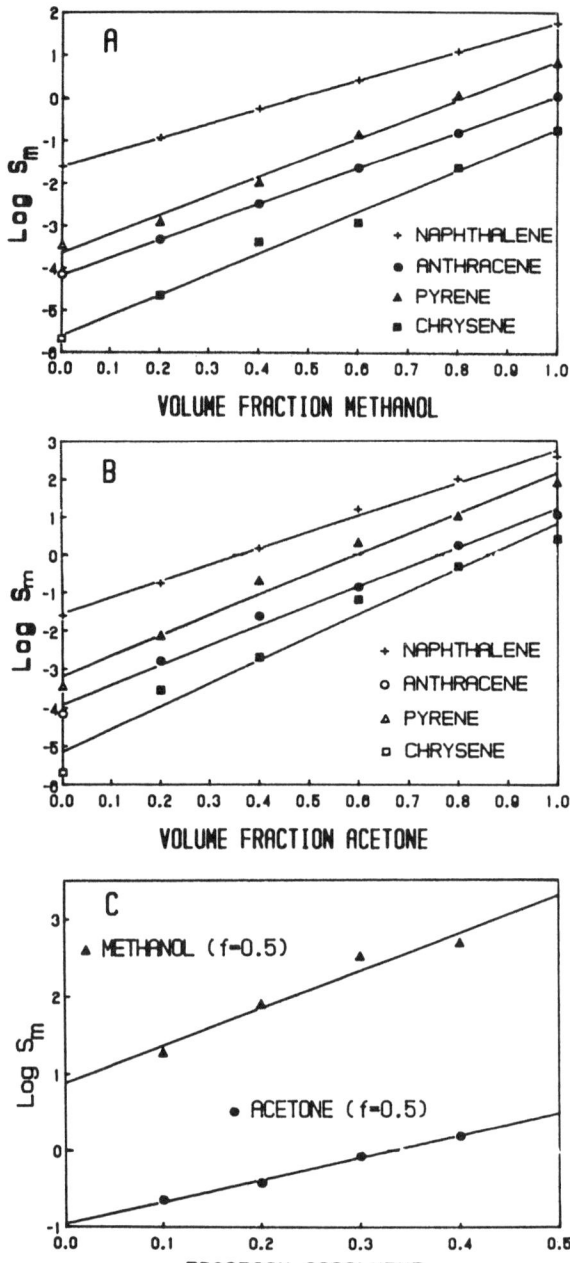

Fig.1A-C. Solubilities for several polyaromatic hydrocarbon chemicals in (**A**) methanol-water and (**B**) acetone-water mixed solvent systems. Solubility data are presented in (**C**) for ternary mixed solvents (methanol-acetone-water) where one cosolvent volume fraction remains fixed at 0.5 as that of the second cosolvent varies with water

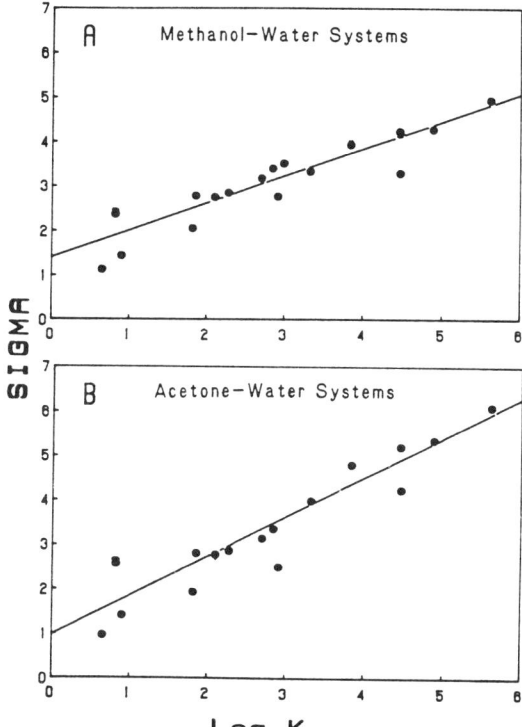

Fig. 2A,B. The linear relationship between σ and log K_{ow} for several hydrophobic organic compounds in (**A**) methanol-water and (**B**) acetone-water systems are shown

relationship was also observed by Rubino and Yalkowsky (1987a) for a number of drugs in several mixed solvents.

Note from Eq. (2) that σ_s, the slope of the log-linear plot of HOC solubility versus cosolvent composition, is directly proportional to log K_{ow} (Fig. 2 A,B). Eq. (2) suggests that, if for a given cosolvent σ_s values are experimentally determined for a few HOC where K_{ow}'s are known, the values of A and B can be determined. In turn, using these values of A and B, σ_s for other HOC in that cosolvent can be determined by measuring or predicting K_{ow} for the HOC (Yalkowsky 1985). Given the values of the constants A and B in Eq. (2) for each CMOS along with the log K_{ow} value for the HOC, the solubility of that HOC in a mixed solvent consisting of any composition of those cosolvents may be calculated (Yalkowsky 1985).

8.2.2 Solubility in Multi-Phasic Solvents

As defined earlier, a multi-phasic solvent consists of water and one or more PMOS, and may or may not contain CMOS. In multi-phasic solvents, the mole-fraction solubilities of the solvents in each of the phases must be known. The influence of the dissolved cosolvents on the solubility of the HOC of interest in each of the liquid phases must also be predicted. Sorensen and Arlt (1979), Lyman (1982), Grain (1982), and Novak et al. (1987) have discussed the thermodynamic basis for liquid-liquid equilibria. Sorensen and Arlt (1979) have presented a large collection of phase diagram data for binary and ternary solvents.

In order to illustrate the problem at hand, let us consider the solubility of a HOC in a simple biphasic solvent – say a mixture of water and a PMOS. In this biphasic solvent, PMOS concentration in the aqueous phase is likely to be at or near its solubility limit. When the PMOS aqueous solubility exceeds 10^{-3} M, the dissolved PMOS may act as a cosolvent in enhancing the HOC aqueous solubility (Munz and Roberts 1986).

A much more complex situation arises, however, if a third solvent, say a CMOS, is added to the biphasic water-PMOS mixture. The presence of even small amounts of the CMOS may significantly increase the PMOS solubility in water. If the CMOS concentration is the same in both binary and ternary mixed solvents, the resulting ternary mixed solvent (water-PMOS-CMOS) is less polar than the binary mixed solvent (water-CMOS) or water alone. Thus, the HOC solubility in the ternary mixed solvent is expected to be greater than either the binary mixed solvent or water. Data are presently unavailable to test whether such effects can be predicted by Eq. (1).

Another feature of the complex solvents must also be recognized. At certain mole-fractions of the CMOS and PMOS, the PMOS solubility enhancement might be sufficiently large that a monophasic, homogeneous mixed solvent may form instead of a biphasic solvent. In some cases complete miscibility of the PMOS in the aqueous phase may result from self-solubilization when the CMOS exceeds a certain critical concentration.

We are unaware of a critical examination of published data for environmentally relevant organic pollutants to evaluate solubility effects discussed above. However, an illustration of such effects can be found in the pharmaceutical literature. Rubino and Yalkowsky (1984, 1985, 1987c) have shown that triacetin (glyceryl triacetate), a solvent relatively insoluble in water, is solubilized in water by propylene glycol or ethanol; the solubilized triacetin then became an excellent cosolvent for several nonpolar drugs. While the deliberate use of an insoluble solvent as a cosolvent can have beneficial effects in drug formulation, similar effects can lead to enhanced aqueous solubility of hydrophobic organic pollutants in water. The environmental consequences of such HOC solubility enhancement are decreased sorption and retardation with the attendant increase in the potential for groundwater contamination.

8.3 Sorption from Complex Solvents

In describing HOC sorption, Rao et al. (1985) noted that the following interactions need to be considered: sorbate-solvent (solubility), solvent-solvent (miscibility), solvent-sorbent, and sorbate-sorbate. We focus here primarily on the first two interactions listed above; the presence of cosolvents may not significantly impact the role of the latter two interactions in determining HOC sorption on natural and synthetic sorbents.

8.3.1 Sorption from Mixed Solvents

Rao et al. (1985) proposed a theoretical approach, based on the predominance of solvophobic interactions, for predicting HOC sorption from mixed solvents. They showed that,

$$\ln K^m = \ln K^w - \Sigma \, \alpha_i \, \sigma_i \, f_i \qquad (4)$$

where
$$\sigma_i \approx (\Delta\gamma_i \, HSA)/(k \, T), \qquad (5)$$

K is sorption coefficient (moles solvent/kg sorbent) with the superscripts w and m indicating values for sorption from water and mixed-solvents, respectively; σ_i is a dimensionless term unique to each solvent-sorbate combination; $\Delta\gamma_i$ is the differential interfacial free energy (J/nm^2) at the solvent-sorbate interface; HSA is the sorbate hydrocarbonaceous surface area (nm^2); k is the Boltzmann constant (J/K); T is temperature (K); f_i is the volume fraction of the i-th cosolvent; and α_i is an empirical constant. The volume fraction of water, f_w, in the mixed solvent is given by $[1 - \Sigma \, f_i]$. As a consequence of the assumption that solubility and sorption coefficient are inversely related, Eqs. (1) and (4) are functionally similar and the σ_i value in both equations is the same (Rao et al. 1985). The solute hydrophobicity is represented by HSA in Eq. (5) and by log K_{ow} in Eq. (2); note that log K_{ow} is directly proportional to HSA. The empirical constants A and B in Eq. (1) are correlated to $\Delta\gamma$ and other solvent properties.

Eq. (4) suggests that with increasing volume fraction (f_i) of a CMOS in a binary mixed solvent, HOC sorption coefficient (K^m) decreases exponentially. This prediction has been experimentally verified for sorption of several HOC by soils from binary mixed solvents (Nkedi-Kizza et al. 1985; Fu and Luthy 1986b; Rao and Lee 1988; Miller et al. 1987). Woodburn et al. (1986) presented data for HOC sorption from ternary mixed solvents in verification of Eq. (4). Rao and Lee (1988) demonstrated the validity of Eq. (4) for HOC sorption from a quinary mixed solvent. For the ternary and quinary mixed solvents, the $\alpha_i \sigma_i$ value needed in Eq. (4) for each solvent-sorbate-sorbent combination was determined from the slopes of ln K^m versus f_i plots using the experimental data for sorption from each binary mixed solvent. Woodburn et al. (1986) and Rao and Lee (1988) have shown that σ_i values can be estimated using the data for HOC retention in reversed-phase liquid chromatographic columns eluted isocratically with mixed solvents.

The slope of the log-linear plot K^m (mole/kg) versus f_i represents the combined effect of both α and σ. Whereas σ_i is a function of solvent-sorbate interactions, the empirical constant α_i appears to reflect solvent-sorbent interactions. The α term should approach unity if the soil organic carbon properties are independent of change in solution phase compositions (Karickhoff 1981). Equation (1) can be used to calculate α_i values from measured σ_i values obtained from solubility data using Eq. (1). Fu and Luthy (1986b) have reported that α may vary from 0.4 to 1.3 depending upon the soil-solute-solvent combinations with an average value of 0.5 ± 0.06. Similar results were obtained reanalyzing the Nkedi-Kizza et al. (1985) data and Miller et al. (1987) data using Yalkowsky's (1986) solubility data. Karickhoff (1981) reported an average α value of 0.83. In most cases the observed value of α is significantly less than unity which means that the effect of CMOS on the decrease of K^m will not be as significant as the effect of CMOS on the increase of solute solubility. This competing effect may be due to the swelling of the organic carbon material by the presence of the CMOS thus increasing solute accessibility into the organic carbon matrix (Freeman and Cheung 1981). These results suggest that while the solvent-solute interactions are predominant, solvent-sorbent interactions do have an impact on HOC sorption.

8.3.2 Sorption from Biphasic Solvents

Extending HOC solubility data to predict HOC sorption by soils from biphasic solvents will require the additional knowledge of sorbent-solvent interactions. The impact of the PMOS on HOC sorption may depend on whether water or the added PMOS is present as a wetting phase. In unsaturated and saturated soils, water is likely to be the wetting phase. For such cases, one needs to only consider the liquid-liquid partitioning of the HOC and the sorption of HOC by the soil directly from the aqueous phase. The aqueous phase will be saturated with the PMOS, and unless the dissolved PMOS concentration exceeds 0.005 molefraction, the cosolute or cosolvent effects of this PMOS on HOC sorption may not be significant (Munz and Roberts 1986).

When the PMOS is the wetting phase and partially or completely coats the sorbent, the net result would be, in effect, to increase HOC sorption because the PMOS films would serve as additional "sinks" (or sorptive volumes) for the hydrophobic sorbate. It must be recognized, however, that various intermediate scenarios are possible depending on the specific sorbent-PMOS interactions as well as on the nature and amount of the CMOS present.

Pesticide sorption from PMOS containing small amounts of water has been studied by a few investigators (Yaron and Saltzman 1972; Hance 1977; van Bladel and Moreale 1974). Their primary focus was to characterize competition between water and pesticides for the sorbent surfaces at low water contents. Rao and Lee (1988) measured sorption of two herbicides, terbacil and atrazine, by Webster soil from biphasic solvents; terbacil sorption was measured using a toluene-water mixture and atrazine sorption was measured from a pentane-water mixture. Measured sorption isotherms for the two herbicides in aqueous solutions and in biphasic solvents were identical. Thus, the presence of the PMOS did not measurably affect herbicide sorption. Miller et al. (1987) have measured atrazine herbicide sorption on two soils from biphasic solvents consisting of water and n-hexane. They reported that with increasing n-hexane content (up to 33.3% by volume of PMOS), atrazine K value was unchanged in one soil but decreased by about a factor of 1.6 in the other soil.

Similar results as discussed above may not be expected for other sorbates or solvents. Rao and Lee (1988) did not consider the effects of the PMOS on the sorbent, which may be significant for other solvents. For multiphasic solvents that are likely to be encountered at waste disposal sites and for viscous PMOS, the sorbent will probably be coated with solvent. If this occurs, the amount of the PMOS associated with the coating and its effect on HOC sorption will need to be considered. No data are available presently to evaluate such effects.

8.3.3 Sorption from Complex Solvents Containing PMOS and CMOS

We are not aware of any published data for sorption of environmentally relevant HOC by soils and aquifer media from complex solvents consisting of water plus CMOS and PMOS. However, as discussed in the previous sections, it should be recognized that the PMOS solubilized into the aqueous phase (assumed to be in contact with the sorbent for this discussion) may be treated as another cosolvent. As the volume fraction of the CMOS increases, the amount of the PMOS solubilized into the predominantly aqueous phase is likely to increase; thus, HOC sorption will continue to decrease in proportion to the volume fractions of the CMOS and the solubilized PMOS.

8.4 Sorption from Aqueous Solutions of Multi-Sorbate Mixtures

Although many of the organic pollutants can be classified as HOC, several important classes of organic pollutants (e.g. phenols, amines, etc.) are hydrophobic ionizable chemicals (HIOC). Because waste disposal sites are characterized by a mixture of several HOC and HIOC, the hydrophobic model may not adequately represent their sorption on soils. Also, the possibility of competitive sorption needs to be considered.

Increasing attention is now being paid to sorption from such multi-sorbate mixtures. Several authors have shown that HOC sorption on to soils and sediments is not significantly influenced by the presence of other sorbates (Karickhoff et al. 1979; Swarzenbach and Westall 1981; Chiou et al. 1983; MacIntyre and deFur 1985; Rao et al. 1986). These results are consistent with the expectation based on the assumption that hydrophobic interactions are dominant in sorption of the HOC (Chiou et al. 1983).

For HIOC, such as phenols and amines, at least four mechanisms must be considered in predicting their sorption. The first mechanism deals with the sorption of the neutral molecular species from the aqueous phase. This is similar to the hydrophobic sorption that has been used to describe the sorption of HOC. A second mechanism of interest is the specific interactions of the dissociated (ionic) species with various functional groups on the sorbent surface. Several models that have been developed for predicting the ion exchange of inorganic ions may be useful in predicting this type of sorption. A third sorption mechanism, known as molecular ion pairing, deals with the transfer of the organic ions with inorganic counterions from the aqueous phase to the organic phase. This type of sorption has been observed by Westall et al. (1985) in his studies of 2,3,4,5-tetrachlorophenol and pentachlorophenol partitioning in octanol-water systems as a function of pH and ionic strength. A fourth mechanism deals with the transfer of the organic ion from the aqueous phase to the organic surface with the counterions remaining in the electric double layer of the aqueous phase. This type of sorption has been described in terms of the Guoy-Chapman model by Cantwell and Puon (1979). While all these mechanisms are plausible, further research is needed to explicitly define the sorption processes involved. The relative contribution of each of these mechanisms will depend upon: (1) the extent of HIOC dissociation as a function of pK_a and solution pH; (2) the ionic charge status of the soil as a function of the pH and the point of zero net charge; and (3) ionic strength and composition of the aqueous phase.

Available data for sorption of organic amines from binary mixtures (Felice et al. 1985; Zachara et al. 1986, 1987) suggest that if the molecular species predominates then competitive sorption effects are not observed, as is the case for HOC. However, if the ionized species are predominant, then competitive sorption effects due to ion exchange involving organic and inorganic cations might be expected (Brown and Combs 1985; Felice et al. 1985; Zachara et al. 1986, 1987). Zachara et al. (1987) used a simplified version of the ideal adsorbed solution (IAS) model (Crittenden et al. 1985) to successfully predict competition observed between quinoline, acridine, and pyridine in acidic subsoil.

We have examined sorption on soils from mixtures of five substituted phenols. The phenols studied were: 2,4-dimethylphenol (DMP); phenol; 2,4-dichlorophenol (DCP); pentachlorophenol (PCP); and 2,4-dinitrophenol (DNP). The pK_a of these weakly acidic compounds are 10.6, 10.0, 7.8, 4.7, and 4.1, respectively. The soils used were: Eustis fine sand (pH = 5.1, OC = 0.78%); Lincoln sand (pH = 6.3, OC = 0.22%); Webster silty clay loam (pH = 7.3, OC = 3.4%); and Kidman very fine sand (pH = 7.7, OC = 0.61%). The differences in the dissociation constants of the phenols and the range in soil pH, soil OC, and soil clay mineralogy provided for a broad spectrum of experimental conditions to study competitive effects for sorption of HIOC from mixtures. Data representing the general trends observed for the sorption isotherms measured using single phenols and quinary mixtures are summarized in Figs. 3 and 4.

For a given soil and pH, the pK_a of the phenol will determine the extent of ionization, and hence, the extent of the competitive sorption from the other phenols. Data illustrating this are presented in Fig. 3. For the sorption of phenol and DMP on Eustis soil the measured isotherms were identical for single- and quinary-sorbate systems (Fig. 3 A,B). Since the pK_a's of both phenol and DMP

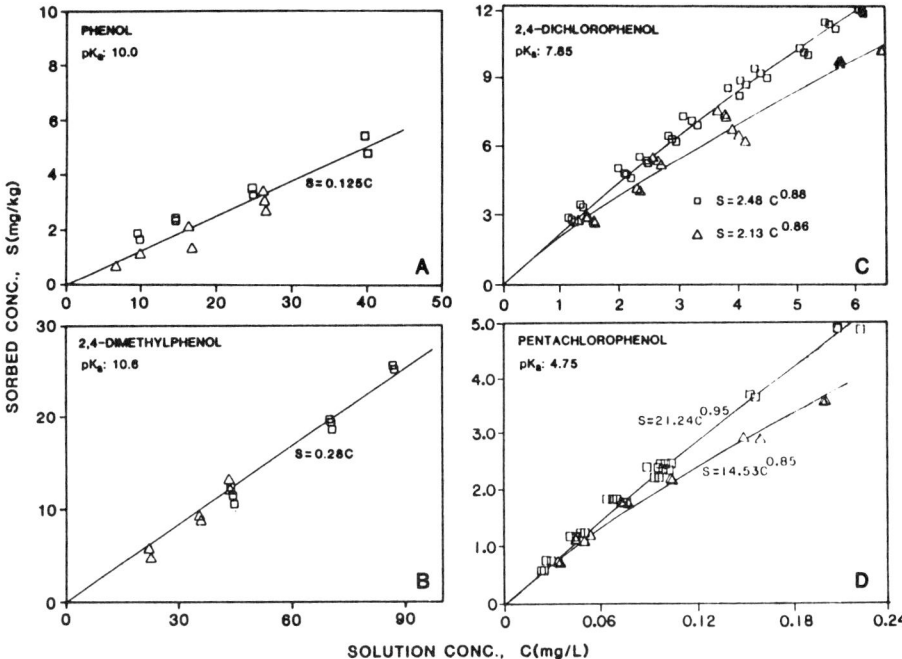

Fig. 3. Equilibrium isotherms sorption of several phenols on Eustis soil. Data for sorption from single-sorbate (□) and multi-sorbate (△) systems are shown

are much greater than the soil pH, the molecular species predominates and no competitive effects from the other phenols in the mixture are expected. Even though essentially all of the DCP is expected to be present as the molecular species (pk_a > pH), the presence of the other phenols seems to have slightly suppressed (by about 14%) the sorption of DCP (Fig. 3C). (Note the percent ionized is calculated using measured bulk pH values which could be significantly higher than those at the soil-solution interface.) The pK_a of PCP is less than the soil pH (4.7 vs 5.1) and about 70% of the PCP will be present in the ionized form. Thus, the observed decrease in PCP sorption (Fig. 3D) may be attributed to competitive sorption of the other phenols; the most likely competing species is DNP ($pK_a = 4.1$) since all other phenols in the mixture exist predominantly in their molecular form at the pH of the Eustis soil. If the predominant sorption mechanism is presumed to be molecular ion-pairing, competition is actually in the solution phase between organic and inorganic ions (i.e. ion selectivity) prior to sorption of the molecular ion pair.

Isotherms for PCP sorption by four soils from single- and quinary-sorbate mixtures are shown in Fig. 4. The pK_a of PCP is less than the pH of all four soils, and about 70% or more of the PCP is expected to exist in the ionized form. Thus, competitive effects are to be expected for all four soils, as seen in Fig. 4. The PCP sorption coefficients for Eustis and Webster soils are essentially the same, which

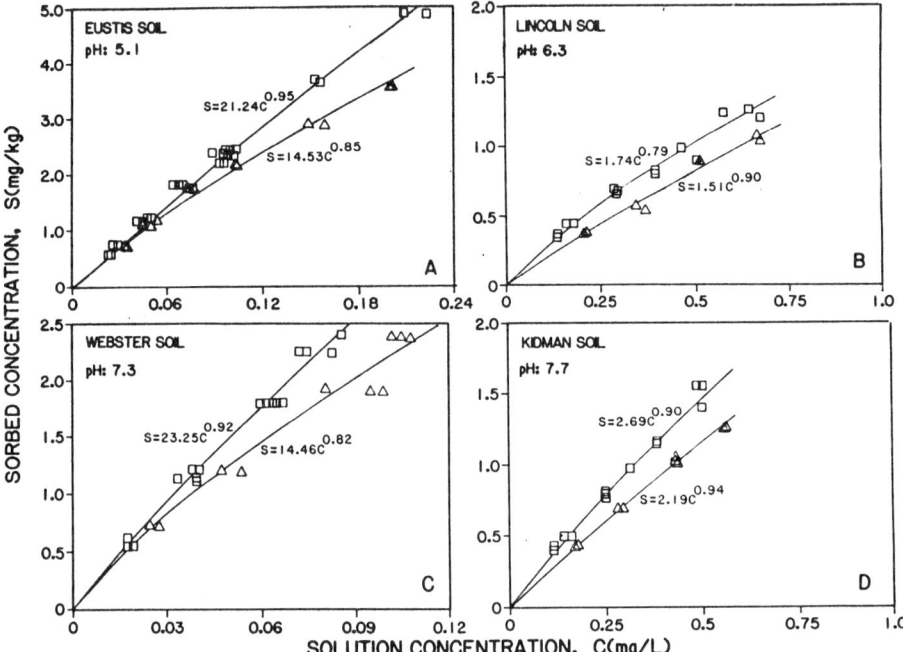

Fig. 4. Equilibrium isotherms for sorption of pentachlorophenol by four soils. Data for sorption from single-sorbate (□) and multi-sorbate (△) systems are shown

further supports that soil OC alone may not be sufficient for predicting sorption of ionizable compounds. PCP is almost completely ionized in the Webster soil-water system, whereas in the Eustis soil-water system 30% of PCP is still molecular. Using an equation to predict the overall distribution ratio (D) of both the ionized and the molecular forms (Schellenberg et al. 1984) the D for PCP was 2.3 and 58 on Webster and Eustis, respectively. Based on these predictions the observed PCP partitioning on Webster was greater than predicted, while partitioning on Eustis was less than predicted. Also, the largest suppression of PCP sorption (30–40%) was observed in these two soils.

Preliminary studies conducted with phenols on the effect of ionic strength showed a logarithmic increase in sorption with increasing $CaCl_2$ concentration (0.001–1.0 N) on all four soils. This effect was the greatest when pH > pK_a and almost negligible when pH < pK_a. Although the soil pH decreases with increasing $CaCl_2$ concentration, the effect of pH was not sufficient to account for the large increase observed in the sorption of phenols. Similar results were obtained in studies with KCl; however, the increase in the sorption of the phenols with KCl concentration was even greater than that observed with $CaCl_2$. These data suggest ion selectivity in the formation of ion pair.

The specific reasons for these various observations on the sorption of the phenols and a quantitative treatment of the sorption data are the topics of ongoing investigations. These data do illustrate the need to consider sorption mechanisms other than hydrophobic association and the importance of knowing the chemical and physical properties of both the sorbent and the sorbates in predicting HIOC sorption in soils.

8.5 Transport in the Presence of Complex Solvents

8.5.1 Transport in Mixed Solvents

On the basis of cosolvent effects on HOC sorption, Rao et al. (1985) have suggested that mobility of HOC in soils in the presence of mixed solvents is enhanced. They showed that the retardation factors for HOC leaching with aqueous and mixed solvents, designated by R^w and R^m, respectively, can be estimated as follows:

$$R^w = [1 + (\rho K^w/\Theta)] \tag{5}$$

and

$$R^m = [1 + (\rho K^m/\Theta)] \tag{6}$$

where K^w and K^m are the HOC sorption coefficients in aqueous and mixed solvents; ρ is the soil bulk density (g/cm^3); and Θ is the volumetric liquid content. Note that K^m value in Eq. (6) decreases with increasing volume fraction of the CMOS in the mixed solvent, as shown in Eq. (4). Nkedi-Kizza et al. (1987) have shown Eq. (6) to be valid for leaching of two herbicides in a soil column eluted *isocratically* with methanol-water mixtures.

Equation (6) is applicable only for the case of isocratic elution, i.e. the CMOS composition in the mixed solvent does not change with either time or position within the soil column. At a disposal site, however, the solvent content and composition is likely to change in a continuous (but predictable) manner with increasing distance away from the site. This is somewhat analogous to the case of gradient elution in HPLC. Appropriate solute transport equations need to be developed to predict such variations in solvent content in order to estimate solvent composition effects on HOC mobility as the pollutant plume migrates away from the disposal or spill site.

8.5.2 Transport in Biphasic Solvents

Experimental and theoretical descriptions of the physics of retention and flow of immiscible solvents in porous media have begun to receive increasing attention (e.g. Baehr and Corapcioglu 1987; Corapcioglu and Baehr 1987; Lenhard and Parker 1987; Kuppusamy et al. 1987; Faust 1985; Parker et al. 1987; Abriola 1984, 1987; Abriola and Pinder 1985a,b). A review of the theory of multiphase fluid flow is beyond the scope of this paper. The impact of multiphase solvents on HOC transport in soils is, however, relevant and will be briefly discussed.

Enfield (1985) and Enfield and Yates (1987) have examined the transport of hydrophobic organic pollutants in soils in the presence of macromolecules (such as dissolved organic carbon) and immiscible solvents. Enfield (1985) considered the fluid phase in the porous medium to consist of water, a PMOS, and a vapor. Assuming steady-state fluid flow conditions and invariant fluid phase composition (i.e. volume fractions of water, PMOS, and vapor are constant), he derived expressions for predicting HOC transport in soils. He suggested that HOC mobility in the presence of a microemulsified immiscible solvent (MES) can be greatly enhanced, especially for strongly hydrophobic pollutants (HOC with $K_{ow} > 10^4$) or when MES concentration is above 1 g/L. West (1984) and Hutchins et al. (1985) arrived at similar conclusions in evaluating the influence of dissolved organic carbon on the enhanced transport of hydrophobic organic pollutants in soil columns. Further experimental evaluation of these findings is warranted.

8.5.3 Transport in Miscible and Immiscible Solvents

We have already seen that the addition of a CMOS to a biphasic solvent (water-PMOS) will solubilize the PMOS into the aqueous phase, which will further increase HOC solubility and will result in an enhancement of HOC transport. As the aqueous solubility of the PMOS is increased with the addition of a CMOS, a *biphasic* solvent could become a single-phase ternary *mixed* solvent (i.e. water-CMOS-PMOS). Such fluid phase changes are more likely to occur at low mole-fraction PMOS and need to be considered in predicting HOC transport in complex solvents. Such might be the case for residual saturation with a PMOS after the free product has been recovered at a waste disposal site.

8.6 Transport of Mixtures of Organic Pollutants

Much of the published data on the transport of organic pollutants in soils has been for a single pollutant. Sorption and degradation of the pollutant are assumed to be independent of other pollutants. The influence of other pollutants present in waste mixtures on sorption, degradation, and transport have generally not been considered. Such effects become particularly relevant in assessing the environmental impact from land treatment of hazardous wastes. The likely effects of pollutant mixtures on sorption have already been discussed. Pollutant transport is influenced not only by the effects of mixtures on sorption, but also by the interacting effects on microbial degradation.

Soil column studies conducted recently in a collaborative effort between the University of Florida and EPA personnel (R.S. Kerr Lab., Ada, OK) permit at least a qualitative assessment of the interacting effects of pollutant mixtures on their movement. A synthetic pollutant waste mixture containing the following chemicals was prepared: three pesticides (atrazine, carbofuran, and diuron); five substituted phenols (phenol; 2,4-dichlorophenol; 2,4-dinitrophenol; 2,4-dimethylphenol; and pentachlorophenol); and four aromatic hydrocarbons (benzene; toluene; p-xylene; and naphthalene). Steady-state water flow conditions and a steady microbial population were established in 150-cm long columns packed with different soils by applying a dilute nutrient solution at a constant flow rate for several weeks. Following this, the nutrient solution spiked with the synthetic waste mixture was continuously applied and the changes in the concentrations of the waste constituents in the column effluent and at several depths were monitored over a period of several months.

On the basis of the variations in pollutant concentration versus pore volumes plots, the observed curves could be grouped into the following four types: (1) the pollutant concentration gradually increased until it was essentially equal to the input concentration; (2) the pollutant concentration increased and levellled off at a value less than that in the input solution; (3) the pollutant concentration increased, reached a maximum (or a plateau), and then declined to undetectable levels; and (4) large fluctuations in the pollutant concentrations were noted. Data representative of each of these types of curves are shown in Fig. 5.

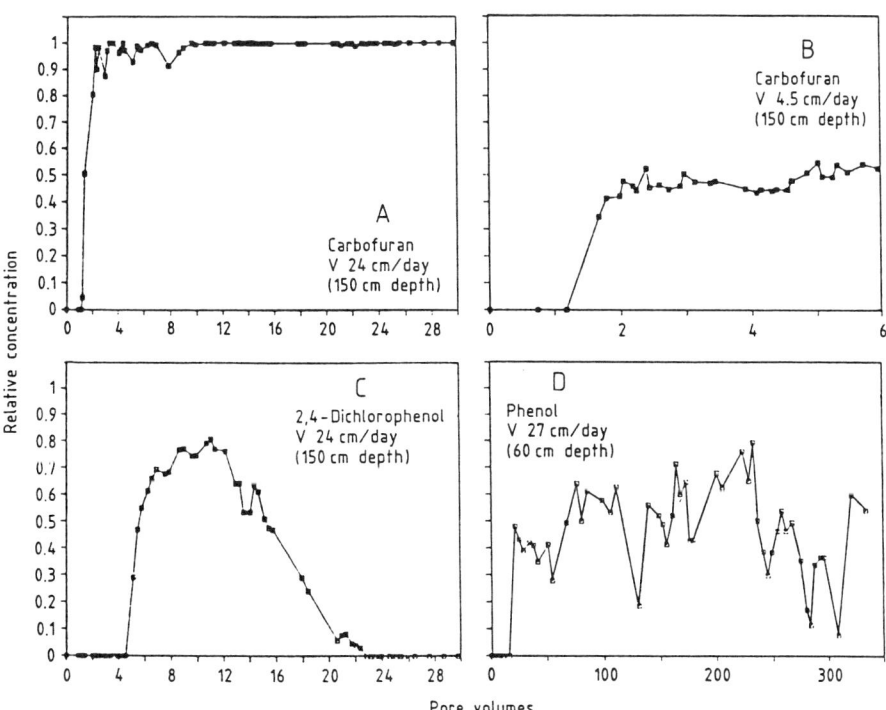

Fig. 5. Changes in the concentrations of organic pollutants in a 150-cm long columns of Lincoln soil to which a synthetic waste mixture was applied continuously. Note that the data shown in **A, B,** and **C** are for concentration changes monitored in the column effluent, while data in **D** are for concentration changes at the 60-cm depth

Type 1 and 2 curves are fairly typical of the results expected when the transport of the organic pollutant is controlled by sorption and a constant degradation rate (i.e. constant microbial activity). The concentration plateau value is determined by the ratio $\xi = (\omega R L/v)$, where ω is the degradation rate coefficient (day^{-1}), R is the retardation factor; L is column length (cm), and v is the average pore-water velocity (cm/day). Note that for larger ξ values, the concentration plateau will be lower, indicating a greater degradation of the pollutant during its transit through the soil column. For the data presented in Fig. 5A and 5B, decreasing the flow velocity (v) is equivalent to a larger ξ value, and the concentration plateaus at a lower value.

Type 3 curve (Fig. 5C) suggests that after a period of time, the degradation rate coefficient (ω) increases and the pollutant concentration declines and reaches a new plateau value determined by the larger ξ value. Among the several explanations for the increase in ω is the possibility that with time a microbial population that can more readily degrade the pollutant evolved, perhaps as a result of adaptation to continued exposure to the pollutant. The pattern of data in Fig. 5D suggests that the ω value might be fluctuating continuously – increasing or decreasing in response to the changes in the concentrations of the waste constituents in the soil column. It is probable that phenol resulting from the degradation of other substituted phenols in the waste mixture also contributes to observed fluctuations.

Prediction of the curves presented in Fig. 5 requires the explicit understanding of the changes in microbial populations and activities as impacted by the composition of the waste mixture. This is an extremely difficult task, but one that needs to be addressed if pollutant migration from waste disposal sites is to be predicted.

Acknowledgements. Research reported here has been supported, in part, by funds from the U.S. Environmental Protection Agency Cooperative Agreements CR-812581, CR-811144, and CR-811035.

References

Abriola LM (1984) Multiphase migration of organic compounds in a porous media. Lecture Notes in Engineering, Vol 8. Springer Berlin Heidelberg New York, 232 p
Abriola LM (1987) Modeling contaminant transport in the subsurface: An interdisciplinary challenge. Rev Geophys 25:125–134
Abriola LM, Pinder GF (1985a) A multiphase approach to the modeling of porous media contamination by organic compounds: 1. Equation development. Water Res Res 21:11–18
Abriola LM, Pinder GF (1985b) A multiphase approach to the modeling of porous media contamination by organic compounds: 2. Numerical simulation. Water Res Res 21:19–26
Baehr AL, Corapcioglu MY (1987) A compositional multi-phase model for groundwater contamination by petroleum products: 2. Numerical solutions. Water Res Res 23:201–213
Bladel R van, Moreale A (1974) Aspects physico-chimiques de la pollution des sols: role et prediction du phenomene d'adsorption. Pedologie XXVII 1:44–66
Cantwell F, Puon S (1979) Mechanism of chromatographic retention of organic ions on a nonionic adsorbent. Anal Chem 51:623–632

Chiou CT, Porter PE, Schmedding DW (1983) Partition equilibria of nonionic organic compounds between soil organic matter and water. Environ Sci Technol 17:227–231

Corapcioglu MY, Baehr AL (1987) A compositional multiphase model for groundwater contamination by petroleum products: 1. Theoretical considerations. Water Res Res 23:191–200

Crittenden JC, Luft P, Hand DW, Oravitz JL, Loper SW (1985) Prediction of multi-component adsorption equilibria using ideal adsorbed solution theory. Environ Sci Technol 19:1037–1043

Enfield CG (1985) Chemical transport facilitated by multiphase flow systems. Water Sci Technol 17:1–12

Enfield CG, Yates SR (1988) Chemical Transport to groundwater. Am Soc Agron Monogr (in press)

Faust CR (1985) Transport of immiscible fluids within and below the unsaturated zones: A numerical model. Water Res Res 21:587–596

Felice LJ, Zachara JM, Schmidt RL, Resch CT (1985) Quinoline partitioning in subsurface materials: Adsorption, desorption, and solute competition, pp 39–41. In: Durham NN, Redlefs AE (eds) Proc 2nd Intern Conf on Groundwater Qual Res. Univ Printing Services, Oklahoma State Univ, Stillwater, OK

Fu JK, Luthy RG (1986a) Aromatic compound solubility in solvent water mixtures. ASCE J Environ Eng 112:328–345

Fu JK, Luthy RG (1986b) Effects of organic solvents on sorption of aromatic solutes on to soils. ASCE J Environ Eng 112:346–366

Grain CF (1982) Activity coefficient, Chap 11. In: Lyman WJ, Reehl WF, Rosenblatt DM (eds) Handbook of Chemical Property Estimation Methods. McGraw-Hill, New York

Hance RJ (1977) The adsorption of atraton and monuron by soils at different water contents. Weed Res 17:137–201

Hutchins SR, Thompson NB, Bedient PB, Ward CH (1985) Fate of trace organics during land applications of municipal wastewater. CRC Crit Rev Env Control 15:355–416

Karickhoff SW, Brown DS, Scott TA (1979) Sorption of hydrophobic organic pollutants on natural sediments. Water Res 13:241–248

Kuppusamy T, Sheng J, Parker JC, Lenhard RJ (1987) Finite element analysis of multiphase immiscible flow through soils. Water Res Res 23:625–631

Lenhard RJ, Parker JC (1987) Measurement and prediction of saturation-pressure relationships in three phase porous media systems. J Contam Hydrol 1:407–424

Lyman WL (1982) Solubility in various solvents, Chapter 3. In: Lyman WJ, Reehl WF, Rosenblatt DH (eds) Handbook of Chemical Property Estimation Methods. McGraw-Hill, New York

McIntyre WG, deFur PO (1985) The effect of hydrocarbon mixtures on adsorption of substituted naphthalenes by clay and sediment. Chemosphere 14:103–112

Miller NA, Wolf DC, Scott HD (1987) Influence of methanol and hexane on soil adsorption of atrazine. Water, Air, Soil Pollution 39:101–112

Munz C, Roberts PV (1986) Effects of solute concentration and cosolvents on the aqueous activity coefficient of halogenated hydrocarbons. Environ Sci Technol 20:830–836

Nkedi-Kizza P, Rao PSC, Hornsby AG (1985) Influence of organic cosolvents on sorption of hydrophobic organic chemicals by soils. Environ Sci Technol 19:975–979

Nkedi-Kizza P, Rao PSC, Hornsby AG (1987) Influence of organic cosolvents on leaching of hydrophobic organic chemicals through soils. Environ Sci Technol 21:1107–1111

Novak JP, Matous J, Pick J (1987) Liquid-Liquid Equilibria. Elsevier Amsterdam

Parker JC, Lenhard RJ, Kuppusamy T (1987) A parametric model for constitute properties governing multiphase fluid flow in porous media. Water Res Res 23:618–624

Rao PSC, Lee LS (1988) Sorption of organic chemicals by soils from multi-sorbate and multi-solvent mixtures. In: Battelle DNL, Richland WA (eds) Proc 24th Hanford Life Sci Symp pp 457–471

Rao PSC, Hornsby AG, Kilcrease DP, Nkedi-Kizza P (1985) Sorption and transport of hydrophobic organic chemicals in aqueous and mixed solvent systems: Model development and preliminary evaluation. J Environ Qual 14:376–383

Rao PSC, Nkedi-Kizza P, Davidson JM (1986) Abiotic processes affecting the transport of organic pollutants in soil. In: Loehr RC, Malina JF Jr (eds) "Land Treatment: A Hazardous Waste Management Alternative". Water Res Symp 13:63–72 Univ Texas, Austin, TX

Rubino JT, Yalkowsky SH (1984) Solubilization by cosolvents: 2. Phenytion in binary and ternary solvents. J Paren Sci Technol 38:215

Rubino JT, Yalkowsky SH (1985) Solubilization by cosolvents: 3. Diazepan and benzocaine in binary solvents. J Paren Sci Technol 39:106–111

Rubino JT, Yalkowsky SH (1987a) Cosolvency and cosolvent polarity. J Pharmacol Res 4:220–230

Rubino JT, Yalkowsky SH (1987b) Cosolvency and deviations from log-linear solubilization. Pharmacol Res 4:231–236

Rubino JT, Yalkowsky SH (1987c) Solubilization by cosolvents: 4. Benzocaine, diazepan, and phenytoin in aprotic cosolvent-water mixture. J Paren Sci Technol 41:172–176

Schellenberg K, Leuenberger C, Scharzenbach RP (1984) Sorption of chlorinated phenols by natural sediments and aquifer materials. Environ Sci Technol 18:652–657

Sorensen JM, Arlt W (1979) Liquid-liquid equilibrium data collection. Chemistry Data Series, Vol V Parts 1 and 2. Deutsche Gesselschaft fur Chemisches Apparatewessen, Federal Republic of Germany

Swarzenbach RD, Westall J (1981) Transport of nonpolar organic compounds from surface water to groundwater. Environ Sci Technol 15:1360–1367

West CC (1984) Organic carbon facilitated transport of trace organics in subsurface systems. Ph D Dissertation, Rice University, Houston, TX

Westall JC, Leuenberger C, Schwarzenbach RP (1985) Influence of pH and ionic strength on the aqueous-nonaqueous distribution of chlorinated phenols. Environ Sci Technol 19:193–198

Woodburn KB, Rao PSC, Fukui M, Nkedi-Kizza P (1986) Solvophobic approach for predicting sorption of hydrophobic organic chemicals on synthetic sorbents and soils. J Contam Hydrol 1:227–241

Yaron B, Saltzman S (1972) Influence of water and temperature on adsorption of parathion by soil. Soil Sci Soc Am J 36:583–586

Yalkowsky SH (1985) Solubility of organic solutes in mixed aqueous solvents. EPA Project Completion Report, CR# 811852–01

Yalkowsky SH (1986) Solubility of organic solutes in mixed aqueous solvents. EPA Project Completion Report, CR# 812581–01

Yalkowsky SH, Roseman TJ (1981) In: Yalkowsky SH (ed) Techniques of Solubilization of Drugs. Dekker, New York

Yalkowsky SH, Rubino JT (1984) Solubilization by cosolvents: 1. Organic solutes in propylene glycol-water mixtures. J Pharmacol Sci 74:416–421

Zachara JM, Ainsworth CC, Felice LJ, Resch CT (1986) Quinoline sorption to subsurface material: Role of pH and retention of the organic cation. Environ Sci Technol 20:620–627

Zachara JM, Ainsworth CC, Cowan CE, Thomas BL (1987) Sorption of binary mixtures of aromatic nitrogen heterocyclic compounds on subsurface materials. Environ Sci Technol 21:397–402

9 Accelerated Degradation of Pesticides

J. KATAN[1] and N. AHARONSON[2]

9.1 Introduction

Our increasing dependence on pesticides in agriculture confronts us with an unsolved dilemma. On the one hand, we are interested in prolonging the duration and persistence of the pesticides at the target site in order to obtain a lasting effect. On the other hand, the long persistence of these toxic substances in the environment further increases their hazards. Much of the extensive research on the fate of pesticides in the environment is related in various ways to these two contradictory objectives. The toxic effects of the accumulating pesticides and their interaction with the living and nonliving components of the environment are a cause of great concern. Thus, the emphasis in studies related to pesticides is placed on problems related to the prolonged persistence of pesticides rather than on those related to a too rapid dissipation.

Degradation of pesticides takes place through biological, chemical and physical processes. However, the microbial — enzymatic processes are considered in many cases as the major natural mechanism for the detoxification of pesticides at agricultural sites. Elucidating the rate and pathways of pesticide biodegradation, under various conditions, is therefore a prerequisite for determining the rational and safe use of pesticides.

In recent years, many cases of a loss of pesticidal effectiveness due to unexpected accelerated degradation were reported. This nondesirable development, in an era where many effective pesticides are suddenly banned due to toxicological considerations, causes a further shrinkage in our arsenal of effective pesticides and therefore deserves great attention.

9.2 Accelerated Degradation of Pesticides

Accelerated or enhanced degradation refers to a phenomenon whereby a pesticide is more rapidly decomposed in the soil, than has been in the past. This is usually accompanied by a rapid loss of pesticidal activity. This phenomenon is

---

[1] Department of Plant Pathology and Microbiology, The Hebrew University of Jerusalem, Rehovot, 76100, Israel
[2] Department of Chemistry of Pesticides and Natural Products, Agric. Research Organization, Volcani Center, Bet Dagan 50250, Israel

linked with previous soil application(s) of the same pesticide or of a structurally similar one. It has been proposed, and demonstrated in many cases, that accelerated degradation is due to the build up and enrichment in soil of microbial populations with improved capacity to degrade the pesticide. This principle has been shown with phenoxy herbicides starting from the late 1940s (Audus 1949; Newman and Thomas 1949) and with additional pesticides in the following years both under laboratory and field conditions, e.g. with endothal (Horowitz 1966). However, it is only in the last few years that accelerated degradation has been connected with remarkable cases of failure of pesticides to control pests in commercial fields. Reports of such failures were presented independently from various countries in different geographical regions, and referred to pesticides belonging to various chemical groups. It is now more commonly accepted than previously that accelerated degradation is a widespread and threatening phenomenon with a potential of great economic importance.

Soils with the capacity for accelerated or enhanced degradation are also referred to as "problem soils" (denoting reduced efficacy in pest control), or "history soils" or "conditioned soils" (denoting previous application(s) of the pesticide(s)). Roeth (1986) suggested the need to distinguish between "self-enhancement"- which refers to enhanced degradation wherein the "historical" pesticide (the one previously applied) and the "challenge" pesticide (which is used for retreating the soil) are the same, and "cross-enhancement" which denotes degradation where the challenge pesticide is different from the historical one. Accepted definitions for accelerated or enhanced degradation will be most helpful for avoiding confusion. "Multiproblem" soils, in which pesticides belonging to two or more chemical groups of pesticides are no longer effective, have also been detected (Kaufman et al. 1985).

The soil is the arena where the populations of the microbial degraders build up upon repeated pesticide applications. Thus, soil-applied pesticides are more likely to be vulnerable to accelerated degradation than foliar-applied materials. However, even the small amounts of pesticides reaching the soil through foliar application might be sufficient to induce accelerated degradation, as shown with the fungicides iprodione (Walker et al. 1986) and benomyl (Yarden et al. 1987). It is possible that contamination of the soil with pesticides from other sources e.g. water, pesticide-treated plant materials, pesticide-treated seeds, will also induce accelerated degradation.

The herbicide EPTC (especially in the form of Eradicane, which contains the antidote) and the insecticide carbofuran were among the pesticides which have been extensively investigated in recent years regarding loss of their efficacy on a large scale (Rahman et al. 1979; Fox 1983; Kaufman and Edwards 1983; Obrigawitch et al. 1983; Rahman and James 1983; Wilson 1984; Gray and Joo 1985; Kaufman et al. 1985; Read 1986; Roeth 1986; Suett 1986; Kaufman 1987 and others). Accelerated degradation was reported with many additional pesticides belonging to various chemical groups including herbicides (e.g. amitrole, butylate, dalapon, diphenamid, MCPA), insecticides (e.g. bendiocarb, carbaryl, isofenphos, lindane, terbufos), nematicides (e.g. ethoprop) and fungicides (e.g.

iprodione, benomyl, metalaxyl) (see references quoted in Kaufman and Edwards 1983; Kaufman et al. 1985; Roeth 1986; Fletcher et al. 1980; Bailey and Coffey 1985; and other papers cited throughout this article). No attempt will be made to cover all pesticides exhibiting this phenomenon but rather to refer to selected examples.

Accelerated degradation might be detected in certain cases as early as after one application – though a longer history of pesticide application or increasing pesticide dosages frequently result in a more pronounced rapid degradation (Fig. 1) (Newman and Thomas 1949; Katan et al. 1984; Kaufman et al. 1985; Read 1986; Suett 1986; Walker et al. 1986; Yarden et al. 1987; Avidov et al. 1988). Accelerated degradation, which is biological in its nature in many cases, may result in either shortening of the "lag period" of the process or an increase in the rate of degradation occurring during the "growth phase" ("fast phase") or in both. This will lead in the end to a more rapid disappearance of the pesticide than in the control soil. There are cases where degradation of a certain pesticide was very rapid even in soils with no known history of application (Kaufman and Edwards 1983). This may have been due to application of other structurally related agrochemicals, to certain agricultural practices, to the crop history, or to the natural properties of the soils. On the other hand, repeated application does not necessarily lead to accelerated degradation of the pesticide, as shown with atrazine, alachlor, metolachlor, cycloate, molinate and other pesticides (Cole 1976; Dowler et al. 1984; Gray and Joo 1985; Harvey 1987). Degradation of alachlor was significantly lower in soils from fields with a continuous alachlor treatment history than in untreated control soils (Aharonson et al. 1987).

A previous application of a pesticide may not be the only means for conditioning the soil to accelerated degradation. Thus, when a soil with a capacity for accelerated degradation of benomyl was mixed with a control soil at the rate of 1:50, the entire soil volume was conditioned to accelerated degradation (Yarden et al. 1987). The implications of this finding might be serious, since foci of accelerated degradation in the field may, in due time, spread and contaminate the adjacent plots.

Rate of degradation usually depends on a variety of factors, e.g. chemical and physical properties of the soil and its microbial composition, crop history, agricultural practices, climatic conditions, presence of pesticides and other agrochemicals. Thus, it is not surprising that accelerated degradation of the same pesticide developed to varying intensities in different soils and under different agricultural situations, as shown with carbofuran (Suett 1986). This might be analogous to a similar phytopathological situation related to the development of soilborne diseases. Certain soils are suppressive i.e. they restrict the development of the pathogen, and consequently of the disease, while other soils are conducive to disease development (Cook and Baker 1983). In many cases, soil suppressiveness to a disease was due to natural biological control mechanisms occurring in the soil. It will be worth studying the possible similarity between mechanisms operating in soils which are conducive to a disease and mechanisms of accelerated degradation.

9.3 Assessment of Accelerated Degradation

Assessing accelerated degradation involves two steps: sampling the soil and evaluating the rate of degradation of the pesticide in question. There are several approaches which can be followed when soil samples have to be collected for the assessment of accelerated degradation. One approach would be to collect samples from fields in which a loss of pesticidal activity is evident or suspected. Special care should be taken when collecting the samples of the corresponding nontreated control soils, since frequently the latter are taken from the nontreated edges of the field, which might differ from the main field in many aspects. Under field conditions the crop is usually treated during the season with several pesticides in addition to the one in question. This may further complicate the picture. A second approach would be to simulate accelerated degradation under controlled laboratory conditions by successively treating soil samples with the tested pesticide. In this case, comparison between the laboratory-treated soil and the corresponding nontreated control soil is more accurate since both soils are from the same source and the only difference between them is pesticide application. However, accelerated degradation obtained in the simulation system cannot be regarded as a proof that such a possibility will also occur under field conditions. Thus, a survey should be made of fields which have been treated with this pesticide. A third approach, which combines the advantages of the above two, would be to simulate accelerated degradation under field conditions. Thus, the pesticide is applied to field plots with a known history of pesticide application. In such a study the control plots are under the same conditions. Adopting simultaneously more than one approach would be very desirable.

The next step after collecting the soil is to add the tested pesticide to it and to follow the dissipation in time of the pesticide in soil (Fig. 1), or the level of pesticidal activity. These studies have to be carried out in parallel in soils with and without a history of pesticide application. A significantly more rapid decline in time in pesticide concentration or in pesticidal effectiveness in the "history" soil denotes an accelerated degradation. One must decide the level of increased degradation that should be classified as accelerated degradation. Treating the soil with a ^{14}C-labeled pesticide and measuring $^{14}CO_2$ evolution is a relatively simple and reliable technique for assessing the rate of biodegradation of the pesticide (Fig. 2) provided that the labeled carbon of the pesticide is released mainly in the form of CO_2 and that the change in the molecule corresponds to changes in bioactivity. Dissipation analyses of the pesticide can be done also by chemical or physical means or by bioassay techniques. Identification of the degradation products of the pesticide enables us to determine whether accelerated degradation also involves a shift in the degradation pathway.

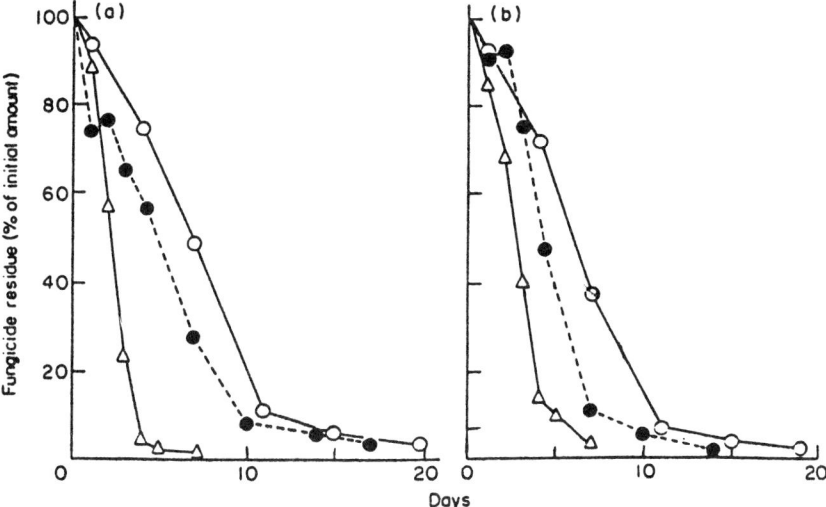

Fig. 1. Degradation of (**a**) iprodione and (**b**) vinclozolin in a clay loam soil. Open circles = first treatment; filled circles = second treatment; opened triangles = third treatment (from Walker et al. 1986)

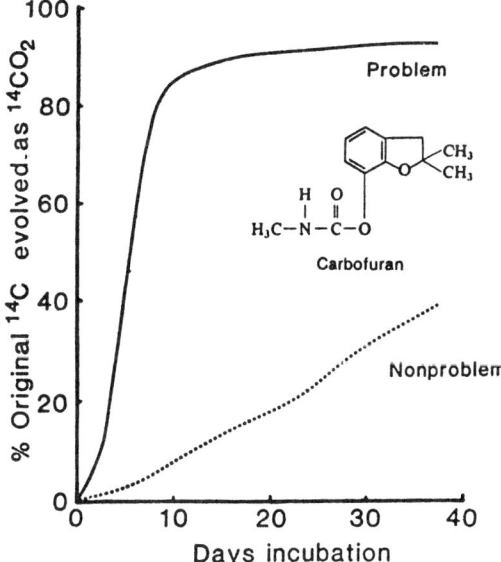

Fig. 2. Evolution of $^{14}CO_2$ during degradation of ^{14}C-carbonyl-carbofuran in carbofuran problem (= with accelerated degradation) and nonproblem (= control) soils (mean of 5 soil pairs). Data from Kaufman and Edwards 1983

9.4 Mechanisms of Accelerated Degradation

It is generally accepted that accelerated degradation is mainly a microbial process related to the enrichment in soil or to changes in populations capable of degrading the pesticide ("degraders"). Indeed, this was shown in a number of cases. The enhanced $^{14}CO_2$ evolution observed in soils with accelerated degradation is regarded as an indication of the microbial nature of degradation. The nullification of accelerated degradation by soil sterilization or by using more specific antimicrobial agents, as shown in many studies (Katan et al. 1984; Read 1986; Roeth 1986; Suett 1986;Kaufman 1987; Racke and Coats 1987), is also indicative of microbial involvement in the accelerated degradation processes. Soil sterilization is usually done by autoclaving, chemosterilization, or gamma irradiation. Drastic sterilization, e.g. by autoclaving, may also cause chemical and physical changes in the soil, and destruction of soil enzymes. Therefore, gamma irradiation which is a less drastic method of sterilization, is more recommended for this purpose. Another approach for the elucidation of the role of microorganisms in pesticide degradation is to use selective chemical agents, e.g. antibiotics, fungicides, bactericides, and other antimicrobial agents, for suppressing certain components of the microbial populations. For example, treating a soil exhibiting accelerated degradation, with the bactericide chloramphenicol, strongly suppressed degradation of isofenphos to a level similar to that observed with the autoclaving treatment whereas the fungicide cycloheximide did not affect degradation (Racke and Coats 1987). This suggests that mainly soil bacteria are responsible for the rapid degradation of this pesticide. However, in such studies we have to take into consideration that the selectivity of such agents is not complete. Morever, the use of a too high dosage of the chemical may distort the picture. In other studies, accelerated degradation was suppressed by fumigation with methyl bromide, solarization and fungicides – all having antimicrobial effects (Avidov et al. 1988; Yarden et al. 1987).

Accelerated degradation may result from an increase in the number of microbial degraders (quantitative change) and or from improved enzymatic activity of the degraders (qualitative change). The degraders may belong to different species or genera, and may differ in their enzymatic capacity or in the metabolic pathways involved in the degradation of the pesticide. Another potential mechanism would be the accumulation in soil of degradative enzymes of microbial origin, provided that the enzymes remain active in soil for long periods. Microbial shifts in the soil or a change in the total biomass do not necessarily denote a comparable change in the populations of the microbial degraders. For example, a higher population of bacteria in soil having accelerated degradation does not necessarily mean that bacteria are responsible for degradation or are more important than fungi or actinomycetes in this respect.

A variety of microbial degraders were isolated from soils with accelerated degradation. They include bacteria e.g. *Acinetobacter, Alcaligenes, Arthrobacter, Bacillus, Micrococcus, Pseudomonas, P. Alcaligenes*; and fungi, e.g. *Alternaria, Bipolaris, Diheterospora, Fusarium, Paecilomyces, Penicillium* and *Verticillium*

(Fisher et al. 1978; Spain et al. 1980; Lee 1984; Yarden et al. 1985; Marty et al. 1986; Racke and Coats 1987; Tam et al. 1987). Usually, the degraders constitute only a small portion of the total microbial population (Lee 1984). Thus, a direct quantitative assessment of microbial degraders in soil is very difficult and laborious, since it involves the assessment of degradation capacity of many individual isolates of microorganisms. Soil perfusion and other enrichment methods, which magnify the populations of the degraders, facilitate this task and may enable us to carry out comparative studies of the populations of the degraders in soil, although these are not direct assessment methods. Thus, using such techniques, with various modifications, the frequency of bacterial degraders (usually obtained in mixed cultures) was higher in soils with accelerated degradation, as compared with the respective nonhistory control soils (Newman and Thomas 1949; Spain et al. 1980; Racke and Coats 1987; Tam et al. 1987). For example, the number of nitrophenol — degrading bacteria in sediment/water cores preexposed to this compound increased four to five orders of magnitude during adaptation (Spain et al. 1980). Although more reports relate accelerated degradation to bacteria than to other groups of soil organisms, no generalization should be made at this stage.

Spain et al. (1980) mentioned three ways by which adaptation, as related to previous exposure of a population to xenobiotics, can occur: (1) Induction or derepression of specific enzymes not present (or present at low levels) in the population before exposure. (2) Selection of new metabolic capabilities produced by genetic changes. (3) Increase in the number of organisms able to catalyze a particular transformation. The third type of change often follows one of the first two. Although it is likely that the degradative enzymes involved are regulated by induction, the involvement of constitutive enzymes should also be considered (Roeth 1986). It is likely that adaptation develops to various levels in a wide range of organisms in the conditioned soil, rather than being restricted to a single species. The selection pressure exerted on the soil organisms upon repeated application of the pesticide will determine, together with other factors, which of the above mechanisms will dominate.

In addition to the buildup in population of microbial degraders, repeated applications of the same pesticide may also create physical or chemical conditions in the soil favorable for the degradation or stimulation of the microorganisms. Desorption of diphenamid from soils with a history of application of this herbicide (in which accelerated degradation was observed) was higher than that from comparable control soils (Kaufman et al. 1985). This would indicate that the more rapid degradation of diphenamid could be due, at least in part, to a reduced adsorption potential and, thus, a greater availability to soil microorganisms for degradation.

Waid (1972) referring to the persistence of accelerated degradation, suggested that the genetic capacity of the microflora to degrade resistant molecules is conserved through the action of a very effective mechanism even though it has no apparent survival value in the absence of an appropriate substrate. He suggested that the episomal transfer of genetic information between bacteria may be an effective mechanism to conserve the potentiality of microorganisms to degrade resistant molecules. The involvement of extra chromosomal elements such as plasmids in the degradation of xenobiotics and as transfer agents has been shown in increasing numbers of cases. The ability of *A. paradoxus* to degrade 2,4-D is encoded by a 58 megadalton conjugal plasmid (Fisher et al. 1978). The degradation of EPTC by *Arthrobacter* sp. is mediated by a 50.5 megadalton plasmid in the cell. The loss of this plasmid resulted irreversibly in mutants unable to degrade EPTC (Tam et al. 1987). The transfer by conjugation of this plasmid to mutants lacking the plasmid, restored completely their capability to degrade the pesticide. Lee (1984) suggested that the involvement of plasmids in EPTC degradation accounts for the loss of EPTC-degrading ability after prolonged storage of bacteria. He also suggested that accelerated degradation may be caused by the rapid transfer of EPTC-degrading information on plasmids to soil bacteria by conjugation. Transfer and spread of pesticide degrading ability through plasmid-related mechanisms has been suggested also for other pesticides, e.g. 2,4-D (Don and Pemberton 1981; Pemberton and Fisher 1977). The research in relation to pesticide degradation and extra chromosomal elements is very extensive and expanding (Pemberton et al. 1979; Chakrabarty 1982) and, it is hoped, may provide in the future clues to some of the questions raised. It is beyond the scope of this paper to review all the literature in this field.

9.5 Persistence of Accelerated Degradation

The duration of accelerated degradation in soil has very important practical implications. Roeth (1986) noted that the apparent reversion of thiocarbamate − enhanced soils to their normal state occurs over a 1- to 2-year period. Yarden et al. (1987) found no evidence of reduction in degradation potency of the fungicide MBC over a 2-year incubation period under laboratory conditions of a soil that had previously been conditioned to accelerated degradation of the fungicide. Accelerated degradation of MCPA lasted for 5 years without further addition of MCPA (Fryer et al. 1980). Determination of the persistence of accelerated degradation in commercial fields has to be done with care, since the possibility exists that other agrochemicals, with a similar chemical structure, which were used in the same field, were responsible for maintaining the adaptation process. There are several potential mechanisms related to the persistence of accelerated degradation for relatively long periods after the application of the pesticide and its apparent dissipation. Low quantities of the chemical are sufficient in certain cases to induce accelerated degradation, as shown with MBC (Yarden et al. 1987) and with nitrophenol at a concentration as low as 60 µg/liter (Spain et al. 1980). Thus, even minute amounts of the chemical released at later stages from soil sites where it has been adsorbed, or from decaying tissues of plants treated previously (as is the case with systemic pesticides), might be sufficient to maintain a constant induction. It is also possible that the shift in biological equilibrium in the previously treated soil is in favor of the degraders, for reasons such as: the pesticide is more toxic to the competitors of the degraders than to the degraders themselves, or the degraders are also capable of decomposing and using other organic substances and therefore their ability to compete for nutrients in soil is improved. The persistence of accelerated degradation will depend on the nature of both the pesticide and its degraders, on the biotic and abiotic environment in the soil, and on the agricultural practices employed and the rate and composition of the other agrochemicals used. In this regard, mechanisms which enable the degraders to persist in soil long after the target pesticide is apparently dissipated, might be similar to those which enable soilborne pathogens to persist in soil for several years in the absence of their principal host, e.g. formation of resistant resting structures. Although some degraders are able to use the pesticide as an energy source; the nutritional role of the pesticide does not seem to be important as a mechanism for survival for long periods. The role of plasmids in persistence should be further elucidated as discussed above.

9.6 Cross Accelerated Degradation

Adaptation of a microorganism to a certain substrate involves the production of an appropriate enzyme which results in an increased ability to degrade the substrate and, possibly, other homologous molecules. This cross adaptation was shown with a variety of pesticides with similarly structured molecules. Cases of cross degradation were shown with phenoxy herbicides already in the earlier studies. Persistence of 2,4-D was reduced by pretreatment of the soil with 2,4-D and other compounds having similar constituent groups, e.g. dichlorophenol (Newman and Thomas 1949). Soil pretreatment with 2,4-D accelerated degradation of MCPA and pretreatment with MCPA accelerated degradation of 2,4-D (Audus 1951; Torstensson et al. 1975). Extensive studies on cross accelerated degradation were carried out with thiocarbamates. In one such study (Wilson 1984), EPTC degradation was accelerated in soil which had previously been treated with butylate, vernolate or EPTC. Vernolate degradation was increased in soil which had previously been treated with vernolate or EPTC. Butylate breakdown was enhanced to a greater extent in soil with previous exposure to butylate. Cycloate degradation, on the other hand, was not enhanced in soil with prior exposure to butylate, EPTC or vernolate. Similar results were also obtained in other studies (Obrigawitch et al. 1983; Harvey 1987). Degradation of various pesticides was accelerated in soils with a history of application of carbofuran or EPTC (Kaufman and Edwards 1983; Kaufman et al. 1985). Kaufman and Blake (1973) isolated several microorganisms that were each able to rapidly degrade a wide range of acetamide, acylanilide, carbamate, toluidine and urea-based pesticides, thus denoting the potential of cross adaptation. In most cases of cross degradation, however, the rate of degradation of the alternate molecule (cross-challenge pesticide) is slower than that of the molecule to which the microorganisms was initially conditioned (history-pesticide). Audus (1960) suggested two possible explanations. Each pesticide may induce its own enrichment flora with different efficiencies for utilization of structurally similar chemicals, or the microorganisms may be the same but with different inducible enzymes that have broad recognition of similar substrates yet different affinities for each pesticide. A third possibility, proposed by Roeth (1986), is that the same microbe and inducible enzyme are active in degradation with the genetic operon possessing different affinities for each pesticide. In this case the operon itself would undergo the adaptive change and the rate of degradation would be a reflection of enzyme synthesis and concentration. Determining the range of cross accelerated degradation for each chemical group of pesticides, is a prerequisite for planning a pesticide regime for each crop which will avoid or delay development of cross degradation. Such studies closely resemble those related to cross resistance of pests to pesticides.

9.7 Control and Management of Accelerated Degradation

Microbial degraders which are responsible for accelerated degradation are organisms which resemble soilborne pathogens in many aspects. Both groups consist of soilborne organisms which exist in agricultural soils at low densities without causing considerable damage, but increase greatly if the substrate (i.e. the pesticide or the host plant) becomes more available due to repeated pesticide application or frequent cropping. Monoculture is conducive to the buildup of the populations of these organisms. Therefore, the basic approaches for controlling soilborne pathogens and microbial degraders are similar. They include monitoring, prevention and the eradication of the causal agents for curing the infested soil. Physical, chemical, cultural and biological means, alone or in combination, may be used for the management and control of both groups of organisms. Strategies followed for controlling resistance to pesticides are especially relevant in this regard. Control and management of accelerated degradation might be achieved through the following approaches:

(1) Monitoring and surveying soils enable us to detect the development of accelerated degradation at very early stages. Such information can be used in the development of warning systems to avoid the further use of the tested pesticide before it loses its efficiency. Moreover, using control measures already at this stage may prove to be very efficient in preventing the accelerated degradation which might develop at a later stage. Reliable, sensitive and rapid methods for the detection of low levels of accelerated degradation have to be developed further and improved. Special effort should be expended to developing assays for the detection of soils which are conducive to accelerated degradation. Such predictive systems when available, will enable adjustment of pesticide application regimes to various soils with minimal risk of undesirable side-effects.

(2) A variety of approaches might be followed for preventing or delaying the development in soil of accelerated degradation, similar to those taken to prevent or delay the development of biotypes of pests resistant to pesticides. For example:

a) Alternating methods of control. Preference should be given to alternation with nonchemical methods, thus shortening the exposure of soil microorganisms to the pesticide.
b) Alternating pesticides belonging to different chemical groups, in which no cross accelerated degradation is evident.
c) Reducing pesticide dosages through the development of integrated pesticide programs which aim at rational use of pesticides.
d) Improving pesticide application in an attempt to reduce drift and run-off resulting in the contamination of nontarget adjacent fields. When possible, foliar application should replace soil application.
e) Rotating crops and reducing frequency of crops which have similar pesticide regimes. Crop rotation enables the use of chemically unrelated pesticides and different control methods.
f) Certain crops may prevent the development of accelerated degradation. Development of accelerated degradation of diphenamid was suppressed in a soil with a history of corn but was enhanced in a soil with a history of soybean (Katan et al. 1984). Degradation of alachlor was slower in corn-cropped soils than in soybeans cropped-soils (Aharonson et al. 1987).
g) Using pesticide mixtures or a combination of control methods which suppress pesticide degradation. Soil fumigation, solarization and the fungicides TMTD and fentin acetate delayed degradation of benomyl, terbutryn and carbofuran (Avidov et al. 1985; Suett 1986; Yarden et al. 1985) and Fig. 3. Degradation of parathion was inhibited by the fungicide captafol (Ferris and Lichtenstein 1980).

(3) Controlling accelerated degradation in a soil which has already acquired this capacity is the most difficult approach. The final goal is to reclaim the soil and to resume pesticidal effectiveness. This can be achieved in various ways: by eradicating the populations of the degraders, suppressing growth and reproduction of the degraders, creating an environment not favorable for the degraders, and inhibiting the degradative enzymes. There are many potential chemical, physical and biological means for achieving these goals. However, only very few methods which are reasonably effective, inexpensive, simple to apply, reproducible and at the same time free of significant negative side-effects, can be considered for this purpose. The chemical methods include regular pesticides which also suppress microbial activity, extenders which are specially designed antimicrobial agents for suppressing degradation, enzyme inhibitors and soil disinfestation. For example, the performance of Eradicane in soil with a previous history of application of this herbicide was much improved when alachlor was used previously (Rahman et al. 1979). Carbaryl and parathion strongly inhibited the enzyme acylamidase isolated from *F. oxysporum* (Blake and Kaufman 1975). There are many reports on the use of dietholate as an effective extender (Prochnow 1981).

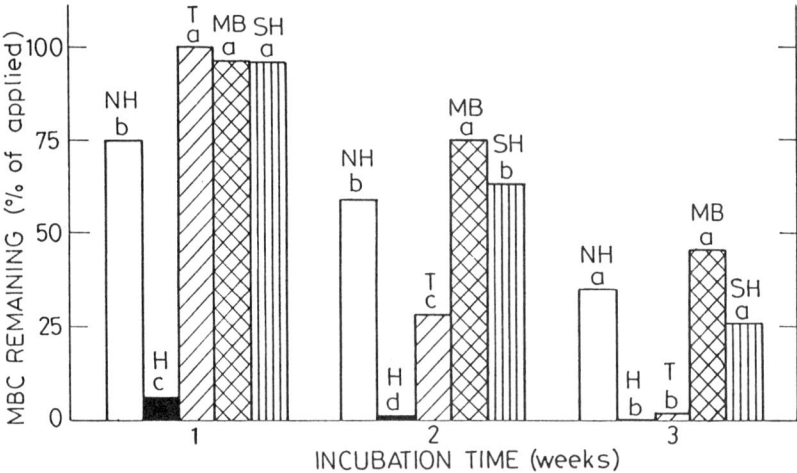

Fig. 3. Curbing accelerated degradation in soil with a history of MBC treatments (H) by the fungicide thiram (T) and disinfestation by fumigation with methyl bromide (MB) or soil solarization (SH), as compared to degradation in soil with no MCB-application history (NH). Within each week, common lower case letters indicate no significant difference ($P = 0.05$) between treatments (from Yarden et al. 1987)

When antimicrobial agents are used in soil, for any purpose, the following should be considered. The antimicrobial agent should not be toxic to beneficial microorganisms such as *Rhizobium* and mycorrhyzae, or to other nontarget organisms; the degraders may develop resistance to this chemical; and its effectiveness will probably depend also on soil properties and on pest population density, as is the case with soil-applied pesticides. It is typical to our unsolved dilemma that persistent pesticides were replaced in the past by nonpersistent ones which, in the turn, have to be supplemented with other toxic compounds to ensure their persistence. The use of biocontrol agents (some of which are produced through biotechnological processes) for controlling pathogens, insects and weeds is approaching, it is hoped, the stage of practical use on a large scale. Adopting this approach, i.e. using such biocontrol agents for controlling the degraders, should also be considered.

9.8 Conclusions and Prospects

Accelerated degradation of pesticides was recognized about 40 years ago but it is only in the last few years that this phenomenon has been considered of significant economic importance, although differing opinions have been expressed (Fox 1983). It has yet to be determined why this phenomenon was only recently detected on a large scale. The possible reasons are: lack of awareness in the past; the confusion of loss of pesticidal effectiveness resulting from accelerated degradation with an homologous phenomenon, i.e. loss of effectiveness resulting from development of resistance to pesticides; the introduction of new pesticides or new formulations, which are possibly more potent in predisposing the soil to accelerated degradation; the frequent application of pesticides which are chemically related; increased monoculture, which leads to the frequent use of the same pesticide and the possible introduction of certain agricultural practices, e.g. irrigation, tillage and pesticide application which, by chance, enhance degradation. Accelerated degradation also poses additional, difficult questions not addressed previously, which might be as serious as those caused by the development of pesticide-resistant pests-which also diminishes the number of effective pesticides available for pest control.

Soils have to be surveyed on a large scale to identify cases of accelerated degradation. The amplitude of this phenomenon and its economic significance should also be evaluated. Detailed records of the pesticide history in these soils are necessary for such studies. Methods for assessing pesticide degradation have to be further developed, improved and adapted for various purposes. Pesticide combinations, formulations, and methods of application, (e.g. slow-release pesticides) and agricultural practices (e.g. reduced tillage, chemigation) have to be reassessed in the light of stimulation or suppression of accelerated degradation. Simulation models, both experimental and theoretical, will be most helpful in such studies. Following molecular and genetic approaches and further elucidation of the role of plasmids and other mobile genetic elements, may reveal additional mechanisms for the conservation, transfer, exchange and intensification of degradation capacity and the possible rise of new recombinants.

Abandoning crop rotation has many negative consequences which include buildup of pest populations, selection of biotypes resistant to pesticides, and outbreak of epidemics of new pests due to shifts in biological equilibria. Accelerated degradation is such an additional undesirable consequence which further emphasizes the need for a more balanced crop rotation. In the long run, such a strategy might prove more economic than the present monoculture system.

Elucidating the mechanisms of accelerated degradation may enable us to develop tools to enhance degradation and decontamination of undesirable agrochemical residues in agricultural sites, industrial wastes, toxic by-products from biotechnological industry and in cases of accidental spillages. Stocks of soils with accelerated degradation capacity, isolates of degraders with superior degradation capacity, or genetically engineered microorganisms might be used for this purpose, with special care, and according to the regulations, bearing in mind the possible consequences of the release of highly potent microorganisms into the environment.

Acknowledgments. Some of the studies carried out by the authors and mentioned in this chapter were supported by the United States-Israel Binational Agricultural Research and Development Fund (BARD). We wish to thank Oded Yarden and A.R. Entwistle for helpful comments and suggestions.

References

Aharonson N, Kaufman DD, Coffman CB, Wilson HP (1987) Acylanilide persistence in continuously cropped soils. North East Weed Sci Soc 41:78–79
Audus LJ (1949) The Biological detoxication of 2,4-D. Plant Soil 2:31–36
Audus LJ (1951) The biological defoxication of hormone herbicides in soil. Plant Soil 3:170–192
Audus LJ (1960) Microbiological breakdown of herbicides in soils. In: Audus LJ (ed) Herbicides and the Soil. Blackwell, Oxford, pp 1–19
Avidov E, Aharonson N, Katan J, Rubin B, Yarden O (1985) Persistence of terbutryn and atrazine in soil as affected by soil disinfestation and fungicides. Weed Sci 33:457–461
Avidov E, Aharonson N, Katan J (1988) Accelerated degradation of dephenamid in soils and means for its control. Weed Sci 36:519–523
Bailey AM, Coffey MD (1985) Biodegradation of metalaxyl in avocado soils. Phytopathology 75:135–137
Blake J, Kaufman DD (1975) Characterization of acylanilide — hydrolyzing enzyme(s) from *Fusarium oxysporum* Schlecht. Pestic Biochem Physiol 5:305–313
Chakarabarty AM (1982) Genetic mechanisms in the dissimilation of chlorinated compounds. In: Chakarabarty AM (ed) Biodegradation and Detoxification of Environmental Pollutants. CRC, Boca Raton, FL, pp 127–139
Cole M (1976) Effect of long-term atrazine application on soil microbial activity. Weed Sci 24:473–476
Cook RJ, Baker KF (1983) The Nature and Practice of Biological Control of Plant Pathogens. APS, St. Paul, 539pp
Don RH, Pemberton JM (1981) Properties of six pesticides degradation plasmids isolated from *Alcaligenes paradoxus* and *Alcaligens eutrophus*. J Bacteriol 145:681–686
Dowler CC, Marti LS, Kuien CS (1984) Degradation rates of butylate and alachlor following repeated field treatments. Proc South Weed Sci Soc 37:323
Ferris IG, Lichtenstein EP (1980) Interactions between agricultural chemicals and soil microflora and their effect on the degradation of ^{14}C- parathion in a cranberry soil. J Agric Food Chem 28:1011–1019
Fisher PR, Appleton J, Pemberton JM (1978) Isolation and characterization of the pesticide – degrading plasmid p JP1 from *Alcaligenes paradoxus*. J Bacteriol:798–804
Fletcher J, Conolly G, Mountfield EI, Jacobs L (1980) The disappearance of benomyl from mushroom casing. Ann Appl Biol 95:73–82
Fox JL (1983) Soil microbes pose problems for pesticides. Science 221:1029–1031
Fryer JD, Smith PD, Hace RJ (1980) Field experiments to investigate long-term effect of repeated applications of MCPA, triallate, simizine, and linoron. II Crop performance and residues 1969–78. Weed Res 20:103–110
Gray RA, Joo GK (1985) Reduction in weed control after repeat applications of thiocarbamate and other herbicides. Weed Sci 33:698–702
Harvey RG (1987) Herbicide dissipation from soils with different herbicide use histories. Weed Sci 35:583–589
Horowitz M (1966) Breakdown of endothal in soil. Weed Res 6:168–171
Katan J, Wilson HP, Kaufman DD (1984) Enhanced degradation of diphenamid and other pesticides in herbicide history soils. Abstr Weed Sci Soc Am :100–101
Kaufman DD (1987) Accelerated degradation of pesticides in soil and its effect on pesticide efficacy. Brit Crop Prot Conf Weeds 2:515–522
Kaufman DD, Blake J (1973) Microbial degradation of several acetamide, acylanilide, carbamate, toluidine and urea pesticides. Soil Biol Biochem 5:297–308
Kaufman DD, Edwards DF (1983) Pesticide/microbe interaction effects on persistence of pesticides in soils. In: Miyamoto J, Kearney PC (eds) Proc Pesticide Chem: Human Welfare and the Environment. Vol 4, Pergamon Press, Oxford, pp 177–182
Kaufman DD, Katan J, Edwards DF, Jordan EJ (1985) Microbial adaptation and metabolism of pesticides. In: Hilton JL (ed) Agricultural Chemicals of the Future. Rowman & Allanheld, New Jersey, pp 437–451

Lee S (1984) EPTC (S-ethyl N,N-dipropythiocarbamate) degrading microorganisms isolated from a soil previously exposed to EPTC. Soil Biol Biochem 16:529–531

Marty JL, Khafif T, Vega D, Bastide J (1986) Degradation of phenyl carbamate herbicides by *Pseudomas alcaligenes* isolated from soil. Soil Biol Biochem 6:649–653

Newman AS, Thomas JR (1949) Decomposition of 2,4-dichlorophenoxyacetic acid in soil and liquid media. Soil Sci Soc Proc 14:160–164

Obrigawitch T, Martin AR, Roeth FW (1983) Degradation of thiocarbamate herbicides in soils exhibiting accelerated rapid EPTC breakdown. Weed Sci 31:187–192

Pemberton JM, Fisher PR (1977) 2,4-D plasmids and persistence. Nature 268:732–733

Pemberton JM, Corney B, Don RH (1979) Evolution and spread of pesticide degrading ability among soil microorganisms. In: Timmis KN, Puhler A (eds) Plasmids of Medical, Environmental, and Commercial Importance. Elsevier, Amsterdam, North Holland, pp 287–299

Prochnow CL (1981) R-33865 herbicide extender for thiocarbamate herbicides. Proc West Soc Weed Sci 34:55–56

Racke KD, Coats JR (1987) Enhanced degradation of isofenphos by soil microorganisms. J Agric Food Chem 35:94–99

Rahman A, James TK (1983) Decreased activity of EPTC + R-25788 following repeated use in some New Zealand soils. Weed Sci 31:783–789

Rahman A, Atkinson GC, Douglas JA, Sinclair DP (1979) Eradicane causes problems. New Zealand J Agric 139:47–49

Read DC (1986) Accelerated microbial breakdown of carbofuran in soil from previously treated fields. Agric Ecosyst Environ 15:51–61

Roeth FW (1986) Enhanced herbicide degradation in soil with repeat application. Rev Weed Res 2:45–65

Spain JC, Pritchard PH, Bourquin AW (1980) Effects of adaptation on biodegradation rates in sediment/water cores from Estuarine and freshwater environments. Appl Environ Microbiol 40:726–734

Suett DL (1986) Accelerated degradation of carbofuran in previously treated field soil in the United Kingdom. Crop Prot 5:165–169

Tam AC, Behki RM, Khan SU (1987) Isolation and characterization of an s-ethyl-N,N-dipropylthiocarbamate – degrading *Arthrobacter* strain and evidence for plasmid – associated s-ethyl-N,N-dipropylthiocarbamate degradation. Appl Environ Microbiol 53:1088–1093

Torstensson NTL, Stark L, Goransson B (1975) The effect of repeated applications of 2,4-D and MCPA on their breakdown in soil. Weed Res 15:159–165

Waid JS (1972) The possible importance of transfer factors in the bacterial degradation of herbicides in natural ecosystems. Res Rev 44:65–71

Walker A, Brown PA, Entwistle AR (1986) Enhanced degradation of iprodione and vinclozolin in soil. Pestic Sci 17:183–193

Wilson RG (1984) Accelerated degradation of thiocarbamate herbicides in soil with prior thiocarbamate herbicide exposure. Weed Sci 32:264–268

Yarden O, Katan J, Aharonson N, Ben-Yephet Y (1985) Delayed and enhanced degradation of benomyl and carbendazin in disinfested and fungicide – treated soils. Phytopathology 75:763–767

Yarden O, Aharonson N, Katan J (1987) Accelerated microbial degradation of methyl, benzimidazol, 2-ylcarbamate in soil and its control. Soil Biol Biochem 19:735–739

Part IV. Petroleum Hydrocarbons

Introductory Comments

The contamination of the land surfaces, the unsaturated zone and groundwater by petroleum hydrocarbons has become one of the major environmental problems of the last few years. Various research groups have recently approached the problem from different angles and the scientific literature has been enriched with contributions on the physico-chemical and biochemical processes governing the fate of petroleum products in porous media and on their transport to the groundwater. Based on the existing understanding of the processes governing the fate of petroleum hydrocarbons in the unsaturated zone, technical procedures for groundwater renovation have been developed and successfully applied.

The papers in this part describe some of the mechanisms governing the interaction, transport and decomposition of petroleum hydrocarbons in the unsaturated zone and the authors' results are discussed in the context of the published literature in their particular field. The included papers concentrate on the principles of various processes and reactions rather than attempting to enumerate all the existing information from the literature.

Abiotic interactions of petroleum hydrocarbons in the unsaturated zone as affected by environmental conditions are discussed by Yaron (Chapter 10). Partitioning of toxic organic compounds between the solid, liquid and gaseous phases of the unsaturated zone, is described and processes such as volatilization, adsorption-desorption and degradation are exemplified. If one considers petroleum contaminants as a mixture of hydrocarbons of varying properties, this paper emphasizes the changes in pollutant composition with space and time during its downward or upward redistribution in the unsaturated zone. The rate and extent of changes in the composition of petroleum-product spills are governed by the properties of each hydrocarbon comprising the original petroleum-product mixture, of the porous media and on environmental conditions. Experimental data are presented in support of the above hypothesis.

The transport of liquid hydrocarbons from the land surface to the groundwater is the topic of an additional chapter in the present part. Rubin and Mechrez (Chapter 11) describe the properties of immiscible liquid petroleum products with regard to their potential transport in a multiphase system which includes also air and water. The properties of the liquids affect their transport parameters. In the process of the simultaneous flow through the porous media, the immiscible fluids affect each other, and the permeability of the multiphase system for one component is affected by the presence of the other components. Various numerical models describing the transport of liquid hydrocarbons in multiphase systems are presented and critically discussed. In the models presented the liquid petroleum products are considered, for convenience, as single components and not as a mixture of hydrocarbons with various properties. Emphasis is placed on the necessity, when modeling, in using transport parameters obtained in the field where they are affected by soil structure as well as physico-chemical and biochemical processes.

The pathways and rates of biochemical induced degradation of petroleum hydrocarbons are discussed by Hoepner et al. (Chapter 12). By comparing the aerobic degradation of hydrocarbons in both marine sediments and soils the authors reach the conclusion that the degradation rate of hydrocarbons is affected less by the properties of the substrate and is determined primarily by environmental conditions. The authors discuss in situ conditions of tidal sediments and present their own research results on the biodegradation of aliphatic compounds. The conclusions reached from basic laboratory experiments are extended to field conditions and management procedures are suggested.

The chapters herein discussed do not cover the entire aspect of petroleum hydrocarbons behavior in the porous media. They do, however, point out various aspects of abiotic and biochemical interactions between petroleum-product components and the porous media, and their redistribution in the unsaturated zone with regard to groundwater contamination.

10 On the Behavior of Petroleum Hydrocarbons in the Unsaturated Zone: Abiotic Aspects

B. YARON

10.1 Introduction

In recent years pollution of the unsaturated zone by petroleum products has aroused concern particularly with regard to ground water quality. The research carried out in this field has dealt mainly with the biological degradation of petroleum products and their residues in soil as well as with their transport through porous media as a liquid immiscible with water or through the gaseous phase.

The aim of this paper is to define the abiotic behavior of petroleum products in the unsaturated zone, and to emphasize the various facets of this behavior and their effect on the ultimate fate of the residues in porous media and ground water. In our approach petroleum products are not considered as homogeneous compounds but as a mixture of hydrocarbons with different physical and chemical properties. Furthermore, the unsaturated zone is not an inert material but can interact with the petroleum products. The properties of the pollutants and of the various components of the porous media will define the dimensions and rate of their interaction. These interactions are affected by the environmental conditions in the unsaturated zone: temperature and moisture content.

The abiotic processes occurring within the porous media are volatilization, dissolution in the water phase, adsorption from the liquid and gaseous phase onto the solid phase, subsequent desorption to the vapor and water phases, surface-induced degradation on the solid phase, and finally transport in the gaseous phase and as a liquid immiscible and miscible with the water phase.

10.2 Properties of the Interacting Materials

10.2.1 Crude Oils and Petroleum Products

The properties of these materials have been described in a series of books and reviews during the last 11 years (e.g. Clark and Brown 1977; Wolfe 1977; Jordan and Payne 1980). Hence, only those properties relevant to the abiotic behavior of the petroleum products in porus media will be summarized. Table 1 (after Clark

---

[1]Department of Organic and Residues Chemistry, Institute of Soils and Water, ARO, The Volcani Center, Bet Dagan, 50250, Israel.

Table 1. Distribution of Hydrocarbons Fractions in Several Crude Oils (after Clark and Brown 1977)

Hydrocarbons	Prudhoe	South Lousiana	Kuwait
Naphtha Fraction, wt%	23.2	18.6	22.7
Paraffins	12.5	8.8	16.2
Naphthenes	7.4	7.7	4.1
Aromatics	3.2	2.1	2.4
Benzenes	0.3	0.2	0.1
Toluene	0.6	0.4	0.4
C_8 Aromatics	0.5	0.7	0.8
C_9 Aromatics	0.06	0.5	0.6
C_{10} Aromatics		0.2	0.3
C_{11} Aromatics		0.1	0.1
Indans			0.1
High-Boiling Fraction wt%	76.8	81.4	77.3
Saturates	14.4	56.3	34.0
n-Paraffins	5.8	5.2	4.7
C_{11}	0.12	0.06	0.12
C_{12}	0.25	0.24	0.28
C_{13}	0.42	0.41	0.38
C_{14}	0.50	0.56	0.44
C_{15}	0.44	0.54	0.43
C_{16}	0.50	0.58	0.45
C_{17}	0.51	0.59	0.41
C_{18}	0.47	0.40	0.35
C_{19}	0.43	0.38	0.33
C_{20}	0.37	0.28	0.25
C_{21}	0.32	0.20	0.20
C_{22}	0.24	0.15	0.17
C_{23}	0.21	0.16	0.15
C_{24}	0.20	0.13	0.12
C_{25}	0.17	0.12	0.10
C_{26}	0.15	0.09	0.09
C_{27}	0.10	0.06	0.06
C_{28}	0.09	0.05	0.06
C_{29}	0.08	0.05	0.06
C_{30}	0.08	0.04	0.07
C_{31}	0.08	0.04	0.06
C_{32} plus	0.07	0	0.06
iso-Paraffins		14.0	13.2
1-ring cycloparaffins	9.9	12.4	6.2
2-ring cycloparaffins	7.7	9.4	4.5
3-ring cycloparaffins	5.5	6.8	3.3
4-ring cycloparaffins	5.4	4.8	1.8
5-ring cycloparaffins		3.2	0.4
6-ring cycloparaffins		1.1	
Aromatics, wt%	25.0	16.5	21.9
Benzenes	7.0	3.9	4.8
Indans and tetralins		2.4	2.2
Dinaphtheno benzenes		2.9	2.0
Naphthalenes	9.9	1.3	0.7
Acenaphthenes		1.4	0.9
Phenathrenes	3.1	0.9	0.3

Table 1. *(Cont.)*

Hydrocarbons	Prudhoe	South Lousiana	Kuwait
Acenaphthalenes		2.8	1.5
Pyrenes	1.5		
Chrysenes			0.2
Benzothiophenes	1.7	0.5	5.4
Dibenzothiophenes	1.3	0.4	3.3
Indanothiophenes			0.6
Polar materials, wt%	2.9	8.4	17.9
Insolubles, wt%	1.2	0.2	3.5

and Brown 1977), for example, shows the main physical characteristics and chemical properties of a crude oil and its distillation products, allowing a comparison of such factors as their boiling point, specific gravity, viscosity and relative proportions of aliphatic, naphthemic and aromatic compounds. The hydrocarbon groups are in themselves a mixture of components with various carbon structures (e.g. *n*-paraffin: C_{11} to C_{32}; iso-paraffins; 1-ring to 6-ring cycloparaffins; aromatics: benzene; toluene; C_8 to C_{11} aromatics, etc.). The relative ratio between the hydrocarbon groups and the composition of each group is defined by both the origin of the crude oil and the distillation procedure. Since the petroleum product is not a homogeneous compound but a mixture of hydrocarbons with various physical and chemical properties its behavior in porous media (e.g. volatilization, dissolution, adsorption) will be controlled not by the ideal behavior of each hydrocarbon participating in the composition of the product, but by the properties of the mixture.

The physical, mineralogical and chemical properties of the *porous media* which may affect the fate of petroleum products are described in most textbooks of soil science or geochemistry. The surface area of the solid phase, as defined by the clay mineralogy and by the soil organic matter, is a major factor in the adsorption of the petroleum hydrocarbons in the unsaturated zone. The porosity of the medium, as well as the air-water ratio in the unsaturated zone, control the transport of the petroleum hydrocarbons in both gaseous and liquid phases. The electrolyte concentration in the soil water may influence the degree of dissolution of the liquid hydrocarbon in the aqueous phase of the unsaturated zone.

10.3 Partitioning Between Gaseous, Aqueous, and Solid Phases

Let us consider a petroleum spill on a land surface. Redistribution of the petroleum hydrocarbon components within the soil volume will occur with time. The highly volatile fraction will be transported in the gaseous phase of the porous media and subsequently lost into the atmosphere, adsorbed onto the solid phase, or dissolved in the ground water. The less volatile hydrocarbons will be transported as a liquid into the porous media. During their transportation through the unsaturated zone, adsorption onto the solid phase and dissolution into the soil water will occur. If the amount of petroleum dispersed within the unsaturated zone is greater than the "retention capacity" of the porous media, the liquid hydrocarbon will reach the ground water. Figure 1 (after Abriola and Pinder 1985a) contains all the elements which may be present in an imaginary contamination scenario. For simplification, only four different regions or phases separated by distinct interfaces are visible: solid (soil), water, gas, and liquid petroleum products. In reality each compound (or group of compounds) of the petroleum products is separated differentially among the above four phases according to its properties. The petroleum hydrocarbons partitioning among the gaseous, aqueous and solid phases of the porous media, their redistribution with depth and their transport to the atmosphere or ground water will be affected by the environmental conditions in the unsaturated zone. Since during their infiltration through the porous media the petroleum hydrocarbons may undergo biological degradation – inducing changes in concentration and composition – the abiotic processes are indirectly affected.

10.3.1 Migration into the Gaseous Phase

An abiotic factor important in determining the fate of petroleum products in the porous media is evaporation. The boiling point ranges of petroleum product components are significant, particularly with regard to losses caused by evaporation. Figure 2, (after Clark and Brown 1977), shows the distillation distribution

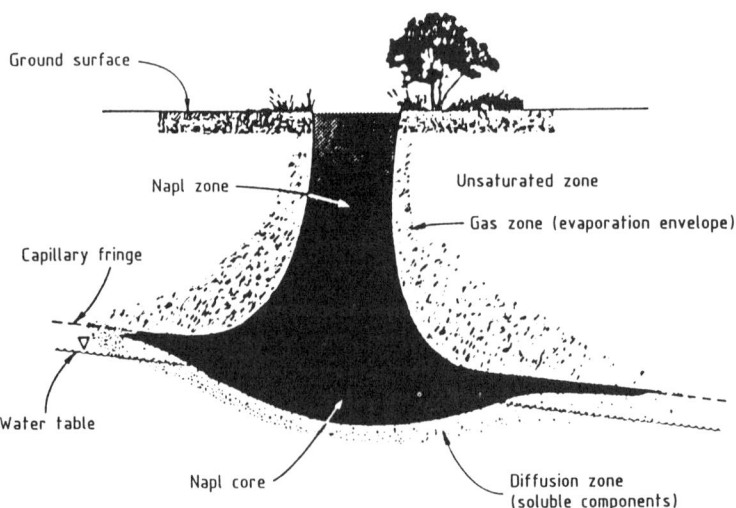

Fig. 1. Schematic representation of higher flow water-immiscible liquid (NAPL) movement through the saturated zone (after Abriola and Pinder 1985a)

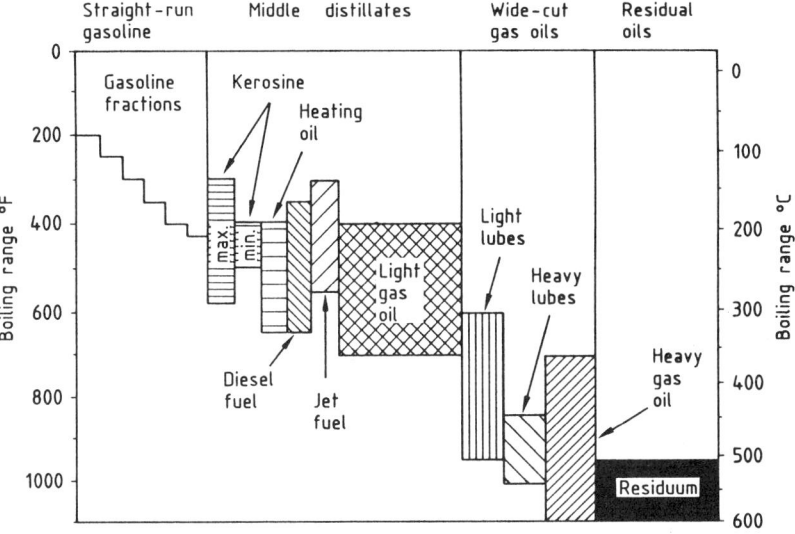

Fig. 2. Fractional distillation distribution of crude and refined oils (after Clark and Brown 1977)

of fractions of crude and refined petroleum. It may be observed that the gasoline fraction is characterized by the lowest boiling point, and is therefore the most volatile of the refined products. Oils, with the highest boiling points, are the least volatile petroleum fractions. The evaporation rate for a specific hydrocarbon is a function of its vapor pressure, which is inversely related to the carbon number and molecular weight. Compounds with molecular weights greater than n-C_{15} will continue to evaporate over long periods of time, although these rates become insignificant after 10^2 hours. As shown previously, each petroleum product is a mixture of hydrocarbons with differing vapor pressures and consequently in each petroleum fraction these will be compounds encompassing a wide range of volatilization properties.

Evaporation from surfaces has been studied mainly in the marine environment. It was observed, for example, that most of the more volatile compounds on water surfaces are lost by evaporative processes within 24–48 h. In Fig. 3 two cases of evaporation of oil from the sea surface are presented. In the case of the south Louisiana crude oil spill (Fig. 3a), the rapid decrease after 42 min of n-C_{12} and n-C_{13} represents the onset of strong winds causing white capping (McAuliffe 1977a), a situation which is not possible in porous media. Similar losses in the aromatic compounds ranging from benzene (C_6) to trimethylbenzene (C_9) are presented in Fig. 3b (McAuliffe 1977b). Changes in the physical properties of oil resulting from the loss of volatile hydrocarbons include increases in the density and kinematic viscosity of the remaining petroleum products. Rostad (1976) showed, for example, that the kinematic viscosity of a Norwegian crude oil increased – due to evaporation of highly volatile compounds – in 48 h from 18.78 to 56.02 (mm^2/sec), and its density from 0.862 to 0.894 g/ml.

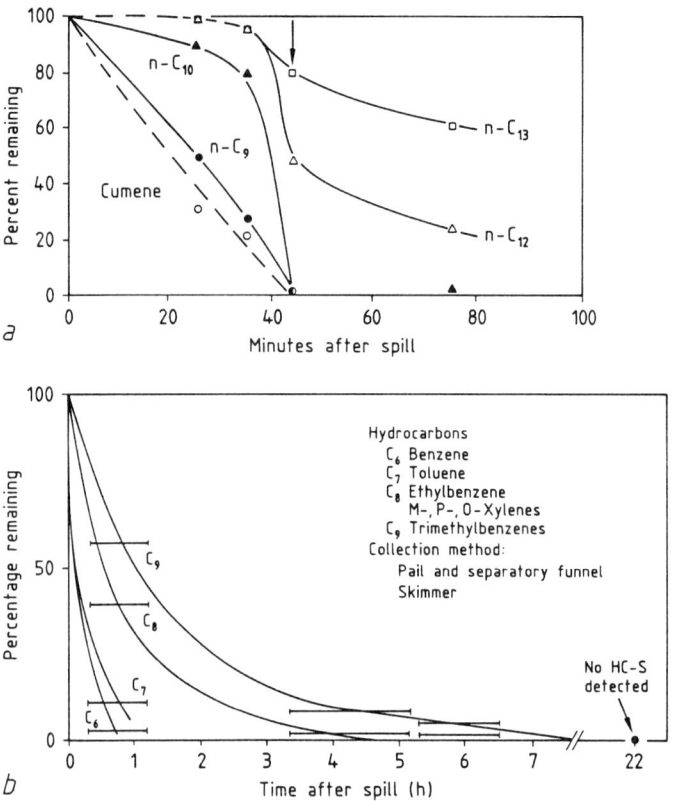

Fig. 3. Percent of hydrocarbons – low-boiling (**a**) and aromatic (**b**) – remaining in surface oil slick due to volatilization (after McAuliffe 1977a,b)

When a petroleum product reaches a land surface its interaction with the soil is affected by the differential volatilization of its compounds. On the one hand, the physical properties (e.g. viscosity, density) of the remaining liquid hydrocarbons are changed, and on the other hand the composition of the gaseous phase will be different from that of the liquid hydrocarbons deposited on the soil. Figure 4 shows the relative composition of a simulated 'kerosene' mixture and the composition of the contiguous gaseous phase obtained by our research group in a controlled laboratory experiment (Yaron et al. 1988). It can be observed that while m-xylene represents only 7% of the total hydrocarbons concentration in the liquid 'kerosene' phase, in the gaseous phase the amount of m-xylene found after 24 h was 40% of the total hydrocarbon content. The presence of each compound in the gaseous phase was found to be in accordance with its vapor pressure. Since the pollution of the unsaturated zone and of the groundwater occurs not only from the liquid phase but also from hydrocarbons in the gaseous phase, it is clear that a difference in composition between the liquid and gaseous phases will lead to a

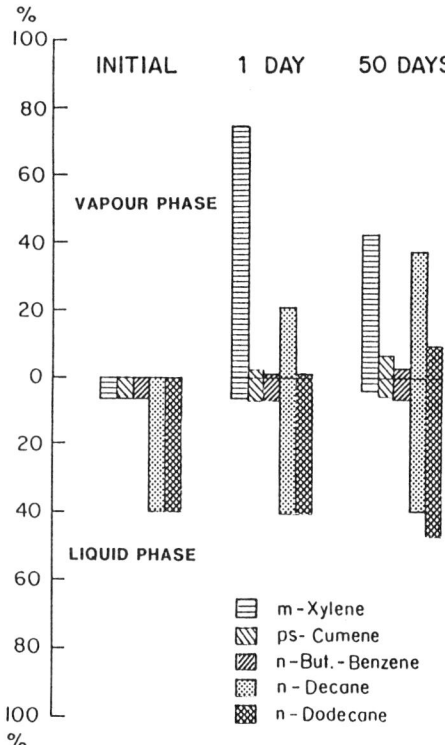

Fig. 4. Relative composition of simulated liquid 'kerosene' and of the surrounding gaseous phase (after Yaron et al. 1988)

different distribution of the pollutant in the porous media. Since the vapor pressures of the hydrocarbons are affected by the temperature, the composition of the saturated 'keresone' atmosphere will also be affected when the ambient temperature changes. Table 2 (from Acher et al. 1987) shows the changes in the composition of a saturated 'kerosene' atmosphere when the temperature was increased from 7° to 34° C. An increase in the concentration of hydrocarbons from both aliphatic and aromatic groups as a result of the temperature increase will affect the pollution hazard in the unsaturated zone.

In summarizing the influence of evaporation on the behavior of petroleum products in the porous media, we can emphasize the following effects: direct reduction of concentration of petroleum products through losses to the atmosphere, changes in the chemical composition of the gaseous phase with regard to the liquid phase, changes in the chemical composition of the liquid phase leading to a change in its subsequent volatilization and dissolution in the water phase, changes the physical properties (e.g. viscosity, fluidity) of the liquid hydrocarbons, thus affecting their infiltration into the porous media.

Table 2. Hydrocarbons present in a saturated 'kerosene' atmosphere at different temperatures (after Acher et al. 1987)

'Kerosene' components	Saturated atmosphere composition L/L			
	Temperature (°C)			
	7	17	27	34
m-Xylene	1.20	2.90	5.00	7.80
ps-Cumene	0.17	0.65	0.98	1.50
n-but.-Benzene	0.04	0.24	0.34	0.62
n-Decane	0.74	2.65	5.10	7.10
n-Dodecane	n.d.	0.08	0.13	0.24

Table 3. Solubilities of selected aliphatic and aromatic petroleum hydrocarbons in seawater and distilled water at 25° C (adopted Sutton and Calder 1974, 1975)

Compound	Solubility in Distilled Water	Solubility in Seawater
Dodecane (C_{12})	3.7 (ppb)	2.9 (ppm)
Tetradecane (C_{14})	2.2	1.7
Hexadecane (C_{16})	0.9	0.4
Octadecane (C_{18})	2.1	0.8
Eicosane (C_{20})	1.9	0.1
Hexacosane (C_{26})	1.7	0.1
Toluene	534.8 ±4.9 (ppm ±S.E.)	379.3 ±2.8 (ppm ±S.E.)
Ethylbenzene	161.2 ±0.9	111.0 ±1.3
o-Xylene	170.5 ±2.5	129.6 ±1.8
m-Xylene	146.0 ±1.6	106.0 ±0.6
p-Xylene	156.0 ±1.6	110.9 ±0.9
Isopropylbenzene	65.3 ±0.8	42.5 ±0.2
1,2,4-Trimethylbenzene	59.0 ±0.8	39.6 ±0.5
1,2,3-Trimethylbenzene	75.2 ±0.6	48.6 ±0.5
1,3,5-Trimethylbenzene	48.2 ±0.3	31.3 ±0.2
n-Butylbenzene	11.8 ±0.1	7.09 ±0.07
s-Butylbenzene	17.6 ±0.2	11.9 ±0.2
t-Butylbenzene	29.5 ±0.3	21.2 ±0.3

10.3.2 Dissolution in the Water Phase

The partitioning of petroleum hydrocarbons, immiscible with water and miscible with water, is of major importance in defining their redistribution in the porous media and their transport to the ground water.

The rate of dissolution for the various components of a petroleum spill depends on rather complex interactions dependent on properties inherent to the oil (i.e. molecular structure of compounds and relative abundance of these components) and the physico-chemical properties of the immediate environment (i.e. salinity, temperature, etc.).

Table 3 (after Sutton and Calder 1974, 1975) lists the solubility of selected petroleum hydrocarbons in distilled water as affected by the carbon number. It can be seen that generally, the solubility in water decreases with an increase in carbon number. The maximum solubility in the paraffins group is exhibited by methane (24 ppm), in the cycloparaffins group by cyclopentane (156 ppm), and in the aromatics by benzene (1780 ppm). Jordan and Payne (1980) showed that the solubility of both chain and ring structures is inversely proportional to the degree of saturation of petroleum hydrocarbons. The addition of a second or third double bond increases solubility proportionally and it has been shown that the presence of a triple bond increases solubility to a greater extent than the presence of two double bonds (McAuliffe 1966). Therefore, the most water-soluble petroleum hydrocarbons will be those with the lowest molar volume and greatest aromatic/olefinic character.

The salinity level of water reduces the solubility of hydrocarbons. To the best of our knowledge, there are no data available on the dissolution of hydrocarbons in soil solution characterized by low electrolyte concentration. The solubility in sea water is about 30–40% less than that in distilled water (Jordan and Payne 1980). For the paraffins, the magnitude of this 'salting-out' effect is directly proportional to the molar volumes, in accordance with the McDevit-Long theory which attributes salting out to the effect of electrolytes upon water structure (Sutton and Calder 1974).

The presence of organic matter in a saline marine environment enhances the solubility due to its surface-active nature. Boehm and Quinn (1973) examined the effect of dissolved organic matter on various hydrocarbons' solubilities, utilizing marine water and NaCl solutions. A 50% to 99% decrease in the amounts solubilized was observed for n-alkanes and isoprenoids when the dissolved organic matter was removed. The aromatics examined (anthracene, phenanthracene and dibutylphthalate) were unaffected by this process.

An additional factor affecting the aqueous solubility of hydrocarbons is the presence of co-solutes. Since the petroleum compounds are mixtures of various hydrocarbons, the presence of the co-solute affects the solubility of a particular hydrocarbon. Eganhause and Calder (1974) pointed out that no general trends are apparent, as solubilities may increase or decrease for various hydrocarbons when other components are added to the system.

10.3.3 Retention by the Solid Phase

Petroleum hydrocarbons are retained in the solid phase of the unsaturated zone as a liquid immiscible with water – trapped in the pores due to the action of capillary forces – or adsorbed on the surfaces. The adsorption occurs as a solute from the water phase or as a gas from the vapor phase.

The trapped hydrocarbons will remain in the unsaturated zone for an indefinite time period (Abriola and Pinder 1985b) serving as a source of contamination which will decrease in magnitude as a result of abiotic processes such as volatilization or dissolution in the water phase. The extent of the 'trapping' process is determined primarily by the physical properties of the vadose zone. If the petroleum product is characterized by a high boiling point and a low solubility it will remain in the unsaturated zone and will not be affected by any abiotic process. In this particular case the porous medium behaves like an inert material and the behavior of the petroleum product depends only on its own properties with no interactions occurring between the two materials.

Adsorption of petroleum hydrocarbons on the soil solid phase and on soil constituents has been the subject of a number of studies over the years. Nathwani and Philips (1977) studied the adsorption of selected hydrocarbons (benzene, o-xylene, toluene, and n-hexadodecane) on a series of Canadian soils. It was found that at concentrations between 1 and 100 ppm the equilibrium distribution of the hydrocarbons studied was described by the Freundlich isotherm. Soils with a higher organic matter content exhibit higher values of the distribution coefficient.

Studies of well-defined clays and organic clay complexes contribute to an understanding of the mechanism which governs the adsorption of aromatic hydrocarbons. Eltantwy and Arnold (1972) presented evidence of interlayer sorption of n-hexane and n-dodecane by Ca-Wyoming montmorillonite. n-Hexane was able to form a single-layer complex with the clay to the extent of 110 mg n-hexane per gram. Similar results were obtained for n-dodecane. These observations strongly suggest that n-alkanes can penetrate the interlayer space of montmorillonite but at a very slow rate. Studies of De Boer and Zwilker (1929); Fripiat (1968) and Pinnavaia and Mortland (1971) further developed the knowledge in this field. It is known today that the nature of the interlayer cation present is apparently of no great influence in the intercalation of n-alkanes. Theng (1974) points out that the intercalation of hydrocarbon species into the expanding layer silicates is essentially a process in which part of the interlayer water associated with the exchangeable cations is replaced by the organic species. In dehydrated systems where the silicate layers of a clay crystal are fully collapsed, intercalation of hydrocarbons is either absent or proceeds only with difficulty. The studies of Clementz (1976) on adsorption of petroleum heavy fractions (Asphaltenes) from solution on montmorillonite surfaces confirmed the above findings.

The previous studies dealt only with the interactions between single hydrocarbon compounds and the soil and soil constituents. Petroleum products are, however, mixtures of hydrocarbons with individual properties competing with each other for the free adsorption sites of porous media. Only in the last few years have petroleum products been considered – in their behaviour in the unsaturated zone – as a mixture of more than one chemical species. Studies of Abriola and Pinder (1985a, b), Curtis et al. (1986) and Corapcioglu and Baehr (1987) included this approach in developing a model for multiphase transport of petroleum products in the unsaturated zone. The research carried out in our laboratory (Acher et al. 1987; Yaron et al. 1988) in the last several years has provided experimental data on the adsorption of petroleum hydrocarbon mixtures from the gaseous phase on soil surfaces, and has added information to the conceptual model quoted above.

Figure 5 (after Yaron et al. 1988) shows the kinetics of adsorption of five components of simulated 'kerosene' from the gaseous phase onto the surfaces of three soils with different clay and organic matter contents (Bet Dagan-clay 12%, OM 0.5%; Gilat-clay 16%, OM 0.6%; Oxford-clay 35%, OM 7.2%). On oven-dry soils the adsorption occurs in the order: Gilat > Oxford > Bet Dagan. Despite the fact that the Oxford soil has a greater clay content than Gilat soil, its adsorption capacity is lower. This fact may be explained by the presence of a high amount of organic matter occupying the free adsorption sites in the Oxford soil. Complementary studies of the adsorption of the same hydrocarbon mixture on peat show very low adsorption values, a fact which confirms this assumption. Figure 6 (Yaron et al. 1988) shows the effect of soil moisture content on the adsorption of petroleum hydrocarbons on Gilat soil. It is seen that an increase in the soil moisture content leads to a decrease of the soil adsorption capacity for all the hydrocarbons studied. When the soil moisture content reaches a value equivalent to 70% of its field capacity (13%), the adsorption of hydrocarbons is almost negligible.

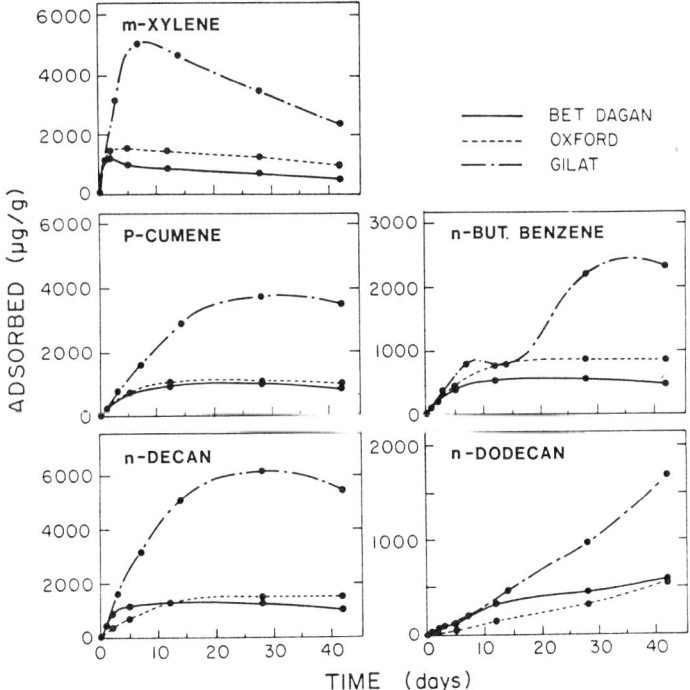

Fig. 5. Kinetics of hydrocarbons adsorbed from a surrounding atmosphere of a simulated liquid 'kerosene' onto the surfaces of three different soils: Bet Dagan (clay 12%, OM 0.5%), Gilat (clay 16%, OM 0.6%), and Oxford (clay 35%, OM 7.2%) (after Yaron et al. 1988)

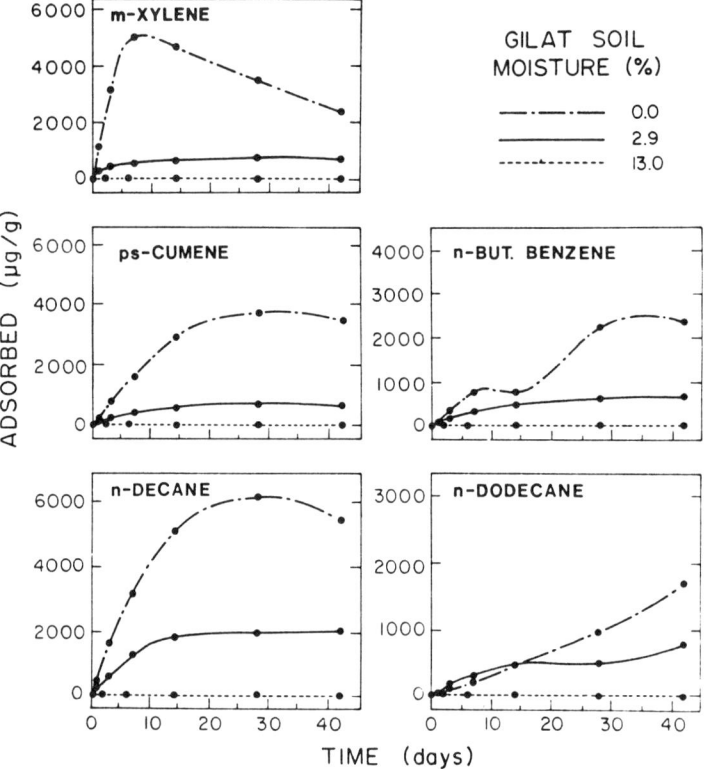

Fig. 6. Effect of initial soil moisture content on the kinetics of hydrocarbon adsorption from a surrounding atmosphere of a simulated liquid 'kerosene' onto the surface of Gilat soil (after Yaron et al. 1988)

The adsorption of both aliphatic and aromatic components of kerosene increases with an increase in temperature. The rise in temperature increases the vapor pressure of the hydrocarbons, thus affecting their concentration in the soil atmosphere, and hence their adsorption. Figure 7 (Acher et al. 1987) shows that on an oven-dried Bet Dagan soil at the highest temperature studied (34° C), adsorption is in the following order: n-decane > pseudo-cumene > m-xylene > t-butylbenzene > n-decane.

It should be emphasized that due to differential volatilization of the 'kerosene' components, the composition of the gas phase is not the same as that of the liquid kerosene, and the adsorption on soil from the gas phase occurs according to the concentration of each component in the gas phase. We do not yet have any data on the adsorption of these kerosene components from the aqueous phase, but it is supposed that the order of adsorption will be affected by their solubility in water and by the competition of water for the adsorption sites on the soil surfaces.

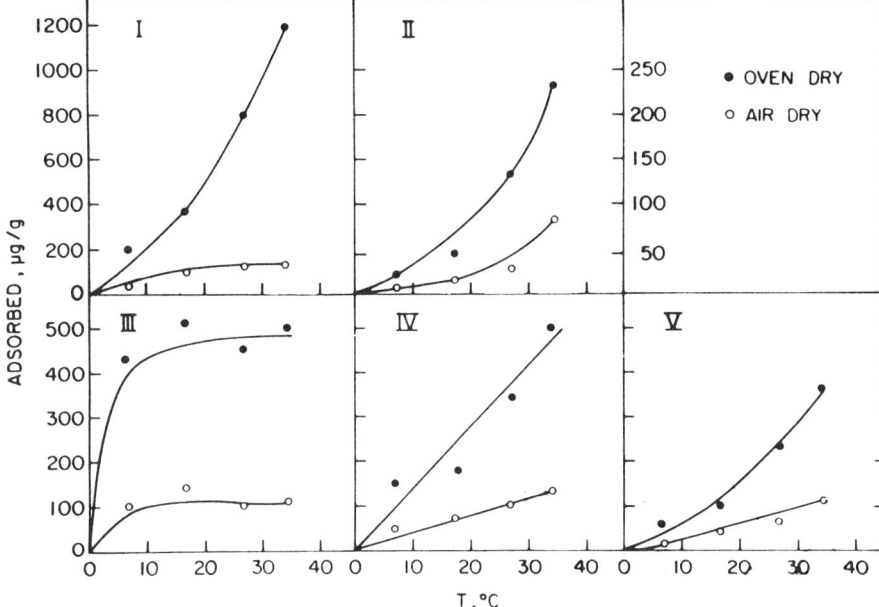

Fig. 7. Adsorption of hydrocarbons onto a Bet Dagan soil from a surrounding atmosphere of a simulated 'kerosene' as affected by soil moisture content: (oven-dry: 0.0%, and air-dry: 0.8%) and temperature: **I**, m-xylene; **II**, n-dodecane; **III**, n-decane; **IV**, ps-cumene; **V**, t-butylbenzene (after Acher et al. 1987)

The desorption of kerosene components from soil surfaces follows the vapor pressure of each component and is affected by the type of soil. Figure 8 (after Yaron et al. 1988) shows the rate of desorption of the five kerosene components studied from the soil surfaces into the gas phase. The highest rate of desorption occurs with m-xylene and the lowest with n-dodecane. For all the components studied, the retention capacities of the soil are of the order Oxford > Gilat > Bet Dagan. The residual amount of hydrocarbons remaining adsorbed on the soil surface 30 days after the soil was polluted is in the following order: n-dodecane > n-butylbenzene > P-cumene > n-decane > m-xylene. Significant amounts of hydrocarbon residues were found in the Oxford soil.

10.4 Surface Degradation

The surface catalytic properties of the porous media are of certain importance in dealing with the fate of hydrocarbons in the unsaturated zone. Theng (1974) has shown that the catalytic activity of clays (which are probably dominant among the minerals present in the unsaturated zone in inducing surface reactions) is often correlated with their acid strength and their ability to act as proton donors

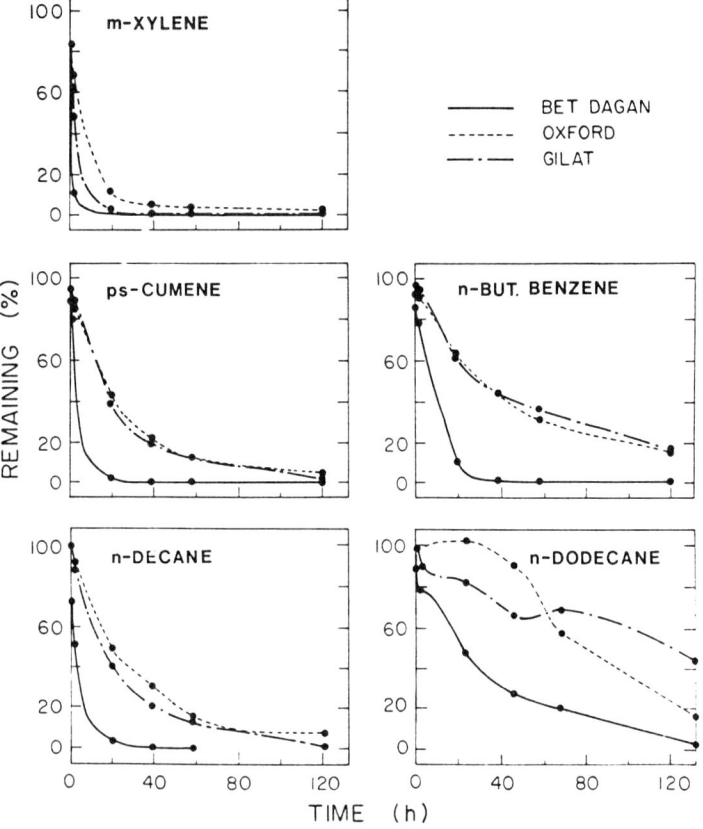

Fig. 8. Rate of desorption of kerosene components into the atmosphere from the surfaces of three oven-dried soils: Bet Dagan (clay 12%, OM 0.5%), Gilat (clay 16%, OM 0.6%), and Oxford (clay 35%, OM 7.2%) (after Yaron et al. 1988)

(Bronsted acids) or electron acceptors (Lewis acids). The first type of acidity is strongly affected by the moisture content and the nature of the compensating cation. At a low moisture content – which characterizes the unsaturated zone – the pH near the surface of the dominant clay minerals may be several units lower than in the bulk solution. There is little information about the conversion of petroleum hydrocarbons on clay surfaces, but the existing knowledge on the behavior of organic molecules from other origins may be used (Mortland 1970; Mingelgrin et al. 1977; Gerstl and Yaron 1981; Saltzman and Mingelgrin 1984). Studies of montmorillonite-petroleum heavy fractions complex formations (Clementz 1976) for example, point out the effect of hydrocarbons on clay structure, but there is no information on the influence of the adsorbent on the adsorbed molecules. It should be pointed out that almost all the information existing on surface interaction with soil materials was obtained from experiments on pure clays.

Despite the fact that the biologically induced degradation of petroleum products in the unsaturated zone is the major degradation process, knowledge of the surface-induced catalysis of the hydrocarbons by soil materials would be useful. The prevailing "dry" conditions which are often characteristic of the unsaturated zone may drastically decrease the biological activity. Under such conditions, surface catalysis of hydrocarbons may become the major decontaminating process and contribute to the renovation of the unsaturated zone.

10.5 Multiphase Transport

Abiotic processes discussed previously, such as dissolution, volatilization and adsorption-desorption, affect multiphase transport and should be considered when dealing with this important problem.

Schwille (1984) reviewed the existing knowledge on migration of organic fluids immiscible with water. He defined the problem of petroleum transport in the unsaturated zone based on the work of Leverett and Lewiss (1941); van Dam (1967); Fried et al. (1979) and on his own research (Schwille 1967, 1975, 1981). Immiscible phase flow was considered to be dependent on the type and composition of the rock formation and on the nature of the fluid, and the physical aspects of its migration were described. Recently, a multiphase approach to modeling the migration of organic contaminants — mainly petroleum products — in porous media, with emphasis on the pollution of ground water was developed (Abriola and Pinder 1985a,b; Pinder and Abriola 1986; Baehr and Corapcioglu 1987; Corapcioglu and Baehr 1987). In these studies the pollution of porous media was considered to entail simultaneous transport of hydrocarbons as unaltered immiscible liquids, as solutes in soil water and as vapor in the air phase. It was also considered that "the chemical composition of the oil phase will differ from the petroleum product introduced to the ground due to the differing susceptibilities of the constituents — e.g. benzene, toluene, xylene, octane and cyclohexane in gasoline — to various phase transfer and chemical transformations" (Corapcioglu and Baehr 1987). This conceptualization of the process was used only for simulation purposes and no experimental results were presented for its support. Recently, Karimi et al. (1987) published data on vapor phase diffusion of benzene in soils, showing that the volatilization flux was greatly reduced by increased soil bulk density and soil water content (Fig. 9).

In a series of experiments carried out in our laboratory to examine the pollution of soil columns by a mixture of hydrocarbons simulating kerosene (Acher et al. 1987), it was observed that during the liquid-kerosene infiltration there is a simultaneous but faster vapor movement producing a penetration front in advance of the liquid front. Hydrocarbon transport in the gaseous phase was two or three times greater than the penetration of the immiscible liquid. Since the chemical composition of the gas phase is different from that of the original 'kerosene' introduced into the soil column (Fig. 4) the pollution

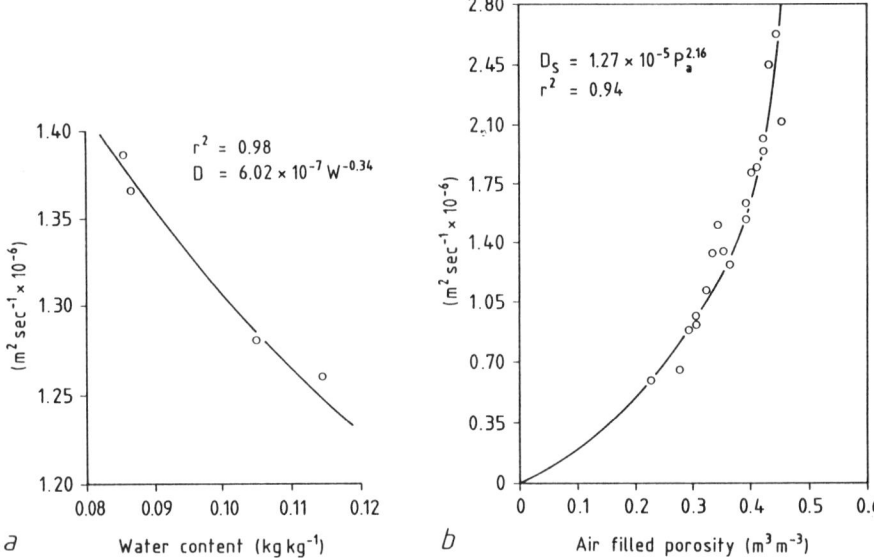

Fig. 9. Effect of soil water (**a**) and air-filled porosity (**b**) on the specific benzene apparent diffusion through a soil layer (after Karimi et al. 1987)

caused by the gas phase will be different from that caused by the immiscible liquid phase.

The initial soil moisture content affects the penetration of hydrocarbons both as an immiscible liquid and as vapor in the gas phase. Figure 10 (Acher et al. 1987) shows the effect of soil moisture on the advancement of a 'kerosone' liquid front in a sandy loam soil. From the slope of the infiltration curves it appears that in the air-dried soil the infiltration rate is 1.5 times faster than in the oven-dried soil. Transport in the gas phase is also affected by the soil moisture content. In oven-dry soil the application of 0.5 cc kerosene to the top of the soil column induced the penetration of vapor to a depth of 32 cm, whereas penetration in an air-dried column (0.8% moisture content) was only 25 mm. Figure 11 (Acher et al. 1987) shows the redistribution over a 46-day period in soil columns of m-xylene and n-dodecane, the most and least volatile 'kerosene' constituents. The redistribution of n-decane is very similar to that of m-xylene, while that of t-butylbenzene and of pseudo-cumene is between that depicted by a and b in Figure 11.

The soil used in column experiments had three different moisture levels: 0.0%, 0.8%, and 12%. The hydrocarbon measurements are plotted as relative values of each hydrocarbon concentration, C/Co – where C is the concentration in each soil layer and Co is the total hydrocarbon concentration in the column – versus the corresponding column depth (cm). In the soil column wetted to field capacity, both aliphatic and aromatic hydrocarbons remained in the upper 200-mm of the column and their redistribution with time did not exceed this depth. In another series of column experiments it could be seen that the gas phase movement preceding the liquid phase front was prevented when the soil moisture

On the Behavior of Petroleum Hydrocarbons in the Unsaturated Zone 227

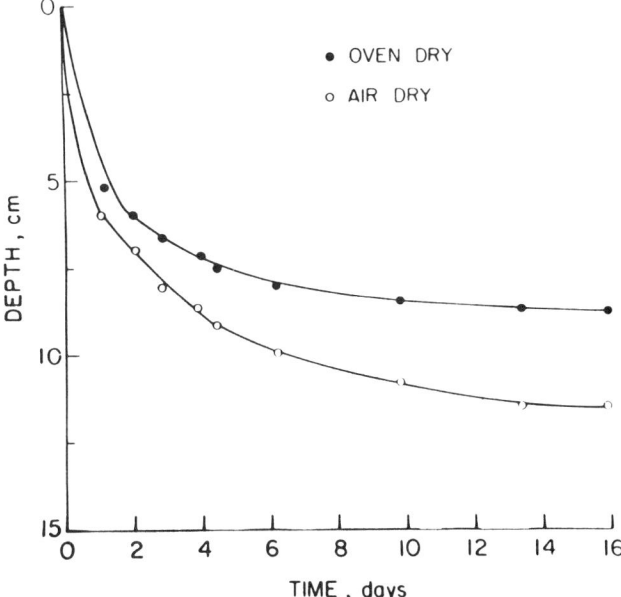

Fig. 10. Advancement of simulated liquid 'kerosene' in a Bet Dagan soil column as affected by initial soil moisture content: oven-dry – 0.0%, and air-dry 0.8% (after Acher et al. 1987)

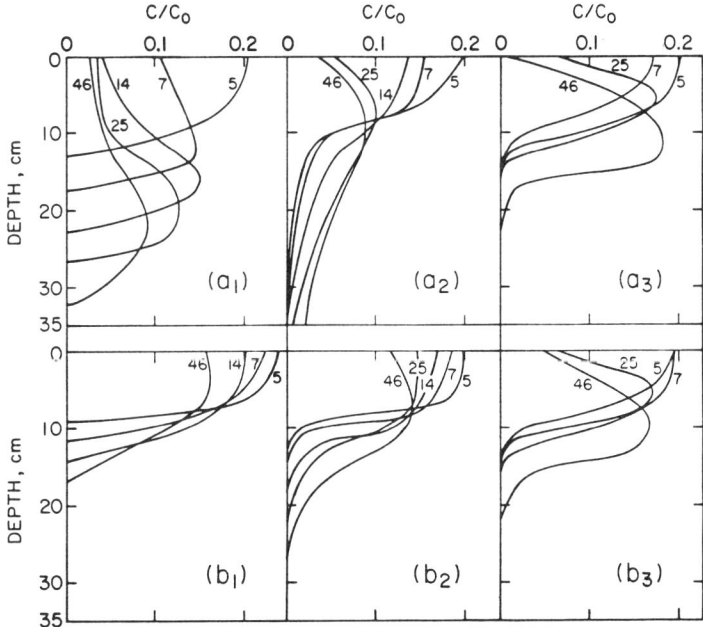

Fig. 11. Redistribution of (**a**) m-xylene and (**b**) n-dodecane from a simulated liquid 'kerosene' in Bet Dagan soil columns with different initial moisture contents: 1–0.0%; 2–0.8%; and 3–12.0%. Numbers on curves represent days after 'kerosene' application to the top of the soil column (after Acher et al. 1987)

content was equal to or more than 4.0%. These results are similar to those of Barbee and Brown (1986), who reported that when xylenes were applied to a sandy loam with a moisture content close to field capacity, the hydrocarbons were retained by the soil and their movement significantly attenuated by the presence of water which clogged the soil pores.

The movement of 'kerosene' in oven- and air-dried soil columns is different. It can be seen that, independent of their properties, the 'kerosene' components move faster and deeper in air-dried than in oven-dried soil columns. This behavior agrees with the adsorption pattern already described above. The extent of redistribution with depth is affected by the hydrocarbon properties. In the case of oven-dried soils, the penetration of the 'kerosene' products follows the order: m-xylene $>$ n-decane $>$ pseudo-cumene $>$ t-butylbenzene $>$ n-dodecane. This behavior is in accordance with the vapor pressure, the vapor concentration, and the adsorption order of the 'kerosene' constituents. In the case of the columns packed with air-dried soil, the redistribution of the "kerosene" components is only slightly affected by their volatility.

The experimental results presented here confirm the concepts adopted by Pinder and Abriola (1986) and Corapcioglu and Baehr (1987) in their modeling of porous media and ground water contamination of the petroleum products.

10.6 Conclusions

Petroleum products, being a complex mixture of both aliphatic and aromatic hydrocarbons, behave in porous media differently from their individual components.

The behavior of petroleum hydrocarbons in the unsaturated zone is affected by a number of abiotic processes, such as: volatilization into the air phase, distribution in the water phase, adsorption-desorption on and from the solid phase, surface-induced degradation, and multiphase transport. The abiotic processes determine a change in the chemical and physical properties of the petroleum product introduced onto the land surface. Both the properties of the unsaturated zone (e.g. constituents of the solid phase, chemical composition of the soil water, etc.) and environmental conditions (e.g. temperature, moisture content, etc.) govern the rate of abiotic processes which affect the fate of petroleum products.

Acknowledgments. This study was supported in part by research grant 5-396-4323 from the Israel National Council for Research and Development and the German Bundesministerium fur Forschung und Technologie.

References

Abriola LM, Pinder GF (1985a) A multiphase approach to the modeling of porous media contamination by organic compounds. 1. Equation development. Water Res Res 21:11–18

Abriola LM, Pinder GF (1985b) A multiphase approach to the modeling of porous media contamination by organic compounds. 2. Numerical simulation. Water Res Res 21:19–26

Acher AJ, Boderie P, Yaron B (1987) Soil and ground water pollution by petroleum products. I. Multiphase migration of kerosene components in soil columns. ARO-Scientific Report to NCDR, 34 pp

Baehr AL, Corapcioglu MY (1987) A compositional multiphase model for groundwater contamination by petroleum products. 2. Numerical solution. Water Res Res 33:201–213

Barbee GC, Brown KW (1986) Movement of xylene through unsaturated soils following simulated spills. Water Air Soil Pollut 29:321–333

Boehm PD, Quinn G (1973) Solubilization of hydrocarbons by the dissolved organic matter in seawater. Geochim Cosmochim Acta 37:2459–2477

Boer JH de, Zwilker C (1929) Adsorption as a result of polarization. Z Physik Chem B3:407–418

Clark RC, Brown DW (1977) Petroleum – properties and analysis in biotic and abiotic systems. In: Mailins DC (ed) Effects of Petroleum in Arctic and Sub-Arctic Marine Environments and Organisms. Academic Press, London, pp 1–89

Clark RC, MacLeod WD (1977) Inputs, transport mechanisms and observed concentrations of petroleum in marine environment. In: Mailins DC (ed) Effects of Petroleum in Arctic and Sub-Arctic Marine Environments and Organisms. Academic Press, London, pp 91–223

Clementz DM (1976) Interactions of petroleum heavy leads with montmorillonite. Clay Clay Min 24:312–319

Corapcioglu MY, Baehr AL (1987) A compositional multiphase model for ground water contamination by petroleum products. 1. Theoretical considerations. Water Res Res 23:191–200

Curtis PG, Robert PV, Reinhard M (1986) A natural gradient experiment on solute transport in sand aquifer 4. Sorption of organic solutes and its influence on mobility. Water Res Res 22:2059–2067

Dam J van (1967) The migration of hydrocarbons in a water-bearing stratum. In: Hepple P (ed) The Joint Problems of the Oil and Water Industries. Inst of Petroleum, London, pp 55–96

Eganhause RP, Calder JA (1974) The solubility of medium molecular weight aromatic hydrocarbons and the effects of hydrocarbon co-solutes and salinity. Geochim Cosmochim Acta 40:555–561

Eltantwy IN, Arnold PW (1972) Adsorption of n-alkanes on Wyoming montmorillonite. Nature Phys Sci 225–237

Fried JJ, Muntzer P, Zilliox L (1979) Ground water pollution by transfer of hydrocarbons. Ground Water V 17:586–594

Fripiat JJ (1968) Surface soils and transformation of adsorbed molecules in soil colloids. Trans 9th Int Congr on Soils Science (Adelaide, S, Australia), Vol 1, pp 679–683

Gerstl Z, Yaron B (1981) Attapulgite-pesticide interactions. Residue Rev 78:69–99

Jordan ER, Payne JR (1980) Fate and weathering of petroleum spills in the marine environment. Ann Arbor Science Publ Ann Arbor, Michigan, 174 pp

Karimi AA, Farmer JJ, Cliath MM (1987) Vapor phase diffusion of benzene in soil. J Environ Qual 16:38–44

Leverett MC, Lewiss WB (1941) Steady flow of gas-oil-water mixture through unconsolidated sands. Trans Aime 142:107

McAuliffe C (1966) Solubility in water of paraffin, olefin, acetylene, and aromatic hydrocarbons. J Phys Chem 70:1267–1275

McAucliffe C (1977a) Dispersal and alteration of oil discharges on a water surface. In: Wolfe DA (ed) Fate and Effects of Petroleum Hydrocarbons in Marine Organisms and Ecosystems. Pergamon, Oxford, pp 19–35

McAuliffe C (1977b) Evaporation and solution of C_2 to C_{10} hydrocarbons from crude oil on the sea surface. In: Wolfe DA (ed) Fate and Effects of Petroleum Hydrocarbons in Marine Organisms and Ecosystems. Pergamon, Oxford, pp 115–132

Mingelgrin U, Saltzman S, Yaron B (1977) A possible model for the surface-induced hydrolysis of organophosphate pesticides on kaolinite clays. Soil Sci Soc Am J 41:519–523

Mortland MM (1970) Clay organic complexes and interactions. Adv Agron 22:75–114

Nathwani JS, Philips CR (1977) Adsorption-desorption of selected hydrocarbons in crude oil and soils. Chemosphere 4:157–162

Pinder GF, Abriola LM (1986) On the simulation of non-aqueous phase organic compounds in the subsurface. Water Res Res 22:109–119

Pinnavaia TJ, Mortland MM (1971) Interlamellar metal complexes on layer silicates. J Phys Chem 75:3975–3962

Rostad H (1976) Behavior of oil spills with emphasis on the North Sea. Report of the Continental Shelf Inst, Trondheim, Norway

Saltzman S, Mingelgrin U (1984) Non-biological degradation of pesticides in the unsaturated zone. In: Yaron et al. (Eds) Pollutants in Porous Media. Springer, Berlin Heidelberg New York, pp 153–161

Schwille F (1967) Petroleum contamination of the subsoil – a hydrological problem. In: Heple P (Ed) The Problem of Oil and of Oil and Water Industries. Elsevier, Amsterdam pp 23–54

Schwille F (1975) Ground water pollution by mineral oil products IAHS-AISH Publ 103:226–240

Schwille F (1981) Ground water pollution in porous media by fluid immiscible with water. Sci Technol Environ 21:173–185

Schwille F (1984) Migration of organic fluids immiscible with water in the unsaturated zone. In: Yaron B et al. (Eds) Pollutants in Porous Media. Springer, Berlin Heidelberg New York, pp 27–47

Sutton C, Calder JA (1974) Solubility of higher molecular weight n-paraffins in distilled water and seawater. Environ Sci Technol 8:654–667

Sutton C, Calder JA (1975) Solubility of alkylbenzenes in distilled water and seawater at 25°C. J Chem Eng Data 20:320–322

Theng BGG (1974) The Chemistry of Clay Organic Residues. Hilger, London 343 pp

Wolfe DA (Ed) (1977) Fate and Effects of Petroleum Hydrocarbons in Marine Organisms and Ecosystems. Pergamon, Oxford

Yaron B, Sutherland PM, Galin Z, Acher AJ (1988) Soil and ground water pollution of petroleum products. II. Sorption-desorption of kerosene vapors as affected by tape of soils and humidity. ARO-Scientific Report to NCDR, 45 pp

11 Transport of Organic Pollutants in a Multiphase System

H. RUBIN and E. MECHREZ[1]

11.1 Introduction

This manuscript concerns the nature and characteristics typical of liquid hydrocarbon (LH) contamination in the saturated as well as the unsaturated zones. LH may be introduced into the soil due to an accident or improper performance of LH carrying installations. The movement of the LH must be quantified in order to assess its impact on the soil and the subsurface water. Most types of LH are immiscible contaminants, namely liquids which at environmental pressures and temperatures exhibit limited solubility with available water.

Fuels like gasoline and kerosene are heterogeneous contaminants being solutions of various types of LH. According to a literature survey carried out by Corapcioglu and Baehr (1985) hydrocarbons, halogenated hydrocarbons and pesticides pose a major threat to groundwater resources in the United States. Probably such conditions exist in all modern industrialized societies.

The multiphase system to which this manuscript refers includes the following two multiphase subsystems, the stationary solid skeleton system and the mobile fluid system. The multiphase characteristics of the solid skeleton of the porous medium represent the variability and heterogeneity of the soil. The fluid multiphase subsystem includes LH, water and air. Although the LH is usually assumed to be an immiscible fluid, sometimes significant volumes of this material may be introduced into the water phase due to solubility of light fractions as well as due to emulsification processes. Some volatile fractions of LH may be transferred into the gaseous phase.

11.2 Classification of LH According to Contamination Criteria

There are several hundred different types of LH compounds. Generally, these compounds are classified according to the following four basic types of LH: (a) paraffins (alkanes), (b) olefins (alkenes), (c) naphthenes, and (d) aromatics. Some examples of these different types of LH are represented in Fig. 1.

---

[1] Department of Civil Engineering, CAMERI-Coastal and Marine Engineering Research Institute, Haifa, Technion City 32000, Israel.

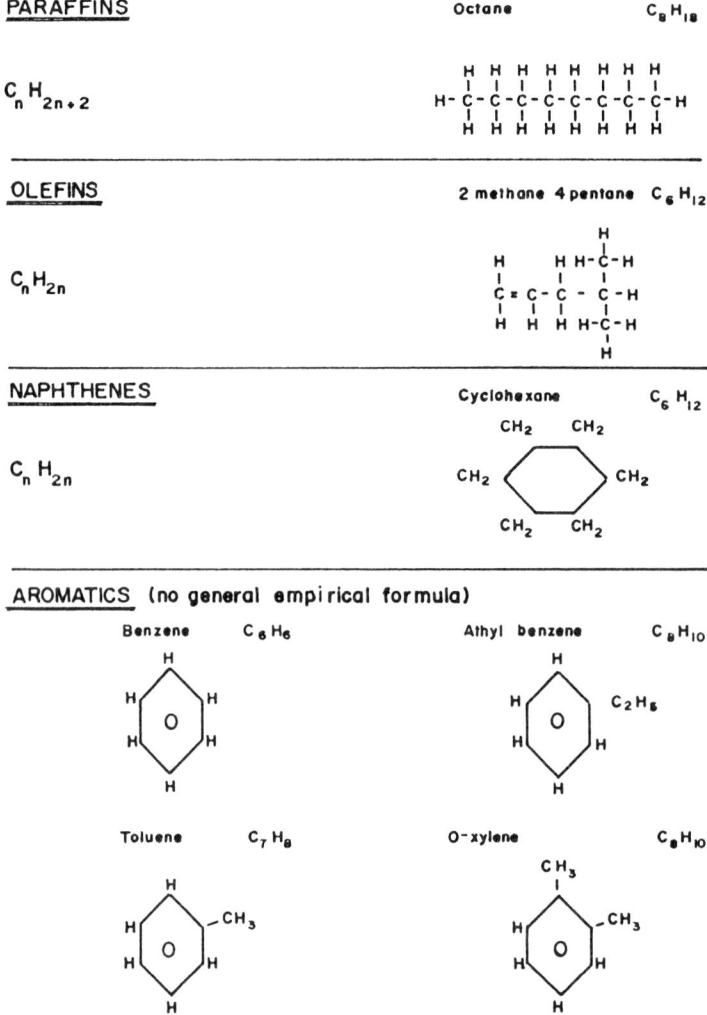

Fig. 1. Examples of the various types of LH

LH compounds are products obtained in the petroleum refinery. Fig. 2 describes a simple refinery (Wittcoff and Reuben, 1980) in which crude oil is separated by distillation into various fractions. The first most volatile fraction obtained by the distillation process consists of methane and higher alkanes. Such gases are not included in the framework of this discussion. The second and third fractions are mixtures of LH being major constituents of gasoline. Such constituents are usually termed as "light distillates". Some of them are proved to be hazardous. Benzene, for example has been determined to be a human carcinogen by the National Cancer Institute of the United States. Burmaster and Harris (1982)

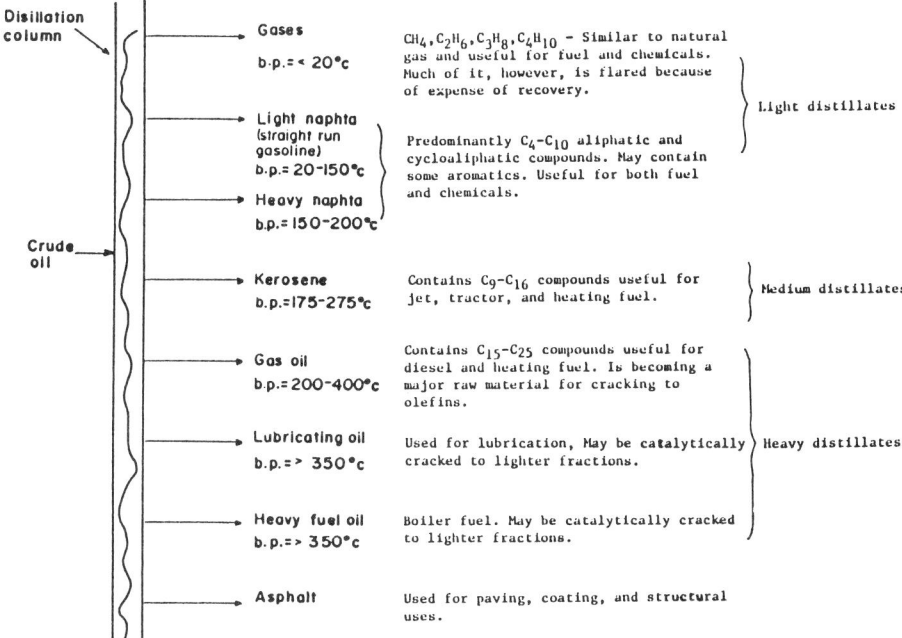

Fig. 2. Crude oil distillation

reported increased incidence of spontaneous abortions and prolonged menstrual bleeding attributed to occupational exposure to benzene, toluene and xylene. These compounds are also designated as hazardous by the Toxic Substance Control Act of the United States. Furthermore, according to McKee et al. (1972) taste and odor of LH can be detected in concentrations as low as 0.005 mg/l. Due to such effects and the wide spread of car engines and gasoline fuel stations, soil and groundwater contamination due to light distillates is reported in many case history studies. Studies of this type have been reviewed by Corapcioglu and Baehr (1985). Referring to the contamination potential of light distillates we have also to consider the possible negative effects of materials such as lead being added to gasoline in order to improve its performance. Light distillates have comparatively low viscosity and high volatility potential. These properties reduce the contamination potential of such distillates. Because of the low viscosity, the residual immobile saturation of light distillates is comparatively small. Due to their high tendency to volatilize their persistance is low, and treatment of the water contaminated by light distillates is comparatively simple.

Returning to the petroleum refining process described in Fig. 2, kerosene represents the fourth fraction of this process. This fuel is used for some tractors, jet aircrafts and for domestic heating. Kerosenes are termed as "medium distillates". Usually the contamination potential of such distillates is assumed to be the highest among LH compounds, due to their effect on people as mentioned with regard to light distillates as well as their average viscosity and low volatility leading to high persistance of these distillates. Groundwater and soil contamination by medium distillates are typical to phreatic aquifers located underneath airports and airbases.

The fifth, sixth and seventh fractions of the refining process are termed "heavy distillates". Although such distillates may sometimes include poisonous compounds, usually their contamination potential is assumed to be comparatively small. Due to their high viscosity they migrate very slowly through the unsaturated zone, and their immobile degree of saturation is comparatively high (Dietz 1979).

Summarizing this section it is possible to classify types of LH according to their environmental contamination potential into the following three different categories: (a) light distillates, (b) medium distillates, and (c) heavy distillates. Medium distillates usually represent LH compounds whose contamination potential is the highest among these three categories.

11.3 Introduction of the Pollutant into the Subsurface Water

The LH may be spilled on the ground surface or at a certain shallow depth of the soil due to an accident or improper performance of an LH carrying installation. Improper performance also includes spillage of LH due to improper codes administrating working procedures.

The LH spill migrates downwards and also spreads horizontally to a limited extent in the unsaturated zone causing contamination of the soil and its environment. Schwille (1975, 1984) described some possible characteristics of LH migration in various types of porous media. He applied some laboratory models, showing that kerosene migration in fissures is extremely different from its migration through sandy soils. Therefore, we may conclude that the heterogeniety of the solid skeleton of the porous medium considerably affects the LH modes of migration in the unsaturated zone. The residual water content in this zone as well as leaching of LH residuals by infiltrating water are very important when considering pollution and restoration of the contaminated soil.

When significant volumes of LH are introduced to the subsurface, the body of the spilled LH penetrates the whole thickness of the unsaturated zone and arrives at the capillary zone of the phreatic aquifer. Here the LH spill is introduced into the water saturated zone. Different authors described differently the introduction of LH into the saturated zone. However it seems that all such differences can be resolved provided that appropriate definitions are applied, and proper physical parameters are used.

According to various texts (e.g. Todd 1980) the capillary fringe is a zone located above the water table through which the saturation decreases along an upward vertical coordinate. According to this description the top of the capillary fringe represents the boundary above which there is no longer any capillary water. However, such a definition can hardly be accurate because of the difficulty to identify the capillary rise in the finest pores of a well graded soil. When considering the LH spill horizontal spreading, major effects are attributed to the lower part of the capillary fringe being saturated with water. This zone is termed in this manuscript as the "saturated capillary zone". It eventually represents the capillary rise in the largest pores of the porous medium. It was found in laboratory experiments (Scheigg 1979, 1980; Mechrez et al. 1986) that in well graded sandy soils the thickness of the saturated capillary zone as explained above is of the order of magnitude of several decimeters. Visual observation cannot indicate what are the real sizes of the saturated capillary zone or the whole capillary fringe. Such observations in laboratory physical models provide information about the location of the groundwater table, where the pressure is equal to the atmospheric one. Above the water table it is possible to observe the size of the saturated capillary zone and a certain wet zone of the soil as shown in Fig. 3. Provided that the LH rate of spillage is smaller than the hydraulic conductivity of saturation, experiments in physical models (Mechrez et al. 1986) indicated that the LH spill

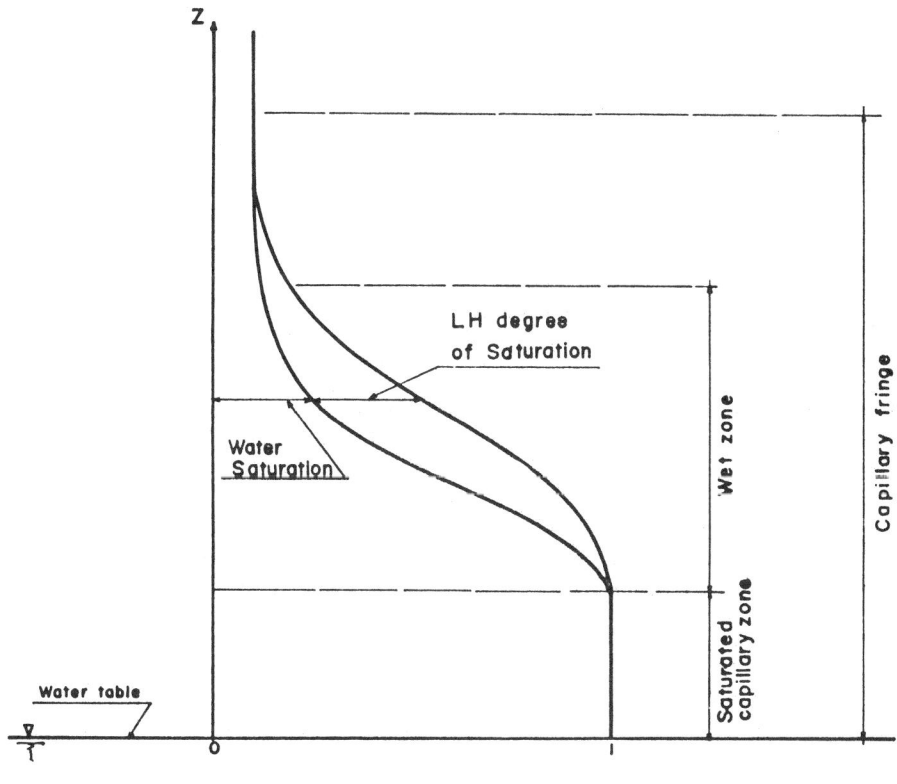

Fig. 3. The capillary fringe, the saturated capillary zone and the LH layer

does not penetrate the saturated capillary zone. It spreads horizontally in a layer of finite thickness whose LH degree of saturation varies as shown in Fig. 3. In the small scale model no LH penetration was observed into the main water body of the aquifer existing below the groundwater table. The saturated capillary zone of about 40cm was an "insulator" between the LH spill and the main water body of the aquifer. Previous studies (Kroszynski 1974) indicated that the water flow in the saturated capillary zone is much slower than that of the main water body although the slope of the top of the saturated capillary zone is almost identical to that of the groundwater table. The slope of the saturated capillary zone and the variability of the LH saturation along the horizontal coordinate provide the potentiometric head gradient leading to the LH spreading. As shown in Fig. 3 the LH degree of saturation varies along the vertical coordinate. Therefore, the triangle of permeabilities shown in Fig. 4 (Leverett and Lewis 1941; Van Dam 1967) indicates that the rate of spreading of the center of gravity of the LH layer should be faster than the rate of spreading of the upper and lower parts of this layer. Eventually the top and bottom of the LH layer are almost immobile. After a long time interval a limited penetration of some LH fingers into the saturated capillary zone can be observed. This phenomenon may be caused by the reduction in the effective surface tension due to the presence of LH. The laboratory experiments mentioned above cannot simply be extrapolated for the description of LH spreading under field conditions. However, provided that the rate of spillage of LH is small, it is possible that the LH would not penetrate the saturated capillary zone, and eventually would not contaminate the main water body of the aquifer. Anyhow many authors referring to ground water contamination by LH claim that due to the LH spillage a lense or pancake of this material is formed on top of the groundwater table as shown in Fig. 5. This figure

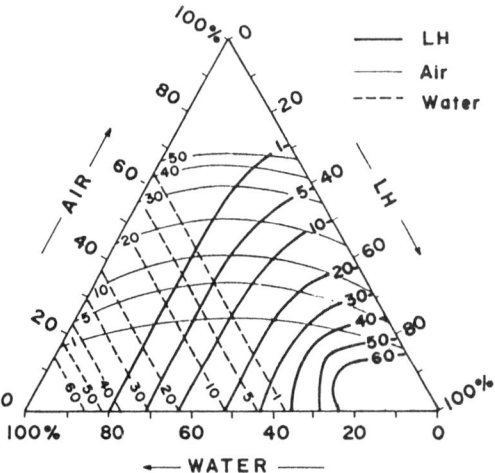

Fig. 4. Relative permeabilities for three phase flow

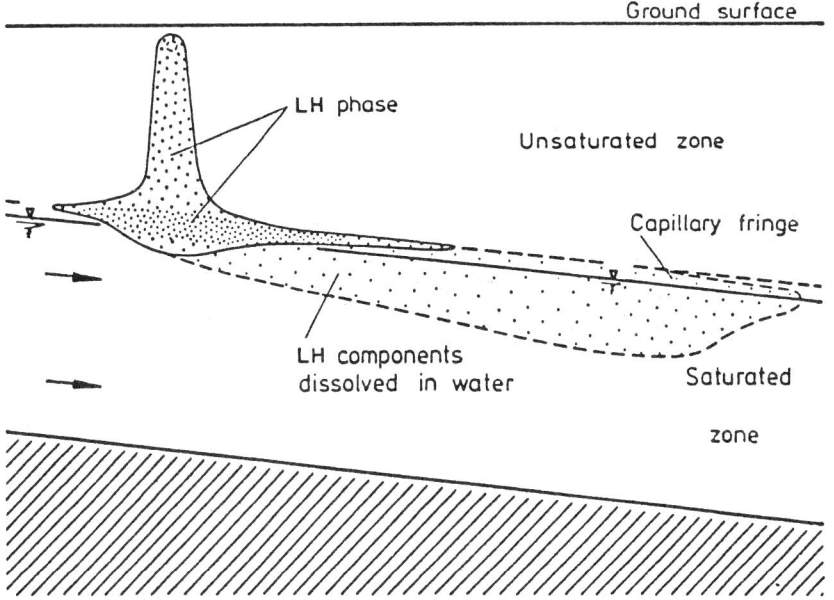

Fig. 5. The classical description of the LH migration in the subsurface

is typical of many reports concerning the contaminations of groundwater by LH. This description, originally presented by Schwille (1975), eventually became a "classical description" of LH spill migration in the subsurface. However, as implied by the next paragraph the classical description may prevail in field conditions only under particular circumstances. It is obvious that if the LH volume is smaller than the residual LH saturation of the soil layer representing the unsaturated zone then the LH would not arrive at the aquifer. However if the LH volume is large, various physical parameters determine whether the LH spill penetrates the groundwater table and creates an LH lense in the main water body of the aquifer. The experiments of Mechrez et al. (1986) imply that the rate of release of the LH from the LH carrying installation is very important in such considerations. We have also to recall that Fig. 5 refers to stationary water table, and no groundwater pumpage is performed. Eventually the groundwater table as well as the location of the saturated capillary zone are subject to annual and seasonal fluctuations. Such ups and downs of the groundwater table may be much larger than the thickness of the LH layer floating above the saturated capillary zone or the LH lense which may be formed in some cases. If the aquifer is a source of potable water then some wells are probably utilized in order to pump water from that aquifer. Pumpage is involved with the water table drawdown. Such a drawdown significantly increases local gradients of the potentiometric head, slopes of the groundwater table and slopes of the saturated capillary zone. Decreases and increases in rates of pumpage may cause large local fluctuations in the physical variables mentioned above. In the following paragraph we consider possible effects of the groundwater table fluctuations on the LH contamination process. Mechrez et al. (1986) performed a series of experiments with a physical model simulating all the phenomena mentioned above on a small laboratory scale. They found that drawdown of the water table leads to downward migration of the saturated capillary zone followed by downward migration of the LH layer. Upward migration of the water table hardly induces any movement of the LH layer, it causes blockage of some LH quantities in some pores of the porous medium due to Jamin effect (Jamin 1860). Some LH quantities are subject to emulsification due to the upward migration of the groundwater table. This is probably a major mechanism of LH introduction into the main water body of the aquifer in semi-arid regions. This mechanism of LH introduction into the main water body of the aquifer was eventually observed in various experimental studies, many of them reviewed by Wooding and Morel-Seytoux (1976). Such studies usually refer to oil production and enhanced oil recovery from oil reservoirs. Mechrez et al. (1986) also simulated the effect of groundwater pumpage on the LH layer floating above the saturated capillary zone. The groundwater pumpage as shown in Fig. 6 caused local depression of the water table and the saturated capillary zone, leading to certain flow of the LH layer towards the pumping well. The final result was an accumulation of the LH spill in a cone surrounding the well. This effect can be termed as "downconing of the LH layer". Under such conditions, by stopping the well pumpage, large quantities of the LH cone are introduced into the main water body of the aquifer, as the groundwater table moves upwards. Improper pumpage procedures may therefore lead to intensive contamination of the groundwater aquifer.

11.4 Migration of LH in the Unsaturated Zone

In the unsaturated zone the presence of three fluid phases should be considered. The fluids are LH, air and water being in mobile or immobile states. The fluids characteristics affecting the flow field include densities, viscosities and interfacial tension between parts of fluids. The porous medium characteristics affecting the flow field stem from its pore size distribution being dependent on the materials comprising the porous medium as well as on their geochemical nature. Geochemical and biochemical activities may lead some properties of the fluids and the porous medium to be time dependent.

Wooding and Morel-Seytoux (1976) developed some dimensionless parameters determining the characteristics of the flow field. Their discussion refers to cases in which the porous medium properties are introduced by a single physical parameter being the characteristic pore size of the medium, which has a constant value. However, even sandy soils show significant time variability of their hydraulic conductivity in certain cases. Kessler and Rubin (1985a,b) measured the hydraulic conductivity of soil consisting of 99% sand particles. They found that the hydraulic conductivity of that soil to water decreases with time,

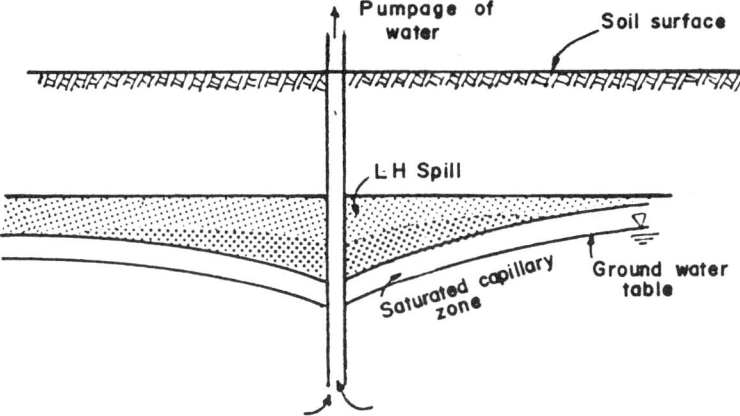

Fig. 6. Downconing of the LH spill due to groundwater pumpage

whereas its hydraulic conductivity to LH increases with time. Probably the 1% of silt and clay particles made these differences. Such particles absorb water and do not absorb LH. Therefore, in certain cases the hydraulic conductivity of soils to LH may be higher than that of these soils to water, even though the LH viscosity is higher than that of water.

Generally, it is possible to identify short time processes and long time processes which should be taken into account whenever LH contamination in the unsaturated zone is considered. The practical implication very often determines which processes should be taken into account. This statement can be represented by some examples. If the rate and amount of vertical migration of the LH spill, which are short time processes, are considered, then the general theory of water infiltration can be useful, and water infiltration data can often be converted in order to quantify the downward migration of the LH spill (Kessler and Rubin 1985a,b; 1987). However, when referring to persistence of the LH in the contaminated soil, which is a long term process then transfer of some constituents between the different phases should be taken into account. We also have to refer to volatility of some constituents, biodegradation, some geochemical reactions etc.

When referring to short time processes the way of introduction of the different fluid phases as well as their initial degree of saturations are of major importance. To exemplify this statement, flow induced by the penetration of LH spill into soil being subject to residual water saturation is involved with effects substantially different from those of flow induced by the penetration of water into soil being subject to residual LH saturation. With regard to the effect of the degree of saturation of the different fluid phases we refer to the three phase permeabilities diagram of Fig. 4. It indicates that only for a narrow range of saturations flow of all three phases is possible. In most practical cases of LH spill penetration into the unsaturated zone the LH is the main phase being subject to flow. In most practical cases of water penetrating into soil being subject to residual saturation of LH, the main flowing phase is water. However the effect of the stagnant fluid phases on the flowing phase should be taken into account. Possible transfer of constituents also should be considered. Such considerations are important with regard to groundwater quality.

In some cases, flow of LH in the unsaturated zone can be induced due to natural or artificial conditions imposed on the main water body of the aquifer. The slope of the groundwater table has a major effect on the horizontal spreading of the LH spill (Scheigg 1979; Mechrez et al. 1986).

11.5 Migration of LH in the Saturated Zone

This section refers to migration of the LH quantities after their introduction into the main water body of the aquifer. Due to the LH introduction mechanisms, LH may be present either as small droplets of emulsion or as solutions of some miscible fractions. Some quantities of LH being blocked in some pores of the porous medium are immobile (Pfannkuch 1984; Wilson and Conard 1984; Conard and Wilson 1986). However, they may gradually release some soluble quantities into the flowing groundwater. With regard to the flow of the LH emulsion, some mechanisms typical to filter flows may lead to transfer of LH quantities into small droplets, large droplets or a continuous LH phase (Spielman and Goren 1970, 1972a,b; Spielman 1977; Spielman and Su 1977; Soo and Radke 1984). Such phenomena should be considered besides the transport of small droplets and miscible fractions of LH due to convection, diffusion and dispersion.

11.6 Modelling Procedures

Various approaches have been developed for the simulation of LH migration in soils and subsurface water. It is possible to classify these approaches into three categories as follows: (a) gross control volumes, (b) differential control volumes referring to the whole flow field, and (c) differential control volumes referring to various parts of the flow field.

11.6.1 Gross Control Volumes

This approach is aimed at developing simple tools for field applications. Dietz (1979) claimed that by semi-quantitative methods based on gross control volumes it is possible to get quick answers, which are usually accurate enough for practical problems like the determination of the depth of LH penetration into the soil, possible shape of the LH lense which may develop on top of the aquifer etc.

Transport of Organic Pollutants in a Multiphase System   241

According to Dietz's method all calculations are based on some simple measurements and assumptions with regard to porous spaces available for LH flow and residual saturation of this fluid. Such measurements and assumptions are generalized into characteristics of large control volumes of the flow field. In certain practical cases the semi-quantitative method suggested by Dietz (1979) can be utilized for the design of some remedial measures needed in cases of LH contamination. However it cannot be applied whenever estimates of time variations of the LH contamination are needed.

Corapcioglu and Baehr (1985) utilized the single cell model (Mercado 1976) in order to get a conservative estimate of the rate at which solubilized fractions of an immobilized immiscible contaminant in the unsaturated zone is leached down and may enter an underlying aquifer. The single cell model can be obtained either by reference to a large control volume or by linearization of the basic conservation equations, and performance of integration with regard to the vertical coordinate and the surfaces bounding the control volume relevant to the contamination process.

11.6.2 Differential Control Volumes Referring to the Whole Flow Field

This approach is coherent with the usually efficient utilization of numerical modelling systems. Basically, the flow field phenomena are determined by the conservation of mass, momentum and energy as well as by the boundary conditions typical to the specific contamination event.

The conservation of mass for each substance present in the flow field yields the following equation relevant to the differential control volume ΔV whose surface is ΔS:

$$\begin{bmatrix} \text{rate of growth in} \\ \text{storage of sub-} \\ \text{stance in } \Delta V \end{bmatrix} = \begin{bmatrix} \text{net flux of sub-} \\ \text{stance into } \Delta V \\ \text{through } \Delta S \end{bmatrix} + \begin{bmatrix} \text{rate of production and destruction of substance} \\ \text{within } \Delta V \text{ (sources} \\ \text{and sinks)} \end{bmatrix} \quad (1)$$

This expression yields several basic equations whose total number is equal to the number of substances being present in the flow field.

The conservation of momentum yields for each fluid phase the following flow equation

$$q_i = - \frac{\rho_i g}{\mu_i} K_{r_i} K_o \nabla \phi_i \quad (2)$$

Where i is a subscript referring to the phase i, ρ_i is the fluid density; g is the gravitational acceleration; μ_i the fluid viscosity; K_r is the relative permeability; K_o is the intrinsic permeability; ϕ_i is the potentiometric head.

The conservation of energy for the differential volume ΔV whose surface is ΔS yields the following expression

$$\begin{bmatrix} \text{rate of change in storage of} \\ \text{thermal energy of all substances and phases in } \Delta V \end{bmatrix} = \begin{bmatrix} \text{net flux of thermal energy by} \\ \text{the flow of the fluid phases} \\ \text{into } \Delta V \text{ through the surface} \\ \Delta S \end{bmatrix} +$$

$$\begin{bmatrix} \text{net flux of thermal energy by} \\ \text{diffusion through solid and} \\ \text{fluid phases of } \Delta V \end{bmatrix} + \begin{bmatrix} \text{rate of release or absorption of thermal energy by} \\ \text{phase change inside } \Delta V \\ \text{(sources and sinks)} \end{bmatrix} \quad (3)$$

Various investigators (Abriola 1983; Abriola and Pinder 1985a,b; Kuppusamy et al. 1987) developed Eqs. (1)–(3) into a single finite differences or finite elements numerical model which can simulate all possible phenomena associated with the contamination of soils and groundwater by LH spills. The investigators showed that in various cases the model developed in such a manner can be very useful. Generally, this type of models is very sophisticated, and needs input of all basic data for all nodal points of the numerical grid. Such a type of model also makes calculations of all possible effects and phenomena in all grid points. The variability of the flow field characteristics reduces the efficiency of the model. In some regions some effects can be ignored. In some regions some data should not be considered at all, etc.

Summarizing, reference to the whole flow field by a single set of basic equations may lead to complicated models subject to low efficiency of the numerical computation, and problems of accuracy, stability and convergence due to the incorporation of large amounts of calculations and nonlinear terms even in regions where they are not needed.

11.6.3 Differential Control Volumes Referring to Various Parts of the Flow Field

According to this approach different mathematical models are developed for different regions of the flow field. All such models are basically based on the basic general expressions of Eqs. (1)–(3). However, in every specific zone of the flow field some simplifications of the general equations are possible by considering only the mechanisms and physicochemical reactions typical to that specific zone. Also reference to specific types of LH may lead to some simplifications. If volatility, for example is not significant then in most practical cases Eq. (3) is not needed for the simulation of the contamination process. Some other simplifications can be introduced into the calculations depending on the type of boundary conditions of the flow field.

Corapcioglu and Baehr (1985) developed a finite differences numerical model for the simulation of the LH contamination process in the unsaturated zone. Their model refers to all possible types of LH and all possible physicochemical and biochemical processes which may take place in the unsaturated zone. Due to its reference to a variety of possible conditions this model is based

Transport of Organic Pollutants in a Multiphase System 243

on the utilization of the basic Eqs. (1)–(3) with no further simplification (except for the reference only to the unsaturated zone).

Kessler and Rubin (1985a,b) referred to the three phase flow diagram of Fig. 4, and claimed that in many practical cases Richards equation (Richards 1931) can be applied in order to simulate the vertical migration and spreading of LH in the unsaturated zone. They considered the use of each of the following basic equations being derived from Richards equation:

The ψ – based model

$$C \frac{\partial \psi}{\partial \psi} = \nabla \cdot (K \nabla z + K \nabla \psi) \tag{4}$$

The θ – based model

$$\frac{\partial \theta}{\partial t} = \nabla \cdot (K \nabla z + D \nabla \theta) \tag{5}$$

and the F – based model

$$C \frac{\partial F}{\partial t} = (\frac{\partial K}{\partial \psi}) \nabla F \cdot \nabla z + K \nabla^2 F \tag{6}$$

where ψ is the capillary pressure head, C is the specific moisture capacity, t is time, z is the vertical coordinate, K is the hydraulic conductivity, θ is the volumetric LH content, D is the moisture diffusivity coefficient, F is an integral of K with regard to ψ.

Kessler and Rubin (1985a,b) claimed that either one of Eqs. (4)–(6) can be utilized for the prediction of LH migration in the unsaturated zone provided that the water degree of saturation is smaller than 10 percent and the time interval is comparatively short. The small water saturation is mentioned as the water is assumed to be an immobile phase which only affects the hydraulic conductivity of the soil for LH. The short time interval is mentioned as geochemical, biochemical and volatility effects are not considered. By the performance of numerical simulations, laboratory experiments and reference to typical spillage phenomena Kessler and Rubin (1987) showed that models based on the use of Eq. (4) can be useful in most practical needs. However, it seems that for every particular case to be simulated, the effect of the boundary conditions and flow conditions in the flow field should be thoroughly studied. Such conditions have variety of implications with regard to the basic model and method to be applied for the simulation process.

Most of the discussion presented in the previous paragraphs refers to the unsaturated zone. However, treatment of a contaminated aquifer is very often involved with LH flow in the unsaturated zone being induced by flow conditions in the aquifer. In such cases the boundary condition represented by the groundwater table and the top of the saturated capillary zone leads to LH movement. The numerical modelling procedure involved in such cases should take into account flow of LH in the unsaturated zone due to a nonlinear boundary condition being determined by flow of water in the saturated zone of the aquifer.

Recently Parker et al. (1987) and Lenhard and Parker (1987) reported on the complexities involved with modelling the three phase flow of air-LH-water. Such a flow is limited to a certain range of saturations. However, the parameters characterizing the system are important even though some fluid phases are eventually immobile.

Fluctuations of the groundwater table lead to emulsification and trapping of LH in the main body of the aquifer. Various authors refer to LH transport after its introduction into the main body of the aquifer. Scheigg (1979); Conard and Wilson (1986) refer to the LH trapping phenomenon. Pistiner (1986) reviewed various studies dealing with modelling of emulsification processes and flow of LH emulsions in porous media (Davies and Rideal 1963; Soo and Radke 1984). In these modelling procedures the basic Eqs. (1)–(3) are utilized in their specific form typical to the saturated zone. The small droplets of LH which are convected by the water flow are assumed to be part of the water phase. The LH phase is partially immobile. Other parts of the LH are represented by a continuous flowing phase.

11.7 Determination of the Flow Field Characteristics

In order to simulate the LH movement in soil and groundwater, mathematical models are usually applied. Models able to simulate two or three phase flow in porous media were reviewed in the previous section of this paper. The utilization of such models requires prior knowledge of functional relationships between fluid pressures, saturations and hydraulic conductivities.

Parameters and coefficients determining physicochemical and biochemical processes should also be known prior to the performance of the numerical simulation.

The lack of experimental techniques to measure the relationships between saturations and partial pressures typical to porous media including water-air-LH phase, very often led investigators to utilize measured characteristics of two fluid phase systems in the simulation process. However, even in two phase systems the determination of the flow field characteristics is involved with complicated tasks associated with some reasonable assumptions.

It is usually assumed that in a two fluid phase system the interfacial pressure difference, namely the capillary pressure, can be represented as a function of the wetting fluid saturation. In the system water-air-LH the wettability of water is higher than that of LH whose wettability is higher than that of air. In the system water-LH the capillary pressure can be expressed as follows

$$P_o - P_w = P_{ow}(S_w) \qquad (7)$$

where subscript o refers to LH, subscript w refers to water, P is pressure, S is saturation.

Various studies (Brooks and Corey 1964; Gardner et al. 1970; Su and Brooks 1976; Van Genuchten 1978) suggest empirical relationships between the capillary pressure and the wetting fluid saturation. However, such relationships depend on the nature of the flow. The monotonic displacement in two phase systems exhibits hysteresis stemming from differences between drainage and imbibition.

Theoretical and empirical formulae (Mualem 1974; Gillham et al. 1976; Scott et al. 1983) are useful for the description of the hysteresis of the curve describing the dependence of the capillary pressure on the wetting fluid saturation. In various cases neglection of the hysteresis phenomenon may lead to substantial errors in the mathematical simulation process.

The traditional approach towards the direct measurement of the hydraulic properties of a two fluid phase system is involved with steady state conditions in the system (Brooks and Corey 1964, 1966; Su and Brooks 1980). Such measurements which yield values of $K(\theta)$ and $\psi(\theta)$ are usually time consuming efforts. In such measurements gamma photon densiometry is often utilized. In this technique a monoenergetic gamma beam is employed in order to determine a single liquid content in a saturated or unsaturated porous medium (Ferguson and Gardner 1962; Fritton 1969; King 1967; Vachaud and Thony 1971; Corey et al. 1971).

Various recent studies concern the development and utilization of transient flow methods which are usually much quicker and simpler than those involved with steady state conditions. Various methods of analysis (Jonson et al. 1959; Morel-Seytoux 1973) imply the use of experimental set-up measuring characteristics of transient flow. Zachmann et al. (1981, 1982) identified the two phase flow parameters by referring to transient flow experiments of gravity drainage. Their interpretation of the experimental results was involved with an estimate of two unknown coefficients in a four parameter model representing the hydraulic properties. Kool et al. (1985) and Parker et al. (1985) estimated the hydraulic properties of the two fluid phase system being determined by a three parameters model by referring to column drainage experiments. Their experiments were initiated by a step change in air pressure at the top of the column.

Direct measurements of hydraulic properties of a three fluid phase system are very complicated. To a limited extent an application of dual-gamma attenuation can be utilized in such measurements being involved with steady state conditions (Ferrand et al. 1986).

In practice, the three phase behavior is usually estimated from two-phase measurements. In the three phase system of water-air-LH the water is the dominant wetting fluid. In such a system it is assumed that the total liquid saturation of water and LH is a function of air-LH capillary pressure. It is also assumed that the water-LH capillary pressure depends on the water saturation (Leverett and Lewis 1941; Corey et al. 1956; Aziz and Settari 1979). These two basic assumptions are represented by the following expressions

$$P_o - P_w = P_{ow}(S_w^{ow}) = P_{ow}(S_w^{aow}) \tag{8}$$
$$P_a - P_o = P_{ao}(S_w^{ao}) = P_{ao}(S_w^{aow} + S_o^{aow}) \tag{9}$$

where superscripts refer to fluid phases present in the porous medium. Therefore air-water capillary pressure is usually estimated to be irrelevant to the three fluid phase system. The assumptions represented in Eqs. (8) and (9) are utilized in various studies concerning three fluid phase flow (Sheffield 1969; Kasemi et al. 1978; Coats 1976, 1980; Eckberg and Sunada 1984). Some studies consider more complicated relationships. Peery and Herron (1969) and Shutler (1969) employed Eq. (8) in order to evaluate S values, but they assumed that LH-gas capillary pressure is a function of the ratio of LH or gas saturation to LH plus gas saturation. The hysteresis effect in the relationships expressed by Eqs. (8) and (9) has not yet been studied.

Studies of Leverett and Lewis (1941) and Corey et al. (1956) indicated that the following expressions for the relative permeabilities in the three fluid phase system can be utilized

$$K_{rw} = K_{rw} (S_w) \tag{10}$$
$$K_{ra} = K_{ra} (S_a) \tag{11}$$
$$K_{ro} = K_{ro} (S_w, S_a) \tag{12}$$

where K_{row} is the LH relative permeability in a two phase LH-water system; K_{roa} is the LH relative permeability in a two phase gas-LH system; K_{rw} is the three phase water relative permeability being equal to the water relative permeability in a two phase water-LH system; K_{ra} is the three phase relative permeability of the gas being equal to the gas relative permeability in the two-phase gas-LH system.

However, experiments with three fluid phases can hardly provide quantification of these expressions. Usually the dependence of the relative permeabilities on the fluid saturations are determined by applying relationships measured in two fluid phase system. The extension of the two phase relationships between saturations and permeabilities being utilized by Leverett (1941) and Leverett and Lewis (1941) led to the diagram of Fig. 4.

Models being able to predict three phase permeabilities from two phase measurements were developed by Stone (1970, 1973). According to his more recent model (Stone 1973)

$$K_{ro} = (K_{row} + K_{rw})(K_{roa} + K_{ra}) - (K_{rw} + K_{ra}) \tag{13}$$

Various investigators (Aziz and Settari 1979; Dietrich and Bondor 1976) introduce some modifications of the models which had been developed by Stone.

Some investigators (Snell 1962; Saraf et al. 1982) studied the effect of saturation history on relative permeability-fluid saturation relationships. Generally, saturation history effect on these relationships is less significant than that effect on capillary pressure-fluid saturation relationships.

The negative sources in Eq. (1) represent phase change and biological degradation of LH. Corapcioglu and Baehr (1985) provided formulae describing this process. Field and laboratory data reported in various studies (Jamison et al. 1976; Raymond et al. 1975, 1976, 1977; Huddleston and Meyer 1979) provide some information concerning the important parameters determining LH biodegradation.

Biodegradation is optimized by controlling conditions of aeration, moisture content and nutrient availability. Such an optimization hardly exists in nature.

However, laboratory tests following the effectiveness of particular microorganisms under various controlled conditions should be carried out for proper quantification of LH biodegradation.

11.8 Practical Aspects

This section concerns the relationships between field activities regarding the "real world" and laboratory – theoretical studies.

Previous sections of this paper imply that usage of any numerical modelling method depends on proper characterization of various characteristics of the flow field. Determination of the flow field characteristics is obtained by the performance of laboratory tests being subject to some theoretical guidelines relevant to test analyses.

Accidental LH spills usually draw much attention. Many theoretical and laboratory studies are relevant to soil and groundwater contamination due to abrupt release of large quantities of LH into the environment. However, accidents in which LH is subject to abrupt spillage of large quantities are quite rare. Most events of soil and groundwater contamination by LH are involved with neglect, uncontrolled disposal and inadequate codes for handling LH. In such cases many years may pass before the groundwater contamination is followed. Then assessment of the contamination scale is a study which should precede any remedial measure. Contamination assessment is involved with data collection related to sources of the LH release, its type, quantities and rates of disposal. Other types of data refer to the soil and groundwater being subject to contamination. Then various modelling approaches can be utilized in order to evaluate and follow the contamination process. Such an evaluation usually finalizes the procedure of the contamination assessment. In the next stage the remedial measures should be applied. Such measures include preventive stages in order to stop the LH release into the environment. Further remedial measures can be taken according to different final goals. Remedial measures at a lowest level of expenditures include preventive efforts directed to avoid further contamination by the LH spreading in the subsurface. More advanced effort may include a certain treatment of the soil being subject to residual LH saturation and a certain treatment of the polluted subsurface water. In the decision making process, regarding the degree and method of the aquifer restoration, the use of quantitative simulations is more than reasonable. Such quantitative simulations provide the basic details needed for the evaluation of the aquifer restoration method being optimized according to environmental safety, economical guidelines and appropriate utilization of groundwater resources.

Acknowledgements. Portions of this research were supported by grants from the National Council for Research and Development Israel and the K.F.K. Karlsruhe, Germany, and a grant from the Water Commissioner Office, Ministry of Agriculture, Israel.

References

Abriola MA (1983) Mathematical modeling of the multiphase migration of organic compounds in a porous medium. PhD Desertation, Civil Engineering, Princeton University, Princeton NJ

Abriola LM, Pinder GF (1985a) A multiphase approach to the modeling of porous media contamination by organic compounds aquation development. Water Res Res 21:11–18

Abriola LM, Pinder GF (1985b) A multiphase approach to the modeling of porous media contamination by organic compound Hydraul Paper No. 3 Colorado State University, Fort Collins, Colorado

Aziz K, Settari A (1979) Petroleum reservoir simulation, Applied Science Publ, London

Brooks RH, Corey AT (1964) Hydraulic properties of porous media, Hydraul Paper No 3, Colorado State University, Fort Collins, Colorado

Brooks RH, Corey AT (1966) Properties of porous media affecting fluid flow. ASCE J Irr Div 92:61–68

Burmaster DE, Harris RH (1982) Groundwater contamination an emerging threat. Technol Rev 84:50–62

Coats KH (1976) Simulation of steamflooding with distillation and solution gas. Soc Petrol Engr AIME J 16:235–247

Coats KH (1980) In-situ combustion model. Soc Petrol Engr AIME J 20:533–553

Conard SH, Wilson JL (1986) Transport and capillary trapping of organic liquids present as a separate phase. EOS 67:945

Corapcioglu NY, Baehr AL (1985) The transport and fate of petroleum products in soils and groundwater, Final Project Rept No CEE-8401438. Civil Engineering, City College of New York, New-York NY

Corey AT, Rathens CH, Henderson JH, Wyllie MR (1956) Three Phase relative permeability. Trans AIME Petrol Eng Div 207:349–351

Corey JC, Peterson SF, Wakat MA (1971) Measurement of attenuation of Cs and Am gamma rays for soil density and water content determinations. Soil Sci Soc Am J 35:215–219

Dam J van (1967) The migration of hydrocarbons in a water bearing stratum. In: Hepple P (ed) The Joint Problems of the Oil and Water Industries. Elsevier, Amsterdam, pp 55–96

Davies JT, Rideal EK (1963) Interfacial Phenomena. Academic Press, London

Dietrich JK, Bondor PL (1976) Three phase oil relative permeability models. 51st Annual Fall Meeting of Soc Petrol Engr AIME, New Orleans, LA

Dietz DN (1979) Pollution of permeable strata by oil components, in Water Pollution by Oil. Applied Science, England, pp 127–139

Eckberg DK, Sunada DK (1984) Nonsteady three phase immiscible fluid distribution in porous media. Water Res Res 20:1891–1897

Ferguson H, Gardner WH (1962) Water content measurement in soil column by gamma ray absorption. Soil Sci Am J 26:11–14

Ferrand LA, Milly PCD, Pinder GF (1986) Dual-gamma attenuation for the determination of porous medium saturation with respect to three fluids. Water Res Res 22:1657–1663

Fritton DD (1969) Resolving time mass absorption coefficient and water content with gamma ray attenuation. Soil Sci Soc Am J 33:651–655

Gardner WR, Hillel D, Benyamini Y (1970) Post irrigation movement of soil water I redistribution. Water Res Res 6:851–861

Genuchten MT van (1978) Calculating the unsaturated hydraulic conductivity with a new closed form analytical model. Civil Engr, Princeton University, Princeton NJ

Gillham RW, Klute A, Heermann DF (1976) Hydraulic properties of a porous medium measurement and empirical representation, Soil Sci Soc Am J 40:203–207

Huddleston RL, Meyer JD (1979) Treatment of refinery oil waste by land farming. AIChE Symp Ser S190 75:327–339

Jamin MJ (1860) Memoire sur lequilibre et le mouvement des liquides dans les corps poreux. CR Hebd Acad Seances Sci Paris

Jamison VW, Raymond RL, Hudson JO (1976) Biodegradation of high octane gasoline. Proc 3rd Intal Biodegradation Symp pp 187–196, Applied Science, England

Jonson EF, Bossler DP, Naumann VO (1959) Calculation of relative permeability from displacement experiments. Trans AIME Petrol Eng Div 216:370–372

Kazemi H, Vestal CR, Shank GD (1978) An efficient multicomponent numerical simulator. Soc Petrol Engr J 18:255–268

Kessler A, Rubin H (1985a) Development of a practical method simulating oil spill migration (OSPIM) in soils, in Hydraulics and Hydrology in the Small Computer Age. ASCE Specialty Conf, Lake Buena Vista, FL, pp 1480–1485

Kessler A, Rubin H (1985b) On the simulation of unsaturated oil flow in soils, in Scientific Basis for Water Resources Management. Proc of the Jerusalem Symp, IAHS Publ no 153, pp 195–205

Kessler A, Rubin H (1987) Relationships between water infiltration and oil spill migration in sandy soils, J Hydrol 91:187–204

King LG (1967) Gamma ray attenuation for soil water content measurements using 241Am, in Proc of a Symp on the use of Isotope and Radiation Techniques in Soil Physics and Irrigation Studies. Intal Atomic Energy Agency, Vienna

Kool JB, Parker JC, van Genuchten MT (1985) Determining soil hydraulic properties from one-step outflow experiments by parameter estimation theory and numerical studies, Soil Sci Soc Am J 49:1348–1354

Kool JB, Parker JC, Genuchten MT van (1987) Parameter estimation for unsaturated flow and transport models. Bull J Hydrol (in press)

Kroszynski UI (1974) The influence of the unsaturated zone upon phreatic flows in porous media. D Sc Thesis, Civil Engineering, Technion, Haifa, Israel

Kuppusamy T, Sheng J, Parker JC, Lenhard RJ (1987) Finite element analysis of multiphase immiscible flow through Soil, paper submitted to Water Res Res 23:625–632

Lenhard RJ, Parker JC (1987) Measurement and prediction of saturation-pressure relationships in air-organic liquid-water-porous media systems. Contam Hydrol (in press)

Lenhard RJ, Dane JH, Parker JC (1987) Measurement and simulation of transient three phase flow in porous media. Water Res Res (in press)

Leverett MC (1941) Capillary behaviour in porous solids. Trans Am Inst Miner Eng 142:152

Leverett MC, Lewis WB (1941) Steady flow of gas-oil-water mixtures through unconsolidated sands. Trans Am Inst Miner Eng 142:107–116

McKee JE, Laverty FB, Hertel RM (1972) Gasoline in groundwater. J Water Pollut Control Fed 44:293–302

Mechrez E, Kessler A, Rubin H (1986) Optical visualization of oil spill penetration into the capillary zone of groundwater aquifers. 4th Intal Symp on Flow Visualization, Assos Nationale de la Recherche Technique, Paris

Mercado A (1976) Nitrate and chloride pollution in aquifers a regional study with the aid of a single cell model. Water Res Res 12:731–747

Mualem Y (1974) A conceptual model of hysteresis, Water Res Res 10:514–520

Morel-Seytoux HJ (1973) A conceptual model of hysteresis. Water Res Res 10:514–520

Parker JC, Kool JB, Genuchten MT van (1985) Determining soil hydraulic properties from one step outflow experiments by parameter estimation experimental studies. Soil Sci Soc Am J 49:1354–1360

Parker JC, Lenhard RJ, Kuppusamy TA (1987) A parametric model for constitutive properties governing multiphase fluid conduction in porous media, paper submitted to Water Res Res 23:618–624

Peery JH, Herron EH (1969) Three phase reservoir simulation. J Petrol Technol 21:211–220

Pfannkuch HO (1984) Determination of the contaminant source strength from mass exchange processes at the petroleum groundwater interface in shallow aquifer systems. Proc Petroleum Hydrocarbons and Organic Chemicals in Ground Water-Prevention Detection and Restoration, National Water Well Assoc Conf Houston, Texas

Pistiner A (1986) Migration of fuel pollutants in groundwater. Civil Engr, Technion, Haifa, Israel

Raymond RL, Hudson JO, Jamison VW (1975) Assimilation of oil by soil bacteria. Refining 40th Midyear Meeting American Petroleum Inst (API), Chicago, Ill

Raymond RL, Hudson JO, Jamison VW (1976) Oil degradation in soil, Appl Environ Microbiol 31:522–535

Raymond RL, Hudson JO, Jamison VW (1977) Extended study of cleanup of oil in soil by biodegradation. Project Rept No 307-77, Sun Tech, Marcus Hook, PA

Richards LA (1931) Capillary conduction of liquids in porous mediums. Physics 1:318-333

Saraf D, Batycky JP, Jackson CH, Fisher DB (1982) An experimental investigation of three phase flow of water-oil-gas mixtures through water-wet sandstone. Calif Regional Meeting of the Soc of Petrol Engr, San Francisco, CA

Scheigg HA (1979) Verdrangungs-Simulation dreier nicht mischbarer Fluide in Porosen Matrix. VAW-Mitteilung No 40

Scheigg HA (1980) Field infiltration as a method for the disposal of oil in water emulsions from the restoration of oil polluted aquifers. Water Res 14:1011-1016

Schwille F (1975) Groundwater pollution by mineral oil product. IAHS-AISH 103:226-240

Schwille F (1984) Migration of organic fluids immiscible with water. In: Yaron B, Dagan G, Goldsmidt J (eds) The unsaturated zone, Pollutants in Porous Media. Springer, Berlin Heidelberg New York, pp 27-48

Scott PS, Farquhar GJ, Kouwen N (1983) Hysteretic effects on net infiltration, in Advances in Infiltration, Proc Natl Conf on Advances in Infiltration ASAE, Chicago, Ill, pp 163-170

Sheffield M (1969) Three phase fluid flow including gravitational viscous and capillary forces. Trans AIME Petrol Div 246:232-246

Shutler ND (1969) Numerical three phase simulation of the linear steamflood process. Soc Petrol Engr J 9:232-246

Snell RW (1962) Three phase relative permeability in an unconsolidated sand. J Inst Petrol 48:80-88

Soo H, Radke CJ (1984) Velocity effects in emulsion flow through porous media. J Colloid Interface Sci 102:462

Spielman LA (1977) Particle capture from low speed laminar flows. Annu Rev Fluid Mech 9:297

Spielman LA, Goren SL (1970) Progress in induced coalescence and a new theoretical framework for coalescence by porous media. Ind Eng Chem 61:10

Spielman LA, Goren SL (1972a) Theory of coalescence by flow through porous media. Ind Eng Chem Fundamentals 11:66

Spielman LA, Goren SL (1972b) Experiments in coalescence by flow through fibrous mats. Ind Eng Chem Fundamentals 11:73

Spielman LA, Su YP (1977) Coalescence of oil in water suspensions by flow through porous media. Ind Eng Chem Fundamentals 16:272

Stone HL (1970) Probability model for estimating three phase relative permeability. Trans AIME Petrol Eng Div 249:214-218

Stone HL (1973) Estimation of three phase relative permeability and residual oil data. J Can Petrol Technol 12:53-61

Su C, Brooks RH (1976) Hydraulic functions of soils from physical experiments and their applications, Water Res Res Inst, Oregon State University, Corvallis, Oregon

Su C, Brooks RH (1980) Water retention measurement for soil. ASCE J Irr Div 106:105-112

Todd DK (1980) Groundwater Hydrology. Wiley, New York, pp 30-36

Vachaud G, Thony JL (1971) Hysteresis during infiltration and redistribution in a soil column at different initial water content. Water Res Res 7:11-127

Wilson JL, Conard SH (1984) Is physical displacement of Residual hydrocarbons a realistic possibility in aquifer restoration. Proc Petroleum Hydrocarbons and Organic Chemicals in Ground Water Prevention Detection and Restoration, National Water Well Assoc Conf Houston, Texas

Wittcoff HA, Reuben BG (1980) Industrial Organic Chemicals in Perspective. Wiley, New York, pp 41-45

Wooding RA, Morel-Seytoux HJ (1976) Multiphase fluid flow through porous media. Annu Rev Fluid Mech 8:233-274

Zachmann DW, Duchateau PC, Klute A (1981) The calibration of the Richards flow equation for a draining column by parameter identification. Soil Sci Soc Am J 45:1012-1015

Zachmann DW, Duchateau PC, Klute A (1982) Simultaneous approximation of water capacity and soil hydraulic conductivity by parameter identification. Soil Sci 134:157-163

12 Biochemical Aspects of Hydrocarbon Biodegradation in Sediments and Soils

T. HÖPNER, H. HARDER, K. KIESEWETTER, U. DALYAN, I. KUTSCHE-SCHMIETEN-KNOP and B. TEIGELKAMP[1]

12.1 Introduction

After pollution of sediments and soils by hydrocarbons, only two restoration possibilities exist in principle: Mechanical removal of the pollutant and biodegradation. Even after mechanical removal, considerable residues of the polluted material usually remain in situ so that even in this case biodegradation is a necessary component of restoration management. At the same time the polluted soil or sediment material has to be cleaned, and among the cleaning procedures biodegradation is one of the methods of choice. To optimize or support biodegradation it is necessary to apply biochemical principles to the process and to determine optimum conditions. Under the pressure of severe oil and groundwater contaminations, many attempts have been made to control and support biodegradation of hydrocarbons in situ by supplying the contaminated zone with reactants (such as nitrate) assumed to be necessary to the process besides or instead of oxygen (Sontheimer et al. 1987). Some of these attempts were not unsuccessful, but the question remained, how far away they were from optimized conditions.

Pathways and velocities are the two primary aspects of biodegradation. Pathways of oxygen dependent degradation of aliphatic and aromatic hydrocarbons have been intensely studied with many microorganisms and the most important substrates and intermediates are well known (Singer and Finnerty 1984; Cerniglia 1984). This aspect will be treated in this report only with respect to conclusions which can be drawn for biodegradation conditions. Velocities, however, are related to the conditions under which the microbiological, biochemical and biotechnological studies have been performed and results cannot be transferred to in situ conditions.

The in situ conditions that appear to be rate determining are availability of oxygen, availability of nutrients, water content, and nature and state of the hydrocarbons (fluid, dissolved, adsorbed to surfaces of particles, vapour). If the other conditions are favourable, rates can be most effectively influenced by oxygen, but in many cases the provision with oxygen is the crucial point of an in situ restoration management. From a practical point of view emphasis is given to (a) biodegradation under oxygen limitation and (b) anaerobic biodegradation.

---

[1]Fachbereich Biologie der Universität, Postfach 2503, 2900 Oldenburg, FRG

In situ examinations of biodegradation are easier in marine sediments than in soils, because the biologically active layers are more easily accessible. In intertidal sediments layers with fundamentally different conditions are lying one on top of the other separated by just short distances in the depth profile: The depth of an aerobic surface layer of only several millimeters or centimeters thickness is followed by an even thinner suboxic layer, and the anoxic zone follows immediately (Fenchel and Riedl 1970). In these sediments it is possible to examine the existing conditions and to verify them separately in laboratory experiments (Höpner et al. 1987). It is much more difficult to exactly describe the conditions in deep soil layers and to observe reactions occurring there. For these reasons experiences obtained in marine sediments are included in this report.

12.2 Biochemical Considerations

The first step of any oxygen dependent biodegradation of an aliphatic or aromatic hydrocarbon is the introduction of a first oxygen atom forming an alcoholic or phenolic derivative. This hydroxylation is catalyzed by an oxygenase or dioxygenase, and molecular oxygen is the source of the introduced oxygen molecule. One atom is incorporated into the substrate, the other is reduced to water. This means that the reaction requires molecular oxygen as well as a reductant. In the case of hexadecane:

$$C_{16}H_{34} + O_2 + NADPH + H^+ \rightarrow C_{15}H_{31}CH_2OH + H_2O + NADP^+ \quad (1)$$

In the case of aliphatic hydrocarbons no biochemical reaction is known in which the molecular oxygen is replaced by another oxygen donor. The alcohol, however, can be converted into an aldehyde and further on into a fatty acid by oxygen dependent reactions as well as by oxygen independent ones, if other electron acceptors are present:

$$C_{15}H_{31}CH_2OH + NAD^+ \rightarrow C_{15}H_{31}CHO + NADH + H^+ \quad (2)$$
$$C_{15}H_{31}CHO + NAD^+ + H_2O \rightarrow C_{15}H_{31}COOH + NADH + H^+ \quad (3)$$

The fatty acid enters the normal degradation pathway which in the case of oxygen availability is

$$C_{15}H_{31}COOH + 23\,O_2 \rightarrow 16\,CO_2 + 16\,H_2O \quad (4)$$

The total degradation follows the equation

$$C_{16}H_{34} + 24.5\,O_2 \rightarrow 16\,CO_2 + 17\,H_2O \quad (5)$$

The degradation of the fatty acid can also be independent of molecular oxygen if other electron acceptors are present. Nitrate is an appropriate reactant, it is converted into gaseous nitrogen (denitrification reaction):

$$C_{15}H_{31}COOH + 18.4\,NO_3^- + 18.4\,H^+ \rightarrow 16\,CO_2 + 9.2\,N_2 + 25.2\,H_2O. \quad (6)$$

A speculative alternative oxygen dependent pathway involves the formation of a fatty acid through a labile hydroperoxide intermediate. This nucleophilic reaction is not likely unless in the case of a ternary carbon. Whatsoever, the result is that at least the first oxygenation step has to be considered to be obligatorily oxygen dependent as long as no biochemical alternative is presented, and that a biodegradation of an aliphatic hydrocarbon cannot be expected in an absolutely anaerobic environment.

Nevertheless, there are speculations on an oxygen independent degradation in the literature (e.g. Atlas and Bartha 1981). According to the above-mentioned question, it is focussed on the existence of an oxygen independent key reaction. In the case of saturated aliphatic hydrocarbons, the most probable candidate is the dehydrogenation of a carbon-carbon bond forming a double bond which subsequently could be hydrated. There are allusions to alkane dehydrogenation under anaerobic conditions: Chouteau et al. 1962 (*n*-hept-1-ene from *n*-heptane by resting cells of *Pseudomonas aeruginosa*), Iizuka et al. 1969 (*n*-dec-1-ene from decane by resting cells of *C. rugosa*), Parkeh et al. 1977 (*n*-alkane-degradation by *Pseudomonas* sp.), Senez and Azoulay 1961 (long-chain hydrocarbons by resting cells and cell-free extracts from *Ps. aeruginosa*) and Traxler and Bernard 1969 (same). In the latter case it was stated (Atlas 1981) that the results had not yet been adequately reproduced. From a biochemical standpoint some doubts remain, and if the reactions occur at all, significant contributions to an ecologically relevant biodegradation rate cannot be expected (a supposition which was pronounced by Atlas and Bartha 1981).

There are also several reports on the anaerobic degradation of hydrocarbons in natural ecosystems (Bailey et al. 1973; Brown et al. 1969a,b; Pierce et al. 1975) suggesting that nitrate or sulfate can serve as electron acceptors in the absence of oxygen. However, in these cases as well, doubts remain because the complete exclusion of oxygen was not fully proven and chemical evidences for hydrocarbon degradation were not sufficiently convincing (Atlas 1981). In contrast to the above works, many reports state that the biodegradation is absolutely dependent on oxygen availability in vitro as well as in situ.

Zeyer et al. (1986) discussed an oxygen-independent degradation of two aromatic hydrocarbon substrates (toluene and 1,3,-dimethyl-benzene) which does not exclude that an unsaturated compound can be degraded anaerobically if aromatic double bonds are sufficiently activated by appropriate ligands.

Regardless of whether biochemical pathways of anaerobic hydrocarbon degradation really exist, "the environmental importance of anaerobic hydrocarbon biodegradation can be discounted. Rapid biodegradation of hydrocarbons does not occur in anaerobic environments" (Atlas 1981). We agree with this statement, but the frequently occurring pollution of anaerobic or oxygen-poor environments by hydrocarbons requires appropriate measures to be taken.

There are sound arguments for examining whether nitrate can be used for groundwater, soil and sediment restoration after hydrocarbon contamination, because the denitrification reaction is the only practical possibility of
- removing an organic contaminant by converting it into carbon dioxide without contaminating the environment with a product of the oxidant, because the oxidant nitrate is converted into gaseous nitrogen.

— supplying the contaminated zone with sufficient amounts of an electron acceptor, because the solubility of nitrate in water is high compared to oxygen. (Attempts to replace oxygen by hydrogen peroxide or to use hydrogen peroxide as an oxygen source are not treated here).

If anaerobic nitrate dependent degradation is not possible in the case of a hydrocarbon, and if oxygen dependent degradation cannot be achieved because oxygen support is not sufficient the question arises: is it possible to combine Eq. (1), with equations of type (6) (regardless of whether the alcohol, the aldehyde or the fatty acid is its substrate), which means supporting the degradation by the small amount of oxygen necessary for the first oxygenation reaction and by the high amount of nitrate necessary for the forthcoming degradation steps? This means degradation by a limiting amount of oxygen in the presence of a second electron acceptor.

Under limited oxygen conditions two theoretical cases can be discussed: (a) According to the amount of the available oxygen the appropriate amount of hydrocarbons is converted into oxygen containing intermediates until the oxygen has been fully consumed, and the intermediates are further degraded by nitrate. (b) The oxygen is consumed and the appropriate amount of hydrocarbon is fully degraded without any participation of the nitrate. While the second case appears to have been achieved in laboratory experiments (this report), results of Battermann and Werner (1984) and Werner (1985) do not exclude the first case. The authors found in practical groundwater restoration, that nitrate addition to oxygen containing flushing water led to additional hydrocarbon decrease. Hence the combined use of oxygen and nitrate, which theoretically should be effective and which seems to be possible in practice, comes into focus. But there is no biochemically and microbiologically explainable methodology.

12.3 Application of Results of Biochemical Considerations

The result of the biochemical considerations is, that support of aerobic biodegradation is the first matter of choice, that anaerobic denitrifying biodegradation cannot contribute significantly, and that combined provision with oxygen and nitrate is worth examining. Principally three approaches are at hand:

a) Denitrifying degradation in the presence of oxygen in a low and constant concentration
b) Denitrifying degradation under intermittant support with low amounts of oxygen
c) Denitrifying degradation maintaining a friable soil or sediment structure which allows the formation of microanaerobic niches.

One objection to proposal a) and b) has already been mentioned: limited oxygen means total degradation of a limited amount of aliphatic hydrocarbon, and does not mean provision of oxygen-containing substrates to denitrification. Case a) may be possible in a mixed homogeneous laboratory setup, but seems difficult or rather impossible in practice. To choose an appropriate oxygen concentration is a matter of balance between yield and rate, because the denitrification rate drops to only 10 to 20% if the oxygen concentration is raised from zero to 1% of the atmospheric pressure (Tiedje et al. 1984).

Case b) means frequent re-adaptation of the bacteria to changing conditions, apart from the metabolic questions mentioned above. Denitrifying bacteria are facultative anaerobes, the various enzymes involved are repressed and derepressed depending on the conditions, which means at first oxygen concentration. Derepression times are in the range of one or several hours. It can be expected therefore, that intermittant oxygen supply means low overall reaction rate (for a review see Knowles 1982).

Case c) is assumed to take place in soils (Stolzy and Flühler 1978; Firestone 1982; Hagin and Tucker 1982). The generalization of the microniche hypothesis is the assumption of *spatial concentration gradients* of oxygen, nitrate, and of substrates and intermediates. The occurrence of spatial gradients can also be expected in the root zone (root channels, Ronen et al. 1984) and in layered limnic and marine sediments (Blackburn 1983; Engler and Patrick 1974; Savant and Dedatta 1982). There are several allusions in the literature to *temporal changes* between saturated and unsaturated conditions, which are of major importance for maintaining denitrification (Ronen et al. 1984), and we suggest that the upward and downward migration of the reduction horizon in intertidal sediments is of comparable importance. The concluding hypothesis is that *spatial and temporal discontinuities* are the key to understand the contradiction between denitrification and presence of oxygen, above all in the case of hydrocarbon degradation.

Summarizing these facts, the experimental development of in situ restoration procedures has to take three points into consideration:

1. To obtain a practically relevant degradation velocity, the participation of molecular oxygen is indispensible.
2. To obtain a practically relevant degradation yield under limited oxygen supply, it is desired to combine the oxygen dependent start reaction with a subsequent nitrate dependent degradation.
3. Oxygen inhibits denitrification, but oxygen makes substrates accessible to denitrification. The art is, to elaborate methods which circumvent or overcome this antagonism.

This report deals with aerobic and anaerobic-denitrifying degradation of hydrocarbons in marine sediments and in soils with special respect to the questions treated above.

12.4 Laboratory Experiments with Fresh Marine Sediment

The aim of the experiments was to examine aerobic biodegradation rates [Eq. (5)] as a first step to follow rates under various conditions of oxygen supply. For these examinations some seawater was added to fresh muddy sediment samples so that they could be shaken. The standard setup contained 55 g sediment (dry weight) and 30 ml water in a 250 ml Erlenmeyer flask. In addition to Statfjord oil, hexadecane was used as a model hydrocarbon. If nutrients (nitrate, orthophosphate) were added, the recommendations of Gibbs (1975) (60 mg nitrate-N and 6 mg phosphate-P per gram hydrocarbon) were followed. Each point on the figures of this report (except the manometric assays) is the result of a single setup. Incubation time was mostly 10, 20, 30 and 40 days at 16° C. Hydrocarbon content was mostly followed by CCl_4 extraction and IR assay, in some cases controlled by capillary gas chromatography.

There was no significant biodegradation in a sandy sediment even after nutrient addition (Fig. 1). However, in muddy sediments degradation of hexadecane (Fig. 2a) or Statfjord raw oil (Fig. 2b) started after a lag phase of 10 to 20 days and attained about 0.2 mg per g per day in the presence of added nutrients, whereas the rate was low without added nutrients. After sterilization the deg-

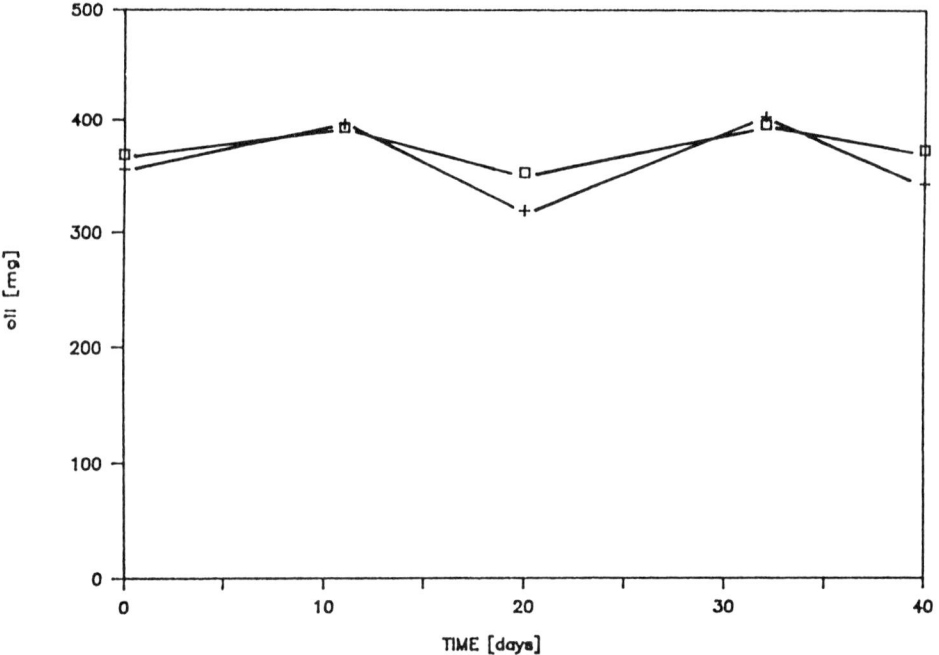

Fig. 1. Behaviour of Statfjord oil in an aerobic shaking culture of fresh sandy sediment - □ - without nutrient addition, - + - with Gibbs' nutrients. For composition of cultures see text

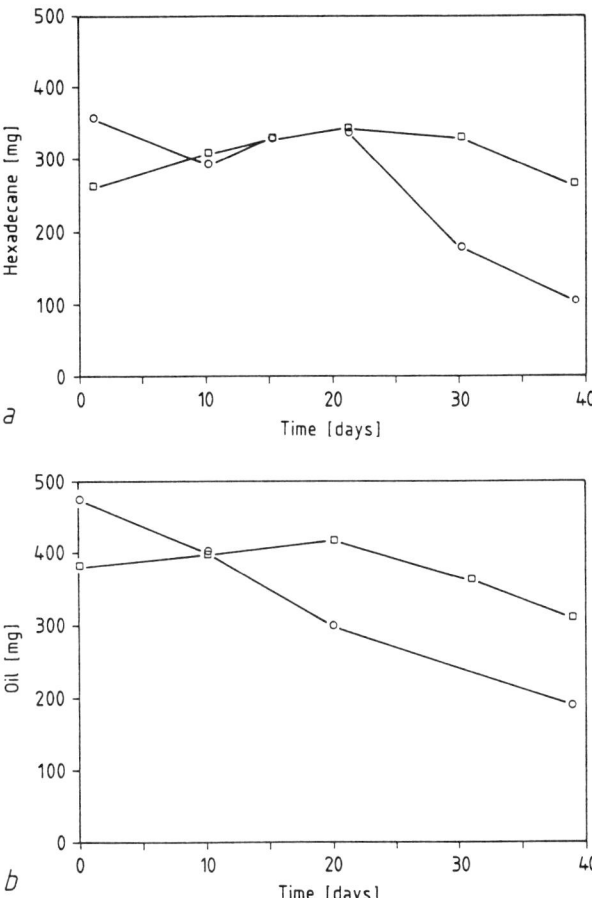

Fig. 2a. Aerobic hexadecane degradation in fresh muddy sediment. -□- without addition of nutrients, -○- with Gibbs' nutrients. **b**: Statfjord oil degradation in fresh muddy sediment. Symbols as in Fig. 2a

radation rate was zero. It was not possible to enhance the degradation rate by adding an adapted inoculum (Fig. 3).

We concluded that the adsorption of the hydrocarbons (and perhaps of the bacteria) to the large surfaces of the silt was the decisive reason for the high degradation rates, because there was no degradation in sand. The effect of nutrients was visible only, if more than 100 mg hydrocarbon were added. Up to this amount the endogenous nutrient content of the mud was sufficient. In the fresh sediment samples a competent population of bacteria seemed always to be present. The observed degradation rate is about 10 mg hydrocarbon per day or about 0.2 mg per g sediment per day.

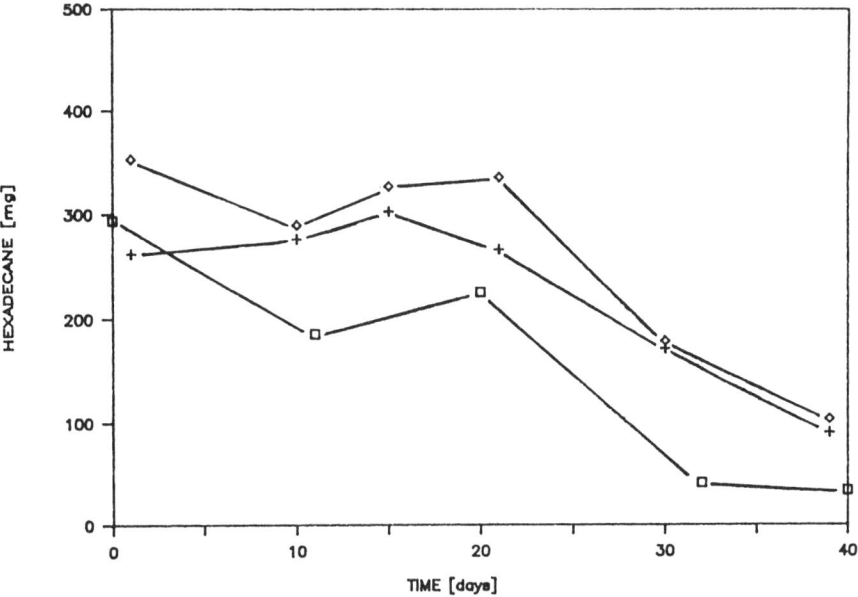

Fig. 3. Hexadecane degradation in the presence of Gibbs' nutrients in fresh muddy sediment (-□-) and in sterilized mud after addition of 10^4 (-+-) and 10^6 (-◊-) bacteria (total colony forming units) per gram sediment

In the progress of the experiments the varying properties of the fresh sediment samples and the different sizes of the lag phases became tedious. So we changed the experimental basis and turned to a standard system with constant and reproducible properties.

12.5 Laboratory Experiments with a Standardized and Inoculated Sediment Mixture

The aim of these experiments was practicability: It would not be possible to work over a long time with sediments of constant property if fresh sediments were sampled for each experiment. Therefore a large quantity of muddy sediment from the site of the above-mentioned fresh samples was air dried, sieved, homogenized and stored. For each shaking culture, 55 g of this material was used. Hexadecane, cetyl alcohol and palmitic acid were added dissolved in pentane. The solvent was removed in a vacuum evaporator. Seawater and distilled water, respectively, were added to attain the previous conditions. To prepare an adapted inoculum, biologically active surface sediment samples were drawn from an indoor mudflat model (2 m^2) which had been kept with its full zoobenthic and phytobenthic populations in a glass house for 30 months. These samples were preincubated with hydrocarbons under the conditions of the degradation experiments and subsamples were used to inoculate the incubation setups.

With this system the biodegradation rates were confirmed (Fig. 4) at about 0.2 mg per g per day. The lag phase phenomenon was no longer observed and the nutrient dependence was confirmed. Oxygen supply was found to be crucial as can be seen in Fig. 4, where the highest rate was obtained under continuous oxygen saturation by vigorous shaking (oxygen concentration not yet controlled), while a low rate was obtained in a closed flask filled with nitrogen. The intermediate result belongs to a closed shaken flask, in which only the approx. 200 ml air content was available. Nutrients had been added as given above. In preparation of the later anaerobic experiments (see Sect. 12.6) oxic degradation rates of hexadecane, cetyl alcohol and palmitic acid were compared (Fig. 5). As expected, the latter two substrates were decomposed more rapidly than hexadecane, suggesting that the first reaction, the hydroxylation, is the rate limiting step of hexadecane degradation (relative rates of substrate incorporation into the bacteria are not considered in this conclusion).

It is well known that in fresh natural sediments, hydrocarbon degradation is possible. Bacterial populations can always be expected because natural or anthropogenic hydrocarbons are always present at least in traces. As we were able to show, the specific surface of the sediment (in our samples about 8 m²/gram) influences the degradation rate. Initially we were surprised that it was so easy to

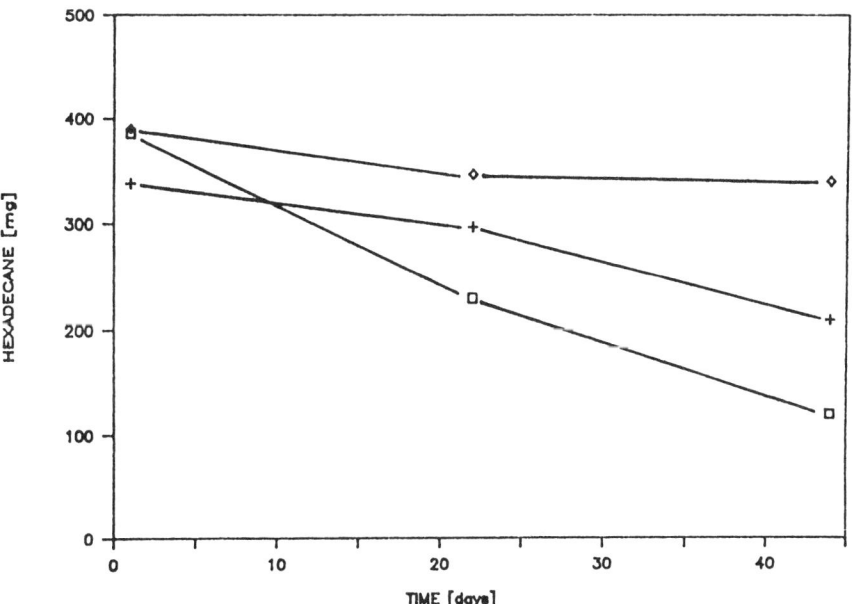

Fig. 4. Aerobic (-☐-), limited (-+-, only air stock in closed flask) and anoxic (-◊-) degradation of 386 mg (1.7 mMol) hexadecane in 55 g standardized and inoculated sediment in the presence of Gibbs' nutrients

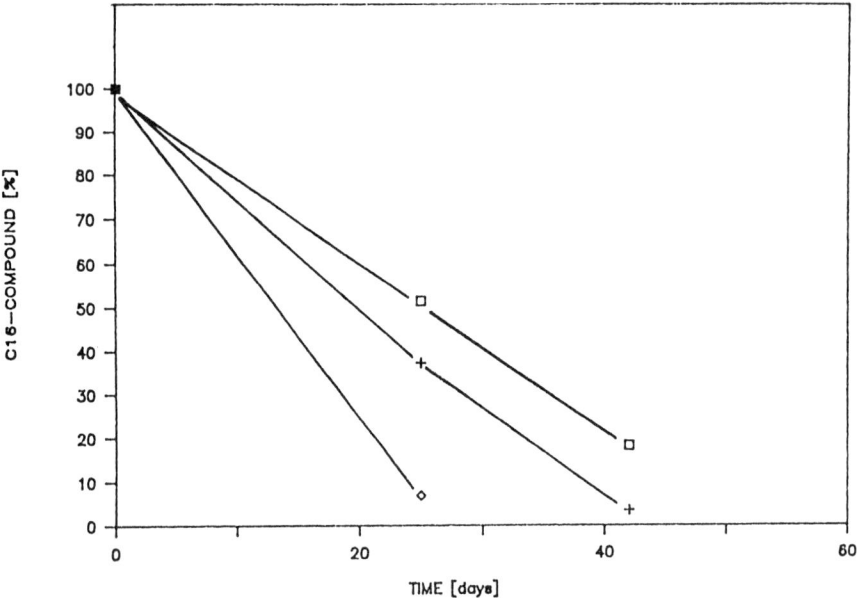

Fig. 5. Aerobic degradation of each 1.7 mMol palmitic acid (-◊-), cetyl alcohol (- + -) and hexadecane (-□-) in 55 g standardized muddy sediment in the presence of Gibbs' nutrients

reproduce the results, which we had obtained in the fresh sediment system, in a standard sediment system. Our laboratory set-up in fact is a small fermenter with the special feature, that we do not work in a homogeneous liquid system, but in a biphasic system in which the water-insoluble substrate is not in the state of emulsified droplets, but in the state of being adsorbed to a solid surface. The ratio between bacterial biomass and degradation rate is in the range of the ratio in an optimized biotechnological fermenter system, as preliminary calculations showed.

The high degradation rates encouraged us to follow biodegradation manometrically by oxygen consumption in a Warburg apparatus. The standard setup was reduced to 4 g sediment. Fig. 6 shows the oxygen consumption rates during an incubation time of 14 days. Each point of this figure is the result of a three hours' observation such as the example given in Fig. 7. The nutrient dependence is clearly visible. Using the data of Fig. 6 a computing model of the whole time course was established (Fig. 8) in which the accumulated oxygen consumption is given in absolute amounts. This set of experiments was done without an adapted inoculum, only a small amount of the oxic sediment from the indoor mud flats was added to the standard sediment. Under this condition the pronounced lag phase (6 days) was to be expected. Because lag-phase phenomena are of practical importance, we studied them in more detail and found, that they could be influenced by nutrient addition (Fig. 9). The decrease of the oxygen consumption rate after 8 days in Figs. 7 and 8 is most likely due to the insufficient mixing effect

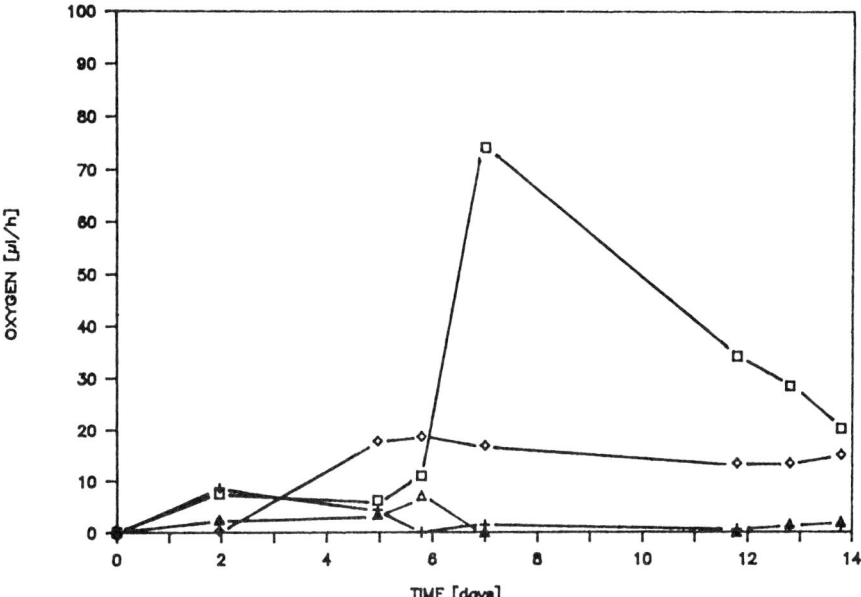

Fig. 6. Oxygen consumption rates during degradation of 28 mg hexadecane in 4 g standardized muddy sediment (manometric assay). -□- in the presence of Gibbs' nutrients. -◊- without nutrient addition. -+- Gibb's nutrients, without hexadecane. -△- without nutrients and hexadecane

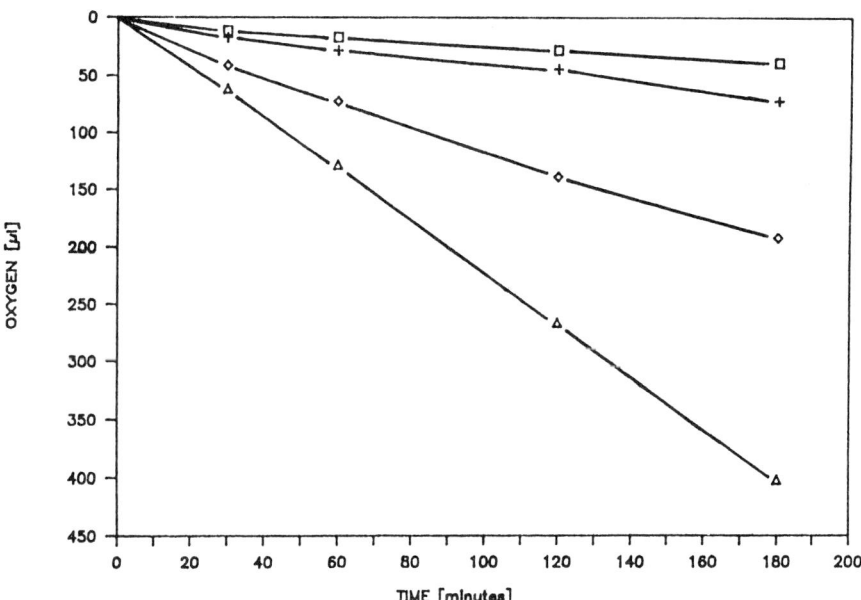

Fig. 7. Oxygen consumption (manometric assay) observed at 5th day after start of degradation of 773 mg hexadecane in 20 g standardized muddy sediment. -△- in the presence of Gibbs' nutrients. -◊- without nutrient addition. -+- Gibb's nutrients, without hexadecane. -□- without nutrients and hexadecane

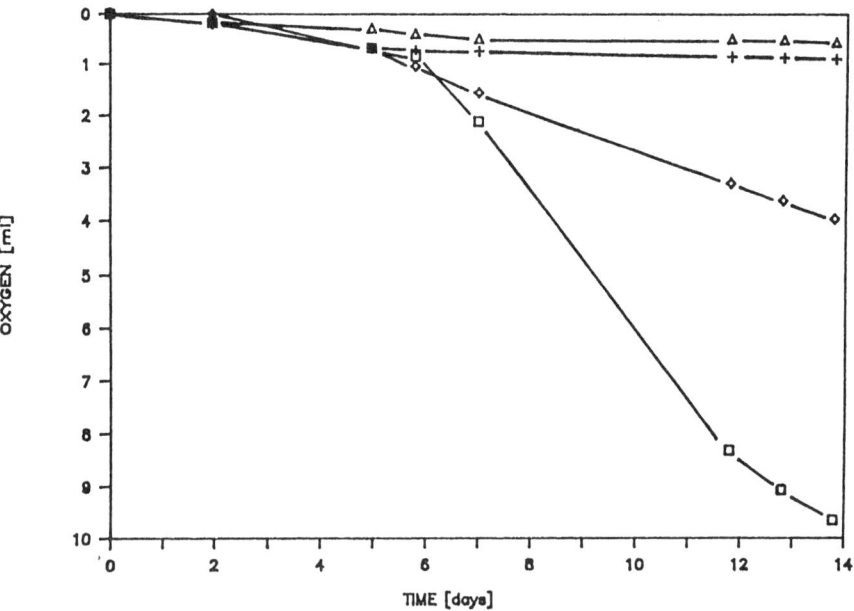

Fig. 8. Oxygen consumption (manometric assay) during degradation of 28 mg hexadecane. For symbols and conditions see legend of Fig. 6. Data calculated from results of Fig. 6

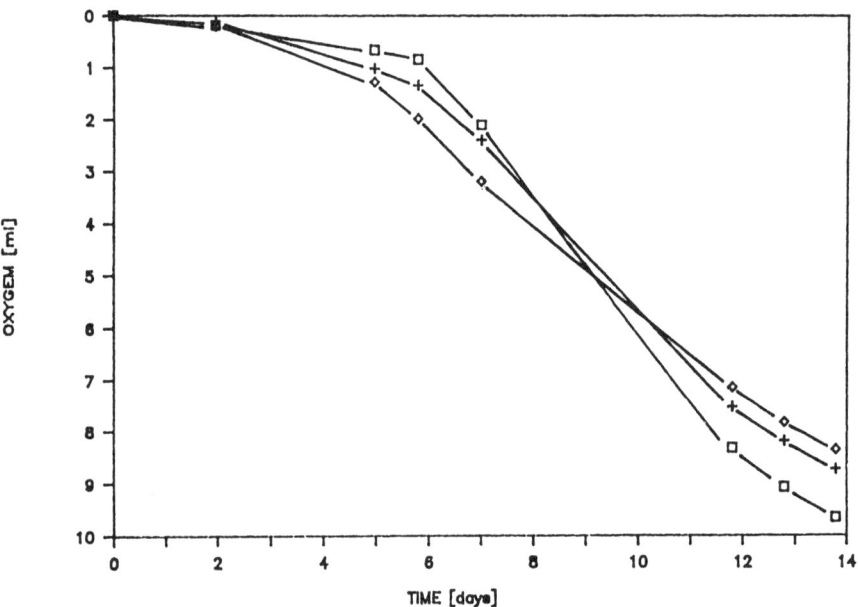

Fig. 9. Effects of different nitrate-N contents on oxygen consumption in the presence of Gibbs' phosphate. -□- 2 fold Gibbs' amounts. - + - 4 fold. -◊- 6 fold. Manometric assay. For other conditions see legend of Fig. 6

of the shaking device of the Warburg apparatus (which is overcome by lowering the sediment amount in the flasks), it was not due to hexadecane exhaustion. The maximum oxygen consumption rate observed corresponds with a hexadecane degradation of 0.18 mg per g sediment per day (total oxidation to CO_2 provided).

12.6 Anaerobic Degradation

We conducted experiments to study anaerobic reaction rates and to find practicable conditions. In these experiments we used at first hexadecane. Up to now we have not experimented with aromatic compounds. Some results are presented here.

As Fig. 4 showed, hexadecane degradation in the "absence" of oxygen is low or zero. Equation (7) could not be verified experimentally.

$$C_{16}H_{34} + 19.6\ NO_3^- + 19.6\ H^+ \rightarrow 16\ CO_2 + 9.8\ N_2 + 26.8\ H_2O. \tag{7}$$

We have to stress the fact that this also means, that there was no or only an immeasurably slow degradation by a sulfate-dependent reaction, because sulfate was always present as a constituent of the seawater (approx. 23 mmol l^{-1}). We were not successful in accelerating degradation by addition of nitrate in the absence of oxygen. So far the biochemical considerations have been confirmed by the experiments. It was also impossible to accelerate the reaction rate by nitrate in the presence of limiting oxygen. From a biochemical or cellular standpoint it is doubtful, whether this acceleration could be expected. Oxygen limitation can mean, that the appropriate amount of the hydrocarbons present is converted to the alcohol, which provides the substrate for the subsequent nitrate dependent degradation. But it can also mean, that a small part of the hydrocarbon is fully degraded. In the case of hexadecane we are presently convinced that the latter case is true, because the preliminary gas chromatographic assays showed no free intermediates of the degradation.

Before drawing final conclusions we wanted to prove that denitrification reactions worked in principle in our standard laboratory system. Our test substrate was glucose. The glucose-dependent denitrification reaction is

$$C_6H_{12}O_6 + 4.8\ NO_3^- + 4.8\ H^+ \rightarrow 6\ CO_2 + 2.4\ N_2 + 8.4\ H_2O. \tag{8}$$

We found a stoichiometric nitrate decay at a rate of 10 μMoles per day in 1 g dry sediment which means 2 μMoles glucose or about 0.4 mg per g sediment per day. The reaction could be followed by monitoring glucose and nitrate decay as well as observing nitrogen formation manometrically. So we were sure that denitrification took place if a suitable substrate was used.

To elucidate the hydrocarbon dependent denitrification step by step we used hexadecane [Eq. (7)] and (jumping across the questionable first oxygenation reaction) in addition its products cetyl alcohol [Eq. (9)] and palmitic acid [Eq. (6)].

$$C_{15}H_{31}CH_2OH + 19.2\ NO_3^- + 19.2\ H^+ \rightarrow 16\ CO_2 + 9.6\ N_2 + 26.6\ H_2O. \tag{9}$$

The three substrates were examined under oxic and anoxic conditions, in the latter case with Gibbs' nitrogen amounts (ammonia or nitrate) as well as with nitrate amounts according to the denitrification stoichiometry. Hexadecane (Fig. 10) showed the already known oxic degradation but no anoxic one. Cetyl alcohol (Fig. 11) was degraded aerobically with a rate of 0.17 mg per g sediment per day. It was also degraded anaerobically according to Eq. (9), the rate being 0.12 mg per g and day. Palmitic acid (Fig. 12) also behaved as expected. The degradation rate following Eq. (4) was 0.3 mg per g sediment and day, following Eq. (6) with 0.2 mg per g and day.

The assumption that the anaerobic degradation is a denitrification reaction like in the case of glucose, is based on the gas chromatographic identification of N_2 and on the stoichiometry of the nitrate reduction. Fig. 13a shows the result of an experiment with a nitrate-palmitic acid ratio according to Eq. (6), while in the experiment of Fig. 13b only one sixth of the stoichiometric amount of nitrate was used. The substrate was fully degraded if the stoichiometric amount of nitrate was present, while the reaction stopped after consumption of the limiting nitrate amount. The reaction rate was about 11 μmol nitrate per g per day (hence it was about the same as with the substrate glucose) or about 1 μmol (0.25 mg) palmitic acid per day in 1 g dry sediment.

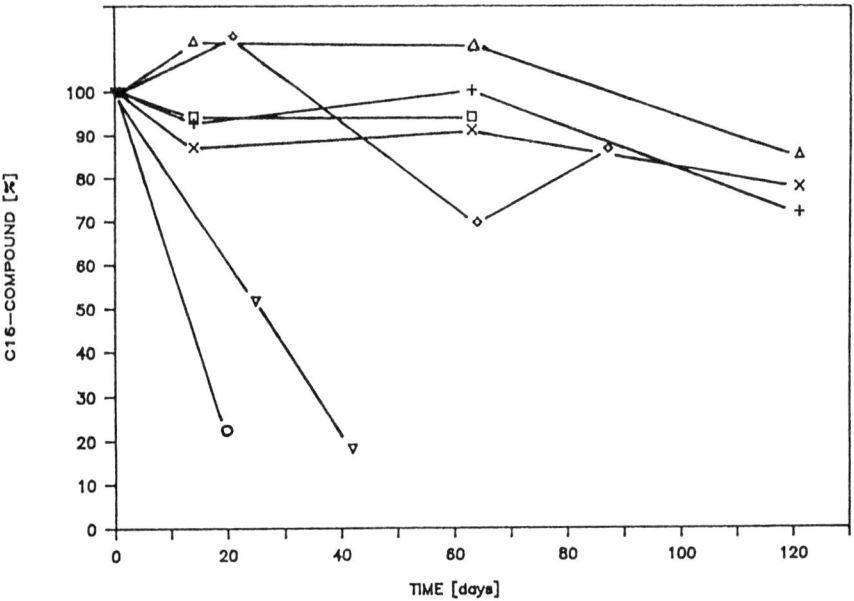

Fig. 10. Anoxic and oxic hexadecane (1.7 mMol) degradation in 55 g standardized muddy sediment in the presence of different nitrogen sources. -○- aerobic, Gibb's nitrate. -∇- aerobic, Gibb's ammonia. -+- anaerobic, Gibb's nitrate. -◊- anaerobic, 2 fold Gibb's nitrate. -△- anaerobic, 5 fold Gibb's nitrate. -×- anaerobic, NO_3- according to the denitrification stoichiometry. -□- anaerobic, Gibb's ammonia. IR-assay of hydrocarbons

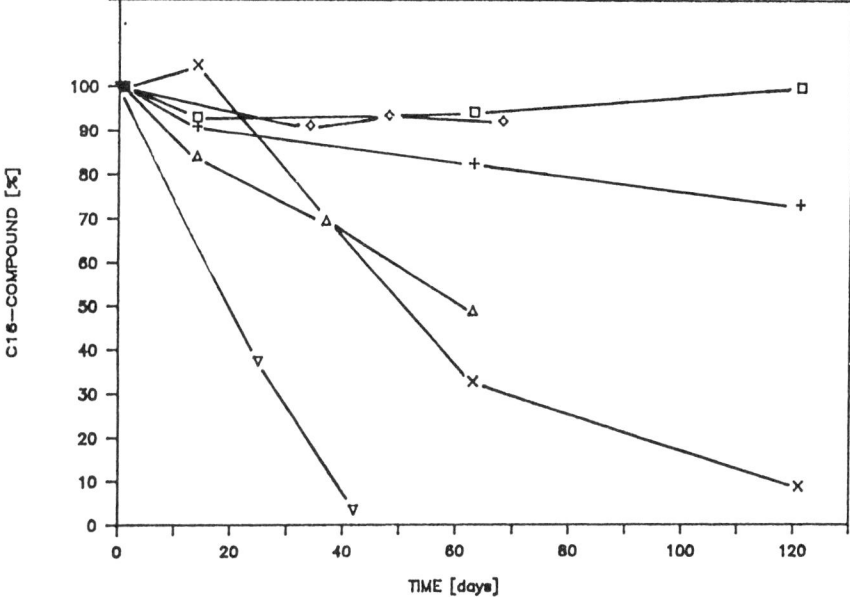

Fig. 11. Anoxic and oxic cetyl alcohol (1.7 mMol) degradation. For symbols and conditions see legend of Fig. 10

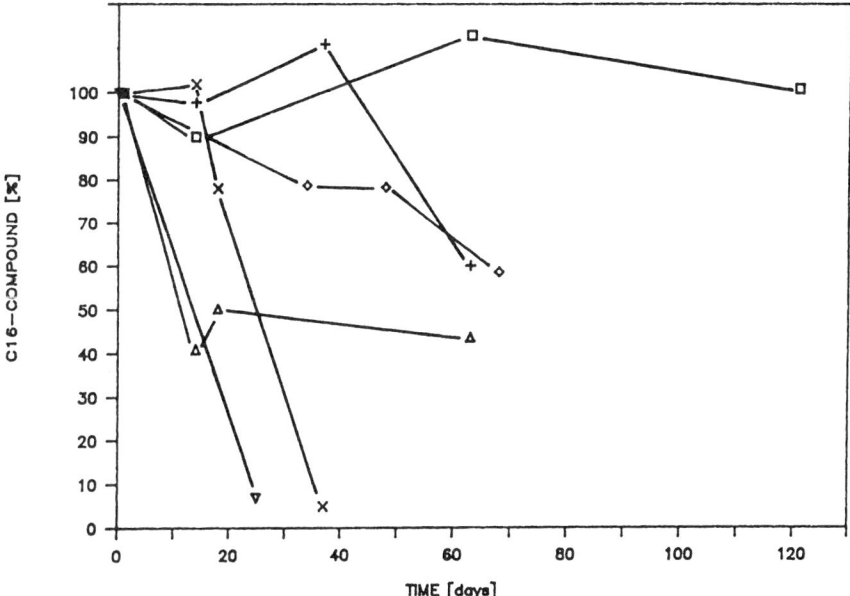

Fig. 12. Anoxic and oxic palmitic acid (1.7 mMol) degradation. For symbols and conditions see legend of Fig. 10

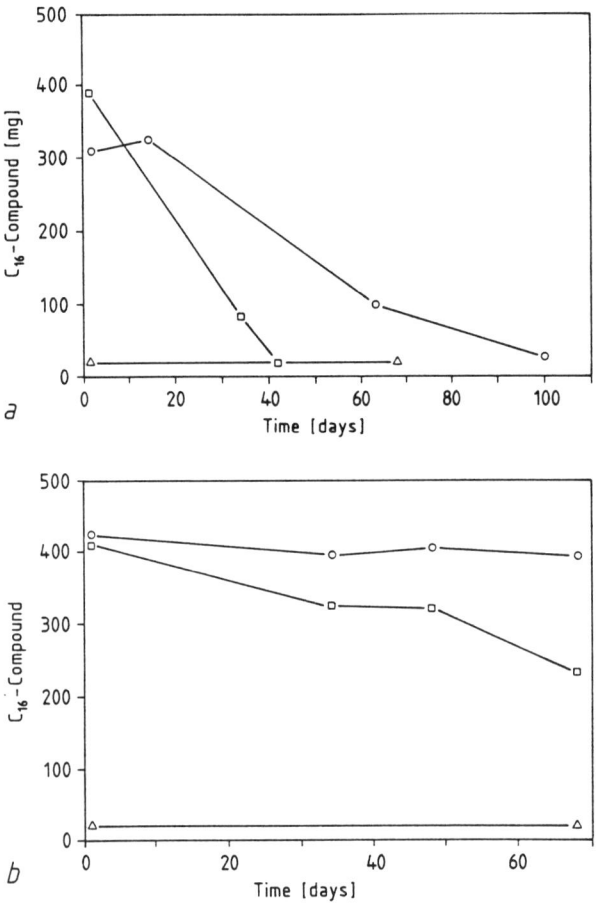

Fig. 13a. Anoxic palmitic acid (-□-) and cetyl alcohol (-○-) degradation in the presence of nitrate according to the denitrification stoichiometry. -△-: no substrate. **b**: same with one 6th of the nitrate in **a**

12.7 Hydrocarbon Biodegradation in Soils

To examine hydrocarbon biodegradation in soil, loess (Wiehengebirge, Osnabrück) was air dried and homogenized, and coarse constituents were removed by sieving. Hexadecane was added dissolved in pentane and the solvent removed in a vacuum evaporator. For water saturated set-ups 55 gram loess and 30 ml tap water were shaken in 250 ml Erlenmeyer flasks. Nutrient amounts are given in the legends of the figures. The set-ups were inoculated from an adapted loess culture. For anoxic incubation air was replaced by nitrogen or argon, the flasks were flushed with nitrogen/argon during the first 10 min of shaking. Soils with 10% water content had a friable structure (aggregates 2–4 mm) with a tendency to aggregate during shaking. Samples were manually mixed every week. Evaporation was minimized by keeping the loosely covered flasks in a water saturated atmosphere and controlled by weighing and water replacement. Loess (150 g) with 3% water content was incubated in closed 1-liter-bottles at an overhead shaking machine. All incubations were at room temperature.

The experiments were concentrated on effects of oxygen supply, of nutrient concentrations and of water content.

Under continuous oxygen saturation (Fig. 14) the initially 1.5 mg hexadecane per g water saturated soil were degraded within 25 days, the maximal rate was 0.1 mg per g soil and day. A nutrient dependency could not be demonstrated (the chosen hydrocarbon amount was so low that the endogenous nutrient content, 3.8 μMol N/g, was sufficient). The nitrogen-flushed setup showed a significantly lower rate.

In loess containing 10% water an initial high rate (about 0.05 mg/day) of degradation was obtained in the presence of Gibbs' nutrients, but the rate decreased and remained very low for a long time (Fig. 15). The reason is not known, but a diffusion limitation is assumed. The initial rate was lower if only the endogenous nutrients were present, and the effect of the inoculum could be easily demonstrated. The degradation rate in the presence of only 3% water was zero or (in other experiments) very low. For comparison results obtained with different water contents are summarized in Fig. 16. In all three cases the Gibbs' nutrients were present.

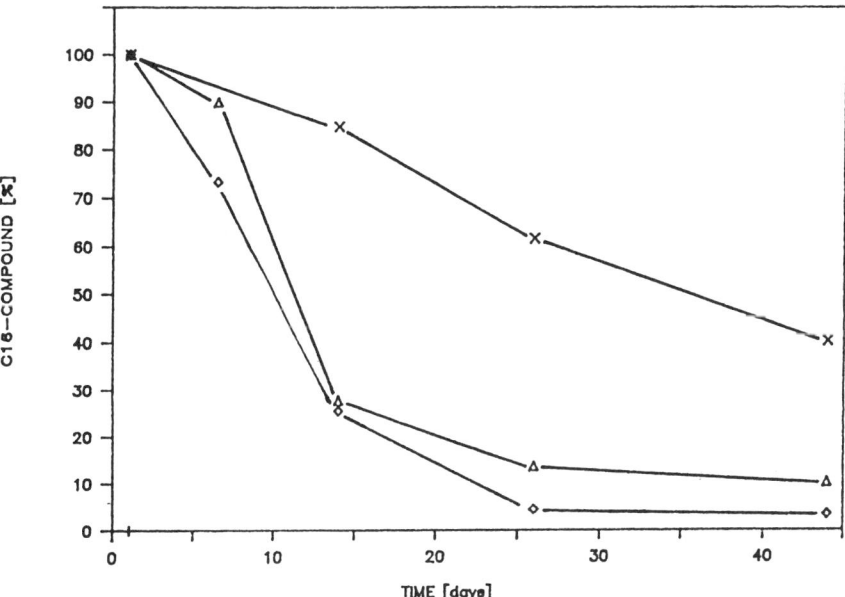

Fig. 14. Degradation of hexadecane (0.3 mMol) in 55 g standardized water saturated loess (= 35% water content) under different oxygen support in the presence of Gibbs' nutrients. -×- anaerobic. -△- limiting oxygen (1.5 mMol in the gas space plus unquantified adsorbed gas). -◊- aerobic. IR-assay of hydrocarbons

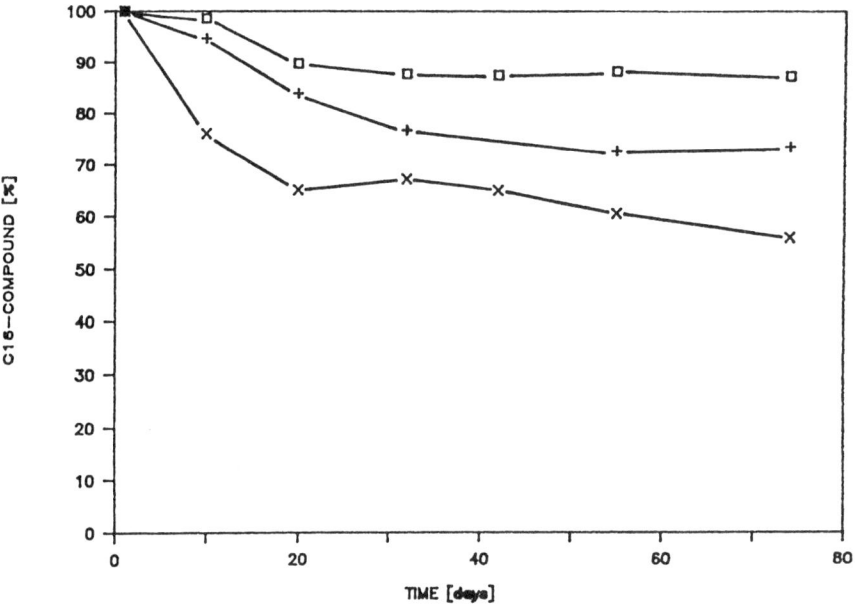

Fig. 15. Aerobic degradation of hexadecane (0.17 mMol) in 10 g standardized loess containing 10% water in the presence of Gibbs' phosphate. -□- without nitrogen and inoculum addition. -+- inoculated, without nitrogen addition. -◊- inoculated, with Gibb's nitrate. IR-assay of hydrocarbons

12.8 Conclusions

The results show that in aerobic water saturated sediment and loess comparable degradation rates (0.1–0.3 mg per g and day) are observed, the kind of substrate and of soil being of minor importance. The rates decreased with decreasing water content and approached zero in the range of 3% water. The anaerobic degradation of palmitic acid proceeded with a comparable rate. We cannot exclude that a step common to these processes, e.g. the diffusion to the microorganisms at the surface of the particles, or the incorporation into the organisms, is rate limiting. As long as a rapid aerobic degradation proceeds, the nutrient support can be rate limiting. In consequence of the rapid oxygen consumption connected with this activity oxygen very soon becomes rate limiting. Under this condition a rate limitation by nutrients is not observed.

It is commonly known that in anoxic sediments hydrocarbon degradation is extremely slow or absent. In the case of the unbranched aliphatic hydrocarbon hexadecane we showed, that the initial hydroxylation and/or the subsequent conversion of the first product into the fatty acid is probably the rate limiting step. Nitrate is a suitable electron acceptor for anaerobic fatty acid degradation. Its conversion into dinitrogen has been derived from the stoichiometry of the reaction, but identification of the gaseous products is still in preparation. Fatty acid degradation using sulfate as electron acceptor cannot be excluded, but, if at all, it proceeds very slowly compared to the denitrification reaction.

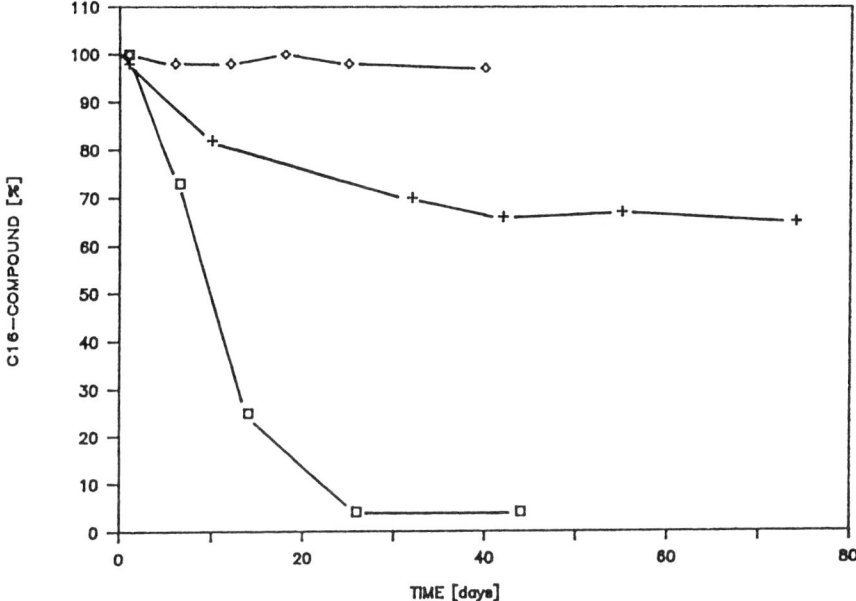

Fig. 16. Aerobic degradation hexadecane in standardized loess of different water content in the presence of Gibb's nutrients. -◊- 3% water, -+- 10% water, -□- 35% water. For experimental conditions of setup with 35% water see legend of Fig. 14, with 10% water see legend of Fig. 15. In the experiment with 3% water 150 g loess contained 2.5 mMoles hexadecane. IR-assay of hydrocarbon

These results are restricted to aliphatic hydrocarbons and it is necessary to examine the behaviour of various aromatic compounds and of raw oil and raw oil products. Nevertheless, the anaerobic experiments confirm the common experiences of the generally low rates and give some insight into the reasons. For practical applications some conclusions can be drawn.

1) In water saturated muddy sediments the apparent hydrocarbon biodegradation rate is limited by oxygen. The first means to support biological degradation is to overcome the oxygen limitation. Mechanical treatment (e.g. mechanical or hydraulic dredging) of oil containing sediments seems to be the first measure which should be examined. Oxygen support to deeper soil layers and to groundwater is technically much more difficult, especially if air saturated water is used as a carrier. The low oxygen solubility is a main argument for replacing this electron acceptor by others (such as nitrate) for which the limit of solubility does not exist.

2) Hydrocarbons in an adsorbed state are more easily degradable than in the state of a separate immiscible layer, of droplets, of an emulsion or of a fluid phase in soils. To support biodegradation, an oil containing sediment or soil should be intensively mixed to achieve a thorough distribution of the oil. This suggestion is in some respects in contradiction to customary opinions. At the same time the treatment is a means to support the sediment or soil with oxygen.

3) When the oxygen limitation has been removed, nutrient limitation is the next bottleneck. Support of biodegradation by nutrients is only effective in the absence of oxygen limitation. Under this condition the biodegradation rate can be enhanced manifold by addition of nutrients. In sediments which are in contact with water it is problematic to keep added nutrients at the area of degradation, it is easier in soil.

4) Biodegradation at or in tidal sediments cannot be supported by spreading out hydrocarbon degrading bacteria. It seems impossible to grow bacteria under the same conditions as present in the material to which they will be added, and bacterial populations which have not been precisely adapted to the conditions of the new medium will die rather than continue to grow.

This experience made in marine sediments remains to be examined in deeper soil layers, where bacterial numbers are much lower than in upper and biologically active sediments.

5) In unsaturated loess the degradation rate is dependent on the water content. Moreover, degradation stops after a relatively short time. The degradation rate drops to zero between 10 and 3% water.

6) The unanswered questions from the field experiments led us to laboratory examinations. It could be shown that laboratory experiments using a standard sediment or soil system provide a hopeful access to the unanswered questions. They demonstrated that the fine grained sediments and soils are very active biocatalytical reaction media for aerobic and anaerobic biodegradation of organic carbon. Degradation conditions can be worked out and optimized in the laboratory, and technical and chemical measures can be conceived to approach and maintain these conditions at the polluted areas. We assume that this kind of combining laboratory and field methods provides access to reasonable oil spill management procedures.

Acknowledgements. This work was supported by the German Federal Minister of Research and Technology (01 ZV 85070 and 01 ZV 85114) and by the Commission of European Communities. The skillful assistance of M. Kanje, B. Kürzel, R. Weinert and I. Wolke is gratefully acknowledged.

References

Atlas RM (1981) Microbial degradation of petroleum hydrocarbons: an environmental perspective. Microbiol Rev 45:180–209

Atlas RM, Bartha R (1981) Microbial Ecology: Fundamentals and Applications. Addison-Wesley, Reading, p 424

Bailey NJL, Jobson AM, Rogers MA (1973) Bacterial degradation of crude oil: comparison of field and experimental data. Chem Geol 11:203–221

Battermann G, Werner P (1984) Beseitigung einer Untergrundkontamination mit Kohlenwasserstoffen durch mikrobiellen Abbau. Gwf-Wasser/Abwasser 125:366–373

Blackburn TH (1983) The microbial nitrogen cycle. In: Krumbein WE (ed) Microbial geochemistry. Blackwell, Oxford, pp 63–89

Brown LR, Phillips WE, Pabst GS, Ladner CM (1969a) Physical, chemical and microbiological changes occurring during degradation of oil in aquatic and brackish water environments. Presented at the American Society of Mechanical Engineers Annual Winter Meeting, Los Angeles, CA

Brown DW, Ramos LS, Fiedman AJ, Macleod WD (1969b) Analysis of petroleum hydrocarbons in marine sediments using a solvent-slurry extraction procedure, pp 161–167. In: Trace organic analysis: a new frontier in analytical chemistry. Special publication no 519. National Bureau of Standards, Washington, DC

Cerniglia CE (1984) Microbial transformation of aromatic hydrocarbons. In: Atlas RM (ed) Petroleum Microbiology. Macmillan, New York, pp 99–128

Chouteau J, Azoulay E, Senez JC (1982) Anaerobic formation of n-hept-l-ene from n-heptane by resting cells of Pseudomonas aeruginosa. Nature 194:576–578

Engler RM, Patrick WH (1974) Nitrate removal from flood-water overlaying flooded soils and sediments. J Environ Qual 3:409–413

Fenchel TM, Riedl RJ (1970) The sulfide system: a new biotic community underneath the oxidized layer of marine sand bottoms. Mar Biol 7:255–268

Firestone MK (1982) Biological denitrification. In: Stevenson FJ (ed) Nitrogen in agricultural soils. Agronomy 22. American Society of Agronomy, Madison, pp 289–326

Gibbs CF (1975) Quantitative Studies on marine biodegradation of oil. I. Nutrient limitation at 14°C. Proc R Soc Lond B Biol Sci 188:61–82

Hagin J, Tucker B (1982) Fertilization of dryland and irrigated soils. Springer, Berlin Heidelberg New York

Höpner TH, Harder H, Kiesewetter K, Teigelkamp B (1987) Hydrocarbon biodegradation in marine sediments: a biochemical approach. In: Kuiper J, Brink WJ van den (eds) Fate and effects of oil in marine ecosystems. Martinus Nijhoff, Dordrecht, pp 41–55

Iizuka H, Ilida M, Fujita S (1969) Formation of n-decene-l from n-decane by resting cells of C. rugosa. Z Allg Mikrobiol 9:223–226

Knowles R (1982) Denitrification. Microbiol Rev 46:43–70

Parekh VR, Traxler RW, Sobek JM (1977) n-Alkane oxidation enzymes of a Pseudomonad. Appl Environ Microbiol 33:881–884

Pierce RH, Cundell AM, Traxler RW (1975) Persistence and biodegradation of spilled residual fuel oil on an estuarine beach. Appl Microbiol 29:646–652

Ronen D, Kanfi Y, Magaritz M (1984) Nitrogen presence in groundwater as affected by the unsaturated zone. In: Yaron B, Dagan G, Goldsmith J (eds) Pollutants in porous media. The unsaturated zone between soil surface and groundwater. Ecological Studies 47. Springer, Berlin Heidelberg New York, pp 223–236

Savant NK, DeDatta SK (1982) Nitrogen Transformations in wetland rice soils. Adv Agron 35:241–302

Senez JC, Azoulay E (1961) Dehydrogenation of paraffinic hydrocarbons by resting cells and cell free extracts of Pseudomonas aeruginosa. Biochim Biophys Acta 47:307–316

Singer ME, Finnerty WR (1984) Microbial metabolism of straight chain and branched alkanes. In: Atlas RM (ed) Petroleum Microbiology. Macmillan, New York, pp 1–59

Sontheimer H, Nagel G, Werner P (1987) Restoration of aquifers polluted with hydrocarbons (this volume)

Stolzy LH, Flühler H (1978) Measurement and prediction of anaerobiosis in soils. In: Nielsen DR, MacDonald JG (eds) Nitrogen in the environment, vol 1. Academic Press, London, pp 363–447

Tiedje JM, Sexstone AJ, Perkin TB, Rewsbech NP, Shelton DR (1984) Anaerobic processes in soil. Plant Soil 76:197–212

Traxler RW, Bernard JM (1969) The utilization of n-alkanes by Pseudomonas aeruginosa under conditions of anaerobiosis. Int Biodeterior Bull 5:21–25

Werner P (1985) A new way for the decontamination of polluted aquifers by biodegradation. Wat Supply 3:41–47

Zeyer J, Kuhn EP, Schwarzenbach RP (1986) Rapid microbial mineralization of toluene and 1,3-dimethylbenzene in the absence of molecular oxygen. Appl Environ Microbiol 52:944–947

Part V. Restoration of the Unsaturated Zone and Groundwater

Introductory Comments

The introduction of toxic species into the unsaturated zone is a particularly insidious form of pollution. This zone is an inhomogeneous solid system. Consequently, exact determination of the distribution of a contaminant throughout the unsaturated zone is often impossible. The most serious danger from pollution of the unsaturated zone is the contamination of the groundwater lying below. Yet, important questions such as how much of a contaminant will reach the groundwater, when will it get there, or for how long will the unsaturated zone serve as a source of contamination of the groundwater are frequently unanswerable. Furthermore, the unsaturated zone is usually more difficult to clean up than is groundwater. The unsaturated zone may lack a continuous liquid phase which can easily be manipulated by the addition of solutes or by controlling its flow through pumping and injection. The complexity of applying cleanup procedures to the unsaturated zone is demonstrated by the fact that cleanup costs of known polluted areas in the United States alone are estimated at billions of dollars.

There are many potential sources for the pollution of the unsaturated zone. Application of agrochemicals, industrial effluents, leaks from storage tanks or pipes, seepage from disposal sites for toxic substances and accidental (or intentional) spills of petroleum-related products and other substances may all result in the contamination of the unsaturated zone. Numerous cases of large-scale pollution of the unsaturated zone have been detected in many parts of the world, and some of these cases are described in this book. The closing of wells and even of water supplies to entire communities is no longer a rare occurrence. Because of the long time that may elapse between the polluting event and the appearance of the pollutant in ground or surface water, the extent of unsaturated zone contamination may be far greater than has already been discovered.

It is apparent that proper management can prevent both ecological disasters and huge expenditures. Proper management requires the ability to estimate the fate of a contaminant present in the unsaturated zone as a function of the properties of the contaminant and of the medium which it pollutes. The ability to predict the fate of a contaminant will make it possible to estimate whether the aquifer is endangered and how long it will take for the pollutant to reach the saturated zone. Malcolm (Chapter 14) describes an experiment which sheds some light on the effect of the water solubility and charge of a contaminant on its retention in low organic matter aquifer material. The author also discusses the effect of pH on the retention. A first order approximation of the extent of retention of an organic molecule is possible from simple data such as the pH of the retaining medium, the water solubility of the contaminant, and its charge at the existing pH.

Once the unsaturated zone is polluted, cleanup may be necessary. Bowman (Chapter 13) reviews procedures for the enhancement of the removal of toxic organic chemicals from the vadose zone. Emphasis is placed on procedures in which the cleanup is attempted within the vadose zone itself. Some of these procedures have been successfully employed while others are innovative ideas which have not as yet been tried in the field. Enhanced volatilization, electrokinetic processes and the introduction of chemicals through the liquid or gas phases are among the procedures described. Cleanup procedures based on biological processes are also discussed. Addition of nutrients to enhance biodegradation by native microorganisms, introduction of microorganisms capable of degrading the pollutant and addition of enzymes are some of the described procedures. In particular, the use of genetic engineering for producing microorganisms capable of degrading specific pollutants is discussed.

It is often easier to dispose of a pollutant after it has already reached the ground or surface water than when it is still in the unsaturated zone. Acher and Saltzman (Chapter 15) demonstrate the use of sensitized photooxidation as a means for removing pollutants and pathogens from water. The photodegradation of a number of classes of pesticides and the disinfection of effluents from bacteria and viruses is discussed. Sontheimer and his coworkers (Chapter 16) describe a procedure for the restoration of an aquifer polluted with hydrocarbons. The procedure which was successfully applied in Karlsruhe, Germany is based on treating the water with ozone as an oxygen source. The increased oxygen content enhances microbial breakdown of the organic pollutants. Another important aspect of the procedure is the control of the flow pattern of the groundwater through the proper location and operation of injection and extraction wells.

Considerable damage has already been caused by pollution of the unsaturated zone. The potential for even more extensive damage from contaminants which have not yet reached water bodies exists. This, and the vast cost of cleanup, has resulted in an intensive research effort into the prevention of contamination of the unsaturated zone and detoxification of areas already polluted. The present section is by necessity only a brief overview of the subject of management of pollution of the unsaturated zone. It serves, however, as a useful introduction to the subject.

13 Manipulation of the Vadose Zone to Enhance Toxic Organic Chemical Removal

R.S. BOWMAN[1]

13.1 Introduction

The vadose zone serves as a medium for the transport of toxic organic chemicals from the soil surface to groundwater. Thus, the vadose zone can itself be considered a source of groundwater pollution; remediation of contaminated groundwater will never be complete as long as mobile contaminants remain in the vadose zone. In addition, even pristine groundwater is threatened with contamination as long as the vadose zone contains xenobiotic materials. Given the long transit times often required for pollutant transport through the vadose zone to groundwater, vadose zone contamination represents an unpleasant legacy for future generations. Clearly, if groundwater pollution is to be avoided or minimized, methods must be applied to remove or control contaminants in the vadose zone.

A spectrum of processes naturally operative in the vadose zone can lead to fixation or degradation of toxic organic chemicals. These natural processes are sometimes insufficient to prevent long-term soil and groundwater contamination by a specific chemical. In this situation, it may be necessary or economically attractive to remove the pollutant from the vadose zone, or to treat it in-situ. This paper reviews physical, chemical, and biological techniques for manipulating the vadose zone to enhance toxic organic chemical removal. Containment methods, such as subsurface barrier formation, are not covered. The future outlook for induced vadose zone chemical removal is discussed.

13.2 Physical Methods

13.2.1 Leaching

Removal of soluble components from the vadose zone by leaching with a solvent (aqueous or organic) is an obvious method of transferring the offending chemical. The objective of vadose zone cleanup is generally to protect groundwater quality; leaching is therefore usually not a viable alternative, since it likely will enhance

---

[1]Dept. of Geoscience, New Mexico Institute of Mining and Technology, Socorro, NM 87801, USA

rather than decrease the threat of groundwater pollution. If the chemical of concern has a significant aqueous solubility, the vadose zone may be cleansed of the chemical by irrigating the soil. Hydrophobic compounds may be removed by leaching with a relatively harmless solvent such as kerosene, followed by extraction of the leachate using recovery wells. The polluting chemical can be recovered from the solvent, if desired, by distillation or adsorption. Such leaching with a non-aqueous solvent will, however, always leave some trace of the solvent in the porous material leached, which may present a long-term pollution hazard in itself. In any case, the addition of organic solvents into a polluted area will not generally be acceptable or desirable.

13.2.2 Enhanced Volatilization

Many of the chemicals of concern which find their way into the vadose zone are liquids which have high vapor pressures in the pure state at ambient temperatures, and/or exhibit high Henry's constants (reflecting their tendency to volatilize) when dissolved in water. These volatile organic compounds (VOC's) partition among pure liquid, dissolved, adsorbed, and vapor phases according to their individual chemical and physical properties and those of the medium in which they are found. This group of compounds includes a wide array of industrial solvents such as trichloroethylene (TCE), tetrachloroethylene (PCE), and chloroform, as well as petroleum fuels such as gasoline. VOC contamination is the most widespread form of groundwater pollution by organic chemicals in the United States (Wilson and Wilson 1985). The U.S. Environmental Protection Agency has estimated that 10–30% of the 3.5 million or more underground tanks used to store chemicals, petroleum products, and other hazardous liquids in the USA may be leaking (Dowd 1984).

Due to their tendency to reside in the vapor phase, VOC's can migrate through the vadose zone via diffusion through the soil gas as well as by transport in the liquid or dissolved state. While such vapor transport allows for more rapid spreading of a contaminant plume and concomitant pollution potential, this property also provides a means of removal of such contaminants from a polluted area. By inducing or enhancing volatilization and providing a sink for the vapor-phase contaminant, the vadose zone can be effectively cleansed, mitigating or preventing groundwater pollution by VOC's.

Little is available in the scientific literature regarding design criteria for enhanced volatilization for VOC decontamination, although Baehr et al. (1989) have developed a one-dimensional model for estimating the rate of removal of a mixture of VOC's from sand. Most of the work in this area has been based on an empirical, site-specific approach applied to the cleanup of actual vadose zone contamination.

The principal of enhanced volatilization is simple. The approach is to reduce the vapor-phase concentration of the VOC's by accelerating the flow of fresh air through the porous medium. This flow can be induced by applying a vacuum in the vadose zone to draw air to an extraction point, by applying air pressure to different points in the vadose zone to induce soil-gas movement, or by a combination of both processes.

A schematic representation of a system to remove VOC's by enhanced volatilization is shown in Fig. 1. Perforated well casing is installed in the contaminated area. With proper initial planning, the same wells used for delineating the contamination plume in the vadose zone and in groundwater can be used for the vacuum extraction at little additional cost. Extraction wells within the plume are connected to a vacuum pump and soil gas is withdrawn. The evacuated gas containing VOC's may be vented directly to the atmosphere or, if this presents an air pollution problem, may be passed through a cleansing system such as activated carbon to remove the contaminants. VOC's may even be recovered by using a condensing system, but generally this will not be economical due to the low concentrations of VOC's in the exhaust gas and the large quantities of gas involved.

The air-inlet wells in Fig. 1 are generally at ambient pressure, although air pressure may be applied to further increase the pressure gradient towards the extraction well. Alternatively, the inlet wells may be omitted entirely, simply allowing air from the surface to be drawn through the vadose zone to the extraction well.

The vacuum extraction system can be useful in decontaminating groundwater as well as in cleansing the vadose zone. The lowered VOC vapor pressure in the vadose zone resulting from vacuum extraction will increase volatilization of contaminants floating on top of groundwater and of contaminants dissolved in the groundwater. To reduce the dissolution and transport of VOC in ground-

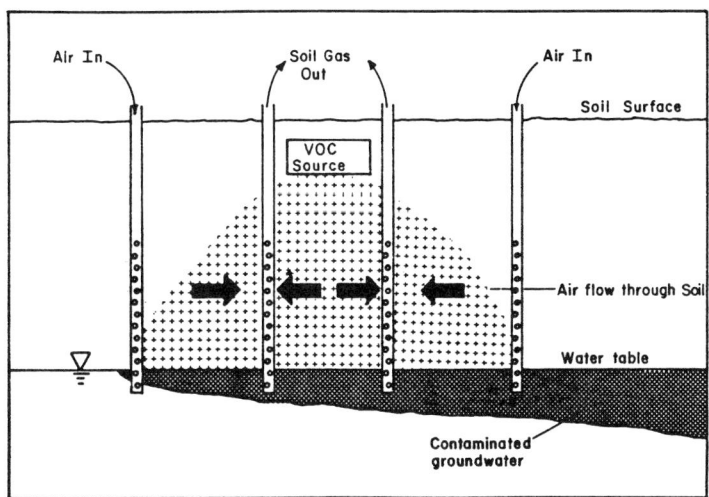

Fig. 1. Schematic diagram of a system to enhance volatilization and removal of volatile organic compounds (VOC's)

water, it is in some instances possible to lower the groundwater level by pumping with a production well, leaving floating VOC stranded in the newly created unsaturated region. The increased VOC-air interface then allows more rapid and efficient removal of pollutant by vacuum extraction.

Two case histories will illustrate the usefulness of this technique. The first example is a system installed for VOC removal at the Twin Cities Army Ammunition Plant in Minnesota, U.S.A. (C. Oster, personal communication 1987). The contaminated vadose zone was below an area which had been used for many years as a waste dump site. The composition and quantity of the VOC contamination was unknown. A diagram of the treatment site is presented in Fig. 2. The slotted polyvinylchloride vent pipes were installed to depths ranging from two to 16 m, packed with gravel, and grouted. The flow rate of exhaust gas ranged from about 3000 ft^3 min^{-1} (1.4 m^3 sec^{-1}) with a single blower at the initiation of the process, to a flow rate of 6200 ft^3 min^{-1} (2.9 m^3 sec^{-1}) with three blowers. Monitoring data indicated that approximately 60,000 lbs (2700 kg) of VOC's were removed from the vadose zone during the first nine months of operation.

Figure 3 shows the rate of VOC and TCE removal as a function of time at the Twin Cities site. The rate of removal decreased as a function of time. The removal rate dropped rapidly after the soil air, saturated with VOC's, was removed, then the rate decreased more slowly as residual chemical was volatilized and transported to the extraction points. The number of exhaust blowers in the extraction system was increased finally to three in order to provide more air flow and maintain a high VOC extraction rate (Fig. 3).

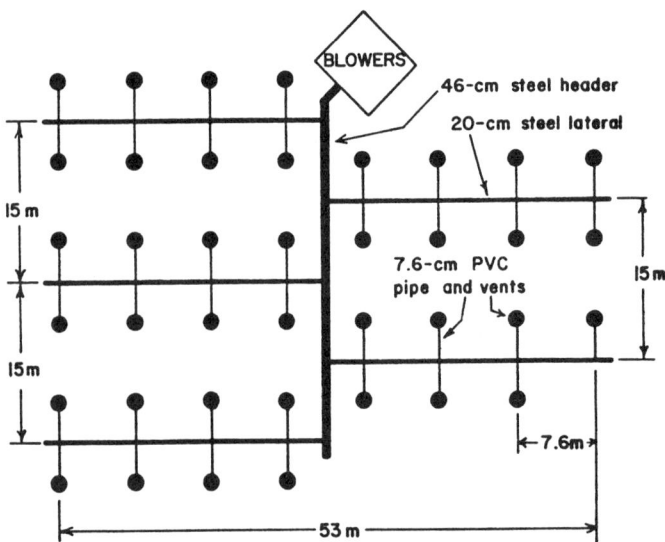

Fig. 2. Schematic diagram of an enhanced volatilization system installed at site D of the Twin Cities Army Ammunition Plant. PVC denotes polyvinylchloride pipe

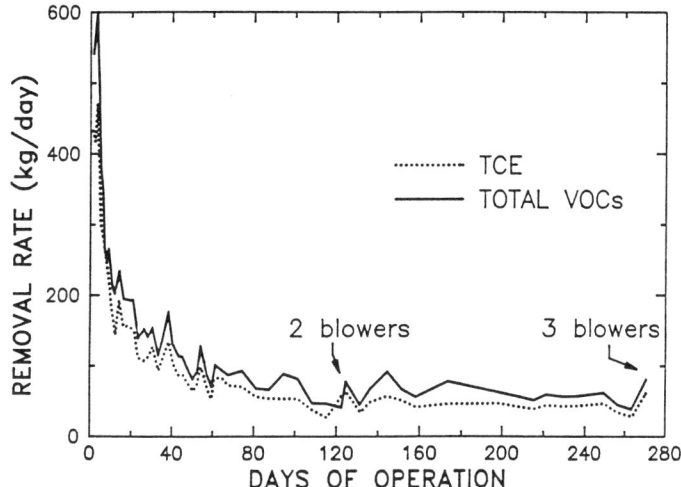

Fig. 3. Removal of hydrocarbons from the vadose zone via enhanced volatilization, at site D of the Twin Cities Army Ammunition Plant

A second example of enhanced in-situ volatilization is provided by Agrelot et al. (1985). In 1982, an underground storage tank in Puerto Rico released approximately 57,000 L of carbon tetrachloride (CCl_4) into the soil. Bore holes drilled in the area following discovery of the leak indicated that approximately 70% of the CCl_4 was contained in the vadose zone, the rest having already arrived at the groundwater, which was at a depth of 100 m. It was determined that enhanced volatilization offered the most promise for recovery of the CCl_4 and the prevention of further contamination of the groundwater, which is a major source of potable water in the area.

Based on pilot studies, a series of vacuum and monitoring wells were installed at depths ranging from 23 to 55 m in the vadose zone. The vadose zone was composed of silty clays of variable thickness (9 to 65 m) overlying permeable limestone. Vacuum was applied to the extraction wells resulting in the development of a pressure field. The observed pressure distribution around one of the extraction wells is presented in Fig. 4.

After 30 months of operation, more than 70% of the spilled volume was removed from the vadose zone by the vacuum extraction system. Cleanup of groundwater during the same period of resulted in recovery of less than 15% of the CCl_4 after extensive pumping of the aquifer. The cost of removing a kilogram of CCl_4 via the enhanced volatilization system was approximately 0.1% of the cost per kilogram for groundwater cleanup.

These case histories illustrate the potential of enhanced volatilization systems to remove VOC's from the vadose zone. All the applications of this technology to date have been in response to groundwater-threatening cases of vadose zone contamination, and have by necessity been empirical in nature. There is a need

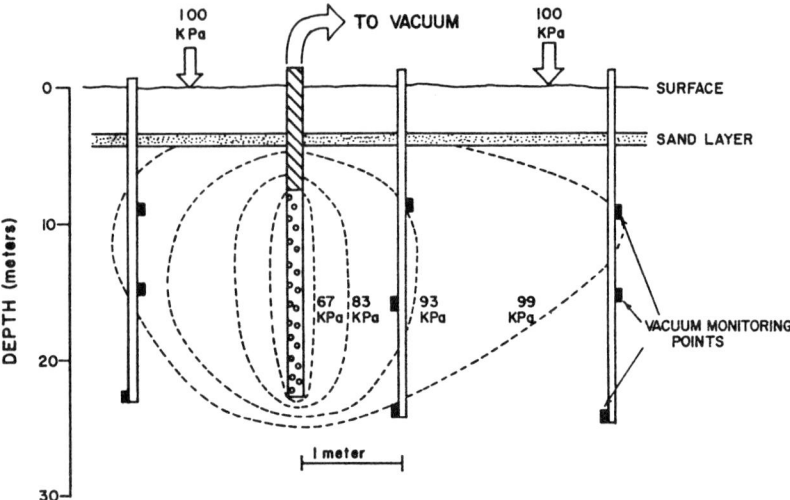

Fig. 4. Pressure distribution in the vadose zone around a vacuum extraction well. Dashed lines are isobars of indicated pressure. Not to scale. (After Agrelot et al. 1985)

for carefully controlled field experiments and model development to allow optimization of well placement and pumping rates for vadose zones of differing texture, structure, and moisture content.

13.2.3 Electrokinetics

Electrokinetic behavior is exhibited by systems in which charged particles and their balancing counter-ions are present. Porous materials such as soils and subsoils containing negatively-charged clays are examples of such systems. Water, particle, and/or electrical current movement can be induced in these systems. The electrokinetic behavior results from the ability of the counter-ions to move relative to the charged particles.

Four types of electrokinetic phenomena can be exhibited by soil-water systems (Mitchell 1976). These are shown diagrammatically in Fig. 5. *Electroosmosis* is the term given to water migration induced by an applied electric field. The field causes the mobile cations to be attracted to the cathode, resulting in a net flow of ions and the associated hydrating water molecules. Due to viscous drag, additional pore water also flows in the direction of the cathode. *Streaming potential* is the opposite of electroosmosis. A flow of water through a bed of charged particles tends to carry with it some of the counter-ions. This results in charge separation and hence a potential difference across the medium. In a system where particles as well as counter-ions are free to move, *electrophoresis* is displayed in response to an applied potential. In electrophoresis, colloidal particles move to the electrode of opposite charge. *Migration potential*, analogous

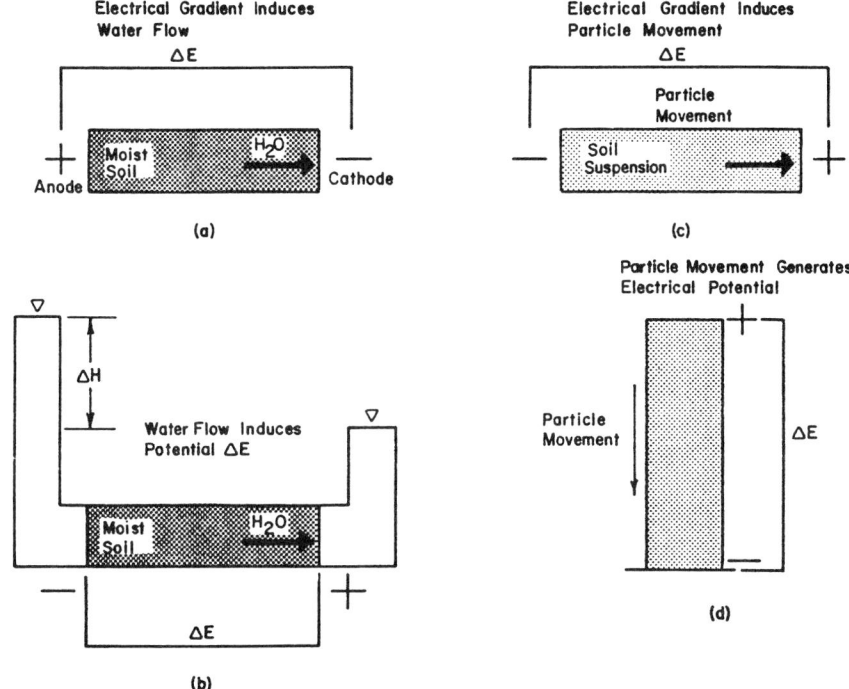

Fig. 5.a-d Electrokinetic phenomena: (a) Electroosmosis. (b) Streaming potential. (c) Electrophoresis. (d) Migration potential (After Mitchell 1976)

to streaming potential in the fixed-bed system, results from differential movement of charged particles and their counter-ions, as in gravitational settling of a colloidal suspension (Fig. 5).

Of these electrokinetic phenomena, electroosmosis would appear to have the greatest applicability to toxic organic chemical treatment in the vadose zone. Using electroosmosis, it may be possible to direct pollutants to specific locations in the vadose zone for treatment or removal.

The application of electroosmosis to contaminant control is just beginning to be investigated. A review of recent work, much of which is unpublished, as well as an extensive bibliography on the use of electroosmosis for dewatering of soils and clays, has been provided by van Zyl and Shakelford (1988). The literature of electroosmosic dewatering will serve as a basis to determine if and how electrokinetics can be used to control pollutants in the vadose zone.

Potential applications of electroosmosis include directing contaminant movement within the vadose zone to avoid groundwater pollution, or to facilitate treatment or removal of the offending materials. A continuous flow system could be created, for example, by injecting water at the anode at the same rate as water is removed at the cathode. Such a system could be used to remove contaminants directly or, alternatively, to transport chemicals or nutrients to adsorbed pollutants in order to transform them chemically or biologically in situ. In addition, it might be possible to construct electrokinetic "barriers", in opposition to prevailing hydraulic gradients, in order to prevent contaminant flow beyond restricted boundaries. Renaud and Probstein (1987) discuss theoretical and practical considerations for the use of electroosmosis in hazardous waste control.

The ability to move water and chemicals about in an unsaturated porous medium would be a compelling advantage of electrokinetics, when used by itself or in combination with other treatment techniques. Further research is needed, however, before electrokinetics can be proven practical for toxic organic chemical removal in the vadose zone. Chemical and physical changes in the porous medium resulting from an applied voltage will need to be quantified. For example, Segall et al. (1980) noted significant changes in leachate chemistry during electroosmotic dewatering of dredged soil, including an increase in pH from 7.8 to 13.4, and a resultant dissolution of soil organic matter. Heavy metal concentrations in the leachate also increased. The impact of such changes to the overall contamination problem would need to be weighed before applying electrokinetic techniques.

13.3 Chemical Methods

Nonbiological degradation of toxic organic chemicals can occur in the vadose zone, and in some situations may be the major mechanism for removal. Saltzman and Mingelgrin (1984) have recently reviewed the literature of nonbiological transformations of pesticides in the unsaturated zone. Hydrolysis, oxidation or reduction, or irreversible sorption by soil organic or mineral matter may singly or in combination serve to remove a contaminant from the vadose zone. Under natural conditions, it is usually difficult to distinguish nonbiological from biological degradation.

Chemical methods to accelerate the detoxification of foreign organics in the unsaturated region are severely hindered by the difficulty of dispersing chemical amendments in a contaminated zone which is more than a few meters in depth. Therefore, most attempts at chemical in-situ treatment have been limited to chemical spill or dump sites where near-surface contaminated soil could be treated and mixed using conventional agricultural implements.

Carpenter and Wilson (1986) recently ranked emerging treatment technologies for decontamination of polychlorinated biphenyl (PCB) -contaminated soils and sediments. Their conclusions are applicable to the treatment of other hazardous halogenated aromatic compounds such as 2,3,7,8-tetrachlorodibenzo-p-dioxin (TCDD) in soils. Of the emerging technologies, treatment with alkaline polyethylene glycol (APEG) shows the most promise for in-situ treatment. APEG is formed by the reaction of polyethylene glycol with an alkali metal such as sodium to form a polyethylene alkoxide. Alternatively, a solution of polyethylene glycol and an alkali metal hydroxide may be used. The degradative effect of the alkoxide is thought to result from the nucleophilic dehalogenation of arylhalides (Kornel and Rogers 1985):

$$\text{AR-X} + \text{RO}^- \rightarrow \text{AR-OR} + \text{X}^-$$
Arylhalide　　Alkoxide　　　Arylether　　Halide ion

The resulting arylether is much more susceptible to degradation by naturally-occurring soil microorganisms than is the parent arylhalide. The APEG process should thus speed the decontamination of arylhalides. One reason APEG is superior to other alkoxides for this purpose is that it is generally miscible with halogenated aromatics. The dehalogenation reaction is much more rapid when alkoxide and arylhalide are in the same phase.

Several laboratory and field studies have been conducted to test the efficiency of APEG conversion of chlorinated aromatic chemicals (Klee et al. 1984; Rogers et al. 1985; Iaconianni 1985). These studies have found that APEG can reduce the concentration of chlorinated aromatics dramatically under the proper conditions. Suitable conditions include low soil moisture content, since APEG activity is lowered in the presence of water. For instance, the TCDD level was reduced by 68% in soil contaminated by 330 ppb of TCDD, after two treatments with APEG over a 21-day period (Klee et al. 1984).

The efficiency of dechlorination by APEG is greater at elevated temperatures (Peterson et al. 1985; Kornel and Rogers 1985). Thus, warming of contaminated soil, perhaps via radio frequency heating, has been suggested as a complement to APEG decontamination (Dev 1986). Heating of the soil would also facilitate moisture removal, further enhancing the effectiveness of the APEG process.

Carpenter and Wilson (1986) estimate that APEG technology will not be developed for application to actual contamination situations before 1989. Even when (and if) the process has been shown effective in the field, it will likely be confined to applications where the contaminant is present in the top few meters of soil.

13.4 Biological Methods

13.4.1 Stimulation of Native Organisms

Many toxic organic chemicals can be metabolized or degraded to some degree by native soil microorganisms. Stimulation or selection of the pollutant-degrading portion of the microbial population can be increased by long-term exposure to the offending chemical, or by providing additional nutrients to promote microbial growth. The observation that long-term use of a pesticide on the same soil often results in increasing rates of degradation and higher pesticide application requirements illustrates how microbial communities can adapt to the presence of a foreign carbon or energy source.

At high concentrations of a xenobiotic chemical in the vadose zone, much of the native microbial consortium may be killed off. The remaining members of the population are of necessity less sensitive to the toxic characteristics of the foreign material, and may display the ability to partially or wholly degrade it. The foreign compound represents a carbon and energy source, and utilization of the compound rapidly depletes the supply of the inorganic nutrients also required for microbial metabolism. These nutrients, along with oxygen, which is required for aerobic degradation, are generally present initially at low levels in the vadose zone. Providing microorganisms with additional nutrients can increase the rate of toxic organic chemical removal.

Enhanced biodegradation of contaminated groundwater by pumping oxygen and nutrients into the contaminated zone is now an established aquifer remediation technique. A recent review of the topic is provided by Wilson et al. (1986). In the basic process, dissolved nutrients in oxygenated water are injected into the groundwater and circulated by pumping water from production wells. The nature and levels of nutrients yielding the optimum degradation rate for the specific contaminants are predetermined by laboratory investigations. Due to the low solubility of oxygen in water, aeration is generally the limiting factor. Recently, Raymond et al. (1986) have received a US patent regarding the use of hydrogen peroxide to enhance biodegradation. Hydrogen peroxide at controlled concentrations can be utilized as an oxygen source by microorganisms.

The same nutrient and oxygen enrichment processes can be applied to vadose-zone contamination, although in this case the nutrients are carried to the organisms by leaching. If the infiltration rate is low, remediation may be a very slow process. Since leaching will also promote transport of mobile contaminants, provisions to pump groundwater to prevent the spread of contamination in the aquifer may also need to be instituted along with the leaching process.

Methane-utilizing bacteria are known to dehalogenate certain chlorinated hydrocarbons which are generally resistant to aerobic degradation. Wilson and Wilson (1985) showed in laboratory studies that degradation of TCE could be enhanced in unsaturated soil columns by passing a stream of air enriched to 0.6% by volume in natural gas (77% methane) over the head of the column. Only 5% of the added TCE was eluted from the methane-treated column, compared to 21 to 28% eluted from a similar column flushed with air alone. In both systems, much of the TCE was removed by enhanced volatilization. The authors showed that the greater removal of TCE from the methane-exposed soil was directly attributable to increased biological degradation. Thus, in vadose zone contamination situations, an enhanced volatilization system might be combined with introduction of natural gas to provide simultaneous physical and biological removal of volatile halogenated organics.

13.4.2 Addition of Altered Organisms

The concept of genetically altering organisms to preferentially degrade specific toxic organic chemicals has caught the imagination of many researchers. However, while stimulation of native organisms to transform toxics is an actively applied technology, the use of organisms specifically designed for subsurface contaminant degradation is still in the laboratory stage.

There is a vast literature on the microbial degradation of toxic organics. In some cases, the enzymes responsible for degradation have been isolated, but in only a few instances have the genes which encode these enzymes been described (Kearney et al. 1987). When the responsible genes have been identified, they have often been found to be clustered on plasmids, which are small fragments of extra-chromosomal DNA. Such plasmids have the potential for incorporation into new or improved organisms using molecular cloning techniques. Bacterial strains have been developed which preferentially metabolize pesticides such as 2,4,5-trichlorophenoxyacetic acid (2,4,5-T) using these procedures (Ghosal et al. 1985).

Many questions remain regarding the potential usefulness of genetically-altered organisms for vadose zone or groundwater removal of toxic organic chemicals. The problem of dispersal of an introduced organism throughout a contaminated region is even greater than is the case for chemical amendments, since microorganisms tend to be sorbed by solid particles. Thus, transport of the organisms from a point of injection via the liquid phase will be retarded.

Assuming that the efficient dispersal of engineered degrading organisms can be accomplished, it remains to be seen if such organisms can effectively compete with native microflora for available oxygen and nutrients. If the introduced organisms do indeed survive and flourish, they may upset the microbial balance once the porous medium has become decontaminated. On the other hand, it may be possible to produce engineered organisms which die off rapidly once contaminant degradation is complete. Kilbane et al. (1983) found a 2,4,5-T-degrading *Pseudomonas cepacia* strain was reduced to undetectable levels after transforming all the 2,4,5-T in a contaminated soil. Subsequent addition of more 2,4,5-T to the soil resulted in a buildup in the degrading bacteria followed by a decline in population again following total 2,4,5-T removal. Such findings suggest that introduced organisms will not compete well with native organisms in the absence of their specific carbon/energy source, but yet may "lie in wait" for the reintroduction of a contaminant. Thus, the potential exists for seeding waste-disposal or pesticide-use areas with genetically altered organisms to prevent the buildup of organic pollutants.

Probably the greatest hurdle in applying genetic engineering to toxic organic chemical removal lies in overcoming justifiable environmental concerns about introducing foreign organisms into the environment. Disagreements among scientists, environmental groups, and the general public have hindered the development of regulations regarding the testing and use of genetically-altered microorganisms, and delayed for several years the field evaluation of organisms such as a bacterium which inhibits frost formation on strawberries (Crawford 1986).

Rather than introduce entire microorganisms into the environment, degradative enzymes mass-produced by genetically-altered bacteria might be used instead. The use of cell-free enzyme preparations to degrade a variety of organic pollutants is well documented (Chakrabarty 1982). Use of enzymes would avoid the need to introduce foreign organisms into the environment, and the enzymes might be more readily dispersed within a porous medium.

Application of biological treatment methods or enhancing biological degradation of pollutants in-situ is hampered by the same sorts of obstacles found in chemical treatment methods. This is primarily related to difficulties in dispersing nutrients and organisms throughout a contaminated area of the vadose zone. Nonetheless, biological treatment processes have received much attention for groundwater remediation and, more recently, for vadose zone remediation.

13.5 Conclusions and Future Research Needs

Manipulation of the vadose zone to enhance toxic organic chemical removal is an emerging technology. With the exception of soil treatment within a few meters of the surface, the major problem with all techniques is one of transport. Either the contaminant must be transported to the site of treatment or removal, or chemical or biological amendments must be transported throughout the contaminated zone for treatment in place. The greatest needs for future research would thus appear to lie in improving physical transport in the vadose zone, and perhaps in developing chemical and biological treatments which utilize gas-phase rather than solution-phase transport. Physical transport could be improved through the optimization of enhanced volatilization systems (for VOC removal), and by using electrokinetics to direct chemical amendments or nutrients to target areas within the vadose zone. Stimulation of methane-utilizing bacteria for dehalogenation of solvents is an example of enhanced removal which utilizes gas-phase transport of an amendment. It may also be possible to develop gaseous chemical reactants for destruction of specific vadose-zone contaminants.

References

Agrelot JC, Malot JS, Visser MJ (1985) Vacuum: defense system for ground water VOC contamination. Proc Fifth Natl Symp on Aquifer Restoration and Ground Water Monitoring, 21–24 May 1985; Columbus, OH, USA

Baehr AL, Hoag GE, Marley MC (1989) Removing volatile contaminants from the unsaturated zone by inducing advective air phase transport. J Contaminant Hydrology 4(1): (in press)

Carpenter BH, Wilson DL (1986) PCB sediment decontamination-technical/economic assessment of selected alternative treatments. EPA-600/2-86/112, National Technical Information Service, Springfield, VA, USA

Chakrabarty AM (1982) Biodegradation and detoxification of environmental pollutants. CRC, Boca Raton, Florida, USA

Crawford M (1986) Regulatory tangle snarls agricultural research in biotechnology arena. Science 234:275–277

Dev H (1986) Radio frequency enhanced in-situ decontamination of soils contaminated with halogenated hydrocarbons. Proc of the Twelfth Annual Research Symp EPA-600/9-86/022 pp 402-412. National Technical Information Service, Springfield, VA, USA

Dowd RM (1984) Leaking underground storage tanks. Environ Sci Technol 18:309A

Ghosal D, You I-S, Chatterjee DK, Chakrabarty AM (1985) Microbial degradation of halogenated compounds. Science 228:135-142

Iaconianni FJ (1985) Destruction of PCB's – environmental applications of alkali metal polyethylene glycolate complexes. EPA/600/2-85/108, National Technical Information Service, Springfield, VA, USA

Kearney PC, Karns JS, Mulbry WW (1987) Engineering soil microorganisms for pesticide degradation, p 591-596. In: Greenhalgh R and Roberts TR (eds) Pesticide science and biotechnology. Blackwell Scientific Publications, London, UK

Kilbane JJ, Chatterjee DK, Chakrabarty AM (1983) Detoxification of 2,4,5-trichlorophenoxyacetic acid from contaminated soil by *Pseudomonas cepacia*. Appl Environ Microbiol 45:1697-1700

Klee A, Rogers C, Tiernan T (1984) Report on the feasibility of APEG detoxification of dioxin-contaminated soils. EPA-600/2-84/071. National Technical Information Service, Springfield, VA, USA

Kornel A, Rogers CJ (1985) PCB destruction, a novel dehalogenation reagent. J Hazard Mater 12:161-176

Mitchell JK (1976) Fundamentals of soil behavior. Wiley, New York, 422 p

Peterson RL, Milicic E, Rogers CJ (1985) Chemical destruction/detoxification of chlorinated dioxins in soils. Proc of the 11th Annual Research Symp EPA-600/9-85/028 pp 106-111. National Technical Information Service, Springfield, VA, USA

Raymond RL, Brown RA, Norris RD, O'Neill ET (1986) Stimulation of biooxidation processes in subterranean formations. US Patent Office. 4,588,506. Patented May 13, 1986

Renaud PC, Probstein RF (1987) Electroosmotic control of hazardous wastes. J Physicochem Hydrol 9:345-360

Rogers CJ, Klee AJ, Kornel A, White JB, Leese KB, Clayton AC (1985) Interim report on the feasibility of using UV photolysis and APEG reagent for treatment of dioxin contaminated soils. EPA/600/2-85/083, National Technical Information Service, Springfield, VA, USA

Saltzman S, Mingelgrin U (1984) Nonbiological degradation of pesticides in the unsaturated zone. In: Yaron B, Dagan G, Goldschmid J, Pollutants in Porous Media. Ecological Studies 47. Springer-Verlag, Berlin

Segall BA, O'Bannon CE, Mathias JA (1980) Electro-osmosis chemistry and water quality. J Geotech Eng Div ASCE 106 (GT10):1148-1152

Van Zyl D, Shackelford CS (1988) Electro-kinetic treatment applications in environmental-geotechnical engineering. Geotechnical News 6:19-25

Wilson JT, Wilson BH (1985) Biotransformation of trichloroethylene in soil. Appl Environ Microbiol 49:242-243

Wilson JT, Leach LL, Henson M, Jones JN (1986) In situ biorestoration as a groundwater remediation technique. Ground Water Review 6:56-64

14 The Relative Importance of pH, Charge, and Water Solubility on the Movement of Organic Solutes in Soils and Ground Water

R. L. MALCOLM[1]

1.1 Introduction

Water solubility and charge of the organic solute and the pH of the natural environment are three dominant factors in the movement of organic pollutants in soils, sediments, and ground-water aquifers. These factors are sometimes overlooked by soil scientists and hydrologists involved in pollution studies who are primarily inorganic chemically oriented. The major objectives of this paper are not only to demonstrate the importance of pH, charge, and water solubility for organic solute movement, but to emphasize that by considering these and a few other factors, a simplistic conceptual model of pollutant movement can be formulated with a limited data base and in a short period of time. Such first approximations of waste movement are imperative today relative to crisis management of pollutant spills, the urgency and relevancy of cleanup procedures, the potential of the spill to contaminate ground waters, and the relative human health hazard of the incident. A series of experiments in the movement of several organic solutes during ground-water recharge of the Ogallala aquifer near Stanton, Texas, by the U.S. Geological Survey will serve to demonstrate the stated objectives of this paper.

14.2 Experimental

14.2.1 Overview of the Injection Site

In experimental investigations several organic and inorganic solutes were systematically added to water used to recharge the Ogallala aquifer through a well. During pulse injections of given solutes, the movements of water and the solutes were monitored in observation wells throughout the saturated aquifer thickness and at intervals of 2, 3, 10, and 15 meters from the injection well. Radial flow of water and solutes from the injection well was postulated. All the observation points were sampled frequently such that a plot of time versus solute concentration was obtained to determine relative movement of solutes.

---

[1]U.S. Geological Survey – WRD, 5293 Ward Road, MS408, Arrada, CO 80002, USA

14.2.2 Properties of the Ogallala Aquifer near Stanton, Texas

At this site the Ogallala aquifer is typically very poorly sorted silty sand and gravel with a few clay lenses. Numerous zones and concretions of caliche occur near the surface, but none is thought to be in the saturated portion of the formation. The saturated aquifer thickness is 12 meters, with a depth of 32 meters from land surface to the water table. The aquifer is unconfined. The cation exchange capacity of typical sands and gravels is less than 1 meq/100 g. Organic carbon content is less than 0.1 percent by weight. The porosity of the aquifer ranged from 25 to 31 percent, the transmissivity from 210 to 300 m^2/d, and the storage coefficient from 3×10^{-2} to 2×10^{-5}. The native water is calcium bicarbonate type with a pH near 7.8 and a specific conductance of 450 µS. Dissolved organic carbon (DOC) of the water is 1 mg C/L (Leenheer et al. 1974). An analysis of the ground water is given in Table 1.

14.2.3 Design and Construction of Injection and Observation Wells

The 0.25-meter injection well was screened throughout the saturation interval. Three groups of three observation wells were located at 2-, 5-, and 30-meter distances from the injection well. In each group, one well was opened and screened through the entire saturated thickness of the aquifer to obtain integrated water samples from the full depth; a second well contained piezometers installed at various depths with fine-grained material between piezometers; and a third well contained point samplers with fine-grained material between the cups similar to those described by Wood (1973), except that the porous cups were replaced by plastic screens. The samplers were emptied by N_2 gas pressure through polyethylene tubing to the surface. Segments of tubing to the various samplers were color-coded black, red, orange, and green from deepest to shallowest, as shown in Fig. 1. In addition to the three groups of wells, water samples were obtained from a well screened from a depth of 41 to 42 meters, located 10

Table 1. Chemical analyses of Ogallala ground water

Constituent	mg/L	Constituent	mg/L
Calcium (Ca)	37	Silica (SiO$_2$)	44
Magnesium (Mg)	23	Strontium (Sr)	1
Sodium (Na)	8	Boron (B)	0.1
Potassium (K)	20	Iron (Fe)	0.005
Chloride (Cl)	19	Fluoride (F)	2
Sulfate (SO$_4$)	28	Bromide (Br)	0.1
On-site alkalinity (HCO$_3$)	116	Manganese (Mn)	0.5
Laboratory alkalinity (CaCO$_3$)	170	On-site pH	7.8
Lithium (Li)	0.06	Specific conductance (µS)	462
Nitrate (N)	0.6	Dissolved solids	288

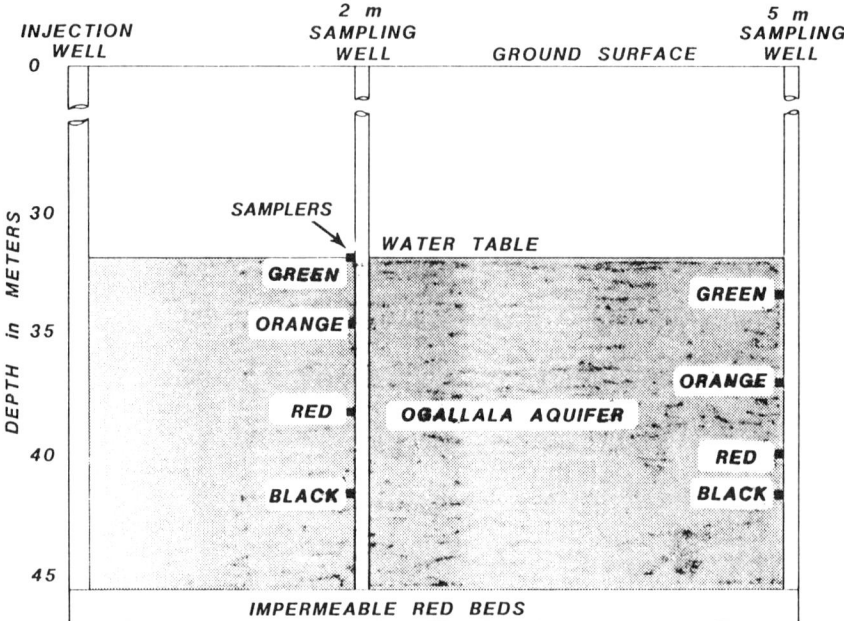

Fig. 1. Diagram of the injection well and the observation well system

meters from the injection well, and a well screened from a depth of 32 to 42 meters, located 15 meters from the injection well.

Water samples in the fully penetrating observation wells at 2, 5, and 30 meters were obtained using small-diameter, gas-operated sample pumps described by Signor (1978). Water samples from the piezometer installed at the same depth as the 5-meter red point sampler and from the wells at 10 meters and 15 meters were sampled using a sampler that consisted of a check valve-equipped copper cylinder connected to the surface by two strings of polyethylene tubing.

14.2.4 Experimental Sequence

Four separate injection experiments were conducted to test the movement of organic and inorganic solutes (Table 2). The first two experiments tested predicted conservative movement of hydrophilic organic solutes as tracers; the last two experiments tested predicted nonconservative movement of organic solutes. Another injection experiment to test the movement of natural organic solutes in playa-lake water (Experiment 3) was reported earlier (Malcolm et al. 1980). Data relating to fluorocarbon and inorganic tracers injected during each experiment also are given in Table 2.

Table 2. Experimental sequence and injection rates

Experiment	Source of Injected Water	Duration and Rate of Water Injection	Tracer	Tracer Concentration (mg/L (or) mg/L as DOC)	Time Period of Tracer Injection
1	Ogallala ground water	*March 1, 1978* 1200–1254: 635 liter/min 1254–2130: 510 liter/min	Ethanol $CBrClF_2$ Iodide	50 0.0007 1.0	*March 1, 1978* 1200–1254 (54-min pulse) 1200–1254 (54-min pulse) 1200–1730 (continuous addition)
2	Ogallala ground water	*March 1, 1978* 2130–2400: 510 liter/min *March 2, 1978* 0000–0400: 510 liter/min 0400–0800: 435 liter/min	Ethanol Benzoate CCl_2F_2	45 5 0.003	*March 1, 1978* 1730–1824 (54-min pulse) *March 2, 1978* 1730–1824 (54-min pulse) 1730–1824 (1-hr pulse)
3	Playa lake-water	*May 24, 1978* 1400–1625: 630 liter/min 1625–1650: 0 liter/min 1650–1920: 490 liter/min 1920–1950: 0 liter/min 1950–2220: 390 liter/min	Natural organic solutes	8.4	*May 24, 1978* 1400–2220 (continuous addition)

Table 2. (Continued)

Experiment	Source of Injected Water	Duration and Rate of Water Injection	Tracer	Tracer Concentration (mg/L (or) mg/L as DOC)	Time Period of Tracer Injection
4	Ogallala ground water	*August 10, 1978* 1300–2400: 342 liter/min	Ethanol	24	*8/10/78* 1300–1700 (4-hr pulse)
			Benzoate	10	1300–1600 (3-hr pulse)
		August 11, 1978 0000–2400: 342 liter/min	Phenol phthalein	2	*8/10/78* 1300–1700 (4-hr pulse)
			Acetate	10	1300–1700 (4-hr pulse)
			CCl_2F_2	6.5	1300–1600 (3-hr pulse)
			Bromide	20	1300–1700 (4-hr pulse)
					1300–1700 (4-hr pulse)
5	Ogallala groundwater	*August 12, 1978* 1710–2400: 342 liter/min 0000–2400: 342 liter/min	Boron	4	*8/12/78* 1710 to 2310 on 8/12/78 (6-hr pulse)
			Ethanol	11.5	1710 on 8/12/78 to 0100 on 8/13/78 (8-hr pulse)
			Aniline	9.5	1710 on 8/12/78 to 0100 on 8/13/78 (8-hr pulse)
			Ethylamine	25	1710 on 8/2/78 to 0100 on 8/13/78 (8-hr pulse)

14.2.5 Solute or Tracer Addition and Observation Well Sampling

The organic solutes tested were ethanol, benzoate, acetate, aniline, ethylamine, and phenolphthalein. They were chosen as the minimum number of organic solutes to evaluate the effect of charge and water solubility on organic solute movement. The inorganic solutes tested were bromide, iodide, and borate. All the solutes except phenolphthalein are small solutes of approximately the same molecular size; therefore, molecular size was not a factor in possible differential movement. Ethanol, acetate, and ethylamine are two-carbon aliphatic compounds differing only in charge and functional grouping at the injected pH of 7.8. Benzoate and aniline are simple aromatic compounds of the same size; one is acidic, the other basic; benzoate is negatively charged at pH 7.8, whereas aniline is uncharged. Ethylamine and aniline are both organic bases; aniline is aromatic and ethylamine is aliphatic; ethylamine is positively charged at pH 7.8, whereas aniline is uncharged. Ethanol, aniline, and phenolphthalein are all uncharged organic solutes at pH 7.8, but differ in water solubility.

Injected tracer concentrations and pulse times are given in Table 2. Each type of solute was introduced into the feedwater stream by metering pumps. A series of injection taps were located approximately 30 meters from the injection well. An injection pump and several bends in the flow line resulted in thorough mixing of the tracers in the feed stream before injection into the well. Injected tracer concentrations were determined on periodic samples taken from a sampling tap at the head of the injection well.

To prevent clogging of the injection well and any unknown complicating interactions between recharge water and the native ground water, Ogallala ground water was used as the recharge water. Ground water used in Experiments 1 and 2 was pumped from a producing well in the Ogallala approximately 95 meters from the injection well. This water was essentially sediment free, but was routed through diatomaceous earth filters before injection. Ground water from the same producing well was used in Experiments 4 and 5, but the water was unfiltered.

Point samplers at the 2-meter observation well were sampled at 5-minute intervals for 1.5 hours after initial tracer injection; 10-minute intervals from 1.5 to 2.0 hours; 15-minute intervals from 2 to 3 hours; 30-minute intervals from 3 to 4 hours; and at 1-hour intervals for the remainder of each experiment. Point samplers at the 5-meter observation well were sampled at 10-minute intervals for the first 2 hours and at the same intervals as the 2-meter well for the remainder of each experiment. The 10-, 15-, and 30-meter wells were sampled at hourly intervals. The red sampling point in the 5-meter well was inoperative during Experiments 1 and 2. All samples were chilled after collection and were kept on ice to prevent organic solute deterioration.

14.2.6 On Site Analyses in the Mobile Laboratory

All samples were analyzed within hours of collection with a Beckman 915 Carbon Analyzer.[2] The standard curve was linear over the DOC range from 2 to 50 mg/L. Phenolphthalein was analyzed at pH 12 at a wavelength of 560 nm on a Beckman Model DB spectrophotometer. Fluorocarbon analyses were performed by gas chromatography using an electron capture detector.

14.2.7 Laboratory Analyses

Benzoate and aniline analyses were performed by direct aqueous injection of 200-μL samples on a Varian 8500 liquid chromatograph fitted with a micro Bondapak C18 column. Benzoate and aniline were detected at 221 nm on a Varichrome spectrophotometer. Sodium phosphate buffered at pH 7 was used as the mobile phase. Ethanol in Experiment 2 was calculated as the difference between DOC and benzoate. Ethylamine was determined by direct chromatographic analyses at Huffman Laboratories, Inc., Wheatridge, Colorado. Bromide, iodide, and borate analyses were performed according to standard methods of chemical analyses of the U.S. Geological Survey (Brown et al. 1970).

14.3 Hypotheses to be Tested

1) The negatively charged inorganic solutes (bromide, iodide) and the neutral solute (borate) were predicted to move conservatively, as fast as the water moved, not be retained by the aquifer solids, and could be used as conservative tracers in hydrologic studies.
2) The small negatively charged benzoate and acetate ions, and ethanol which is uncharged but infinitely soluble in water were predicted to move conservatively, not being retained by the aquifer solids, and could be used as organic conservative tracers in hydrologic studies.
3) By ionization, ethylamine will be a positively charged ion at pH 7.8 and will be strongly sorbed or retained by aquifer solids through cation exchange processes.
4) Aniline will remain an uncharged solute at pH 7.8, but will be slightly retained by aquifer solids because of moderate-to-low water solubility.
5) Phenolphthalein will remain an uncharged solute at pH 7.8, but will be strongly retained or sorbed by aquifer solids because of very low water solubility.

---

[2]The use of brand names in this report is for identification purposes only and does not constitute endorsement by the U.S. Geological Survey.

14.4 Results and Discussion

The complete data results for each solute in all experiments are given in a data report by Bassett et al. (1981). Selected organic data from various experiments are presented in this paper either to support or reject the stated hypotheses. The inorganic and fluorocarbon data will be summarized; detailed discussion of such data is the subject of other reports.

Bromide is a common inorganic tracer and is considered to be a conservative tracer of water movement. The results of this paper also support the conclusion that bromide is a conservative inorganic tracer. Chloride is also touted as a good, conservative inorganic tracer, but the high background concentrations of chloride native to the aquifer precluded its use or testing. Both iodide and borate were slightly retained by the aquifer solids as compared to bromide.

Results of several experiments show strong evidence for the conservative movement of ethanol and benzoate. Representative data for the coincident breakthrough of benzoate and bromide are shown in Fig. 2 for the orange sampling point in the 5-meter observation well during Experiment 4. Evidence for conservative movement of ethanol is its simultaneous breakthrough with benzoate during Experiment 2 at all sampling points in the 2- and 5-meter observation wells (Figs. 3 and 4). In addition, breakthrough curves at each of the points had essentially identical ratios of concentration of observed-to-injected tracer (Co/Ci). Other evidence supporting conservative movement of benzoate and ethanol are the well-defined Gaussian breakthrough curves, the high Co/Ci

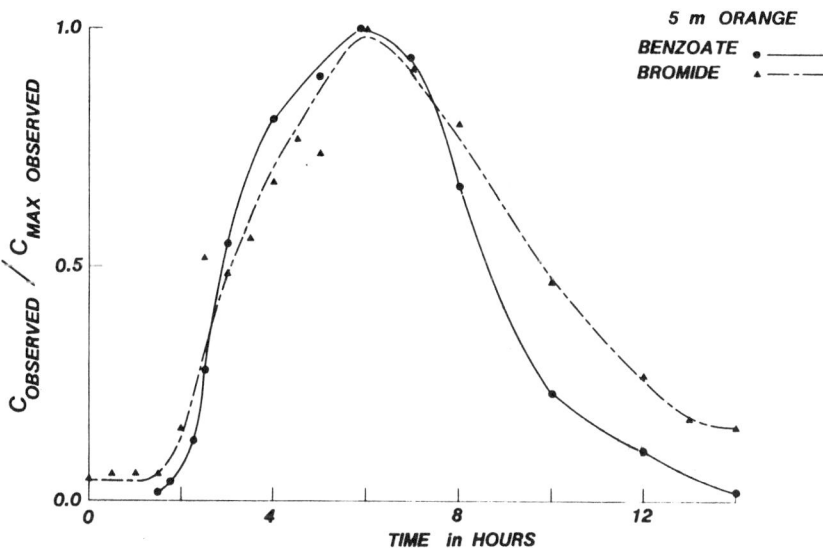

Fig. 2. Breakthrough curves for benzoate and bromide at the 5-meter orange sampling point during Experiment 4

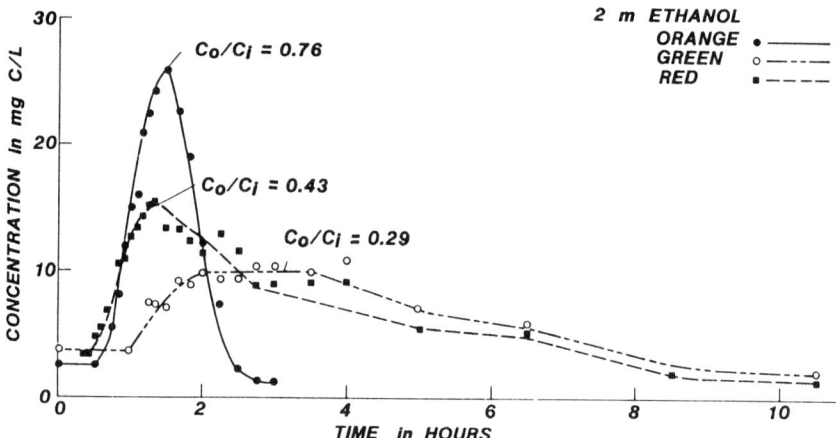

Fig. 3. Breakthrough curve for ethanol at the 2-meter observation well during Experiment 2

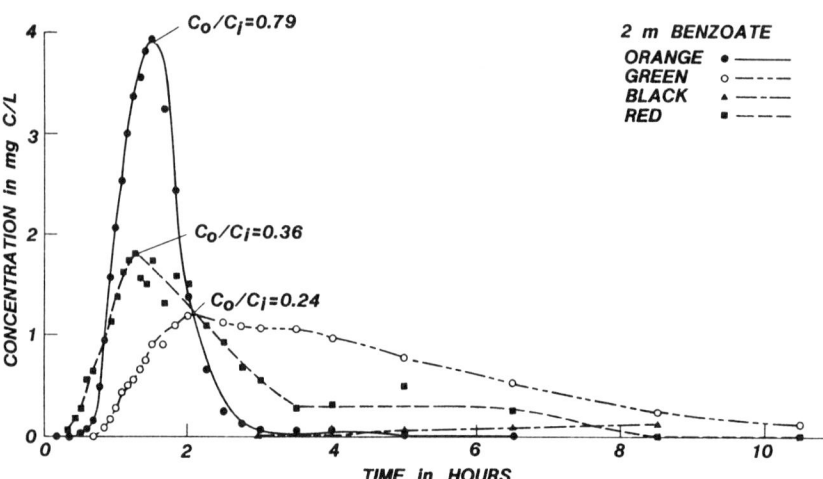

Fig. 4. Benzoate breakthrough curves at the 2-meter observation well during Experiment 2

ratios, and the duplication of breakthrough curves in Experiments 1 and 2. Conservative movement after a short pulse injection is suggested by the sharpness of the tracer breakthrough curve. Breakthrough curves for the more permeable orange and red zones in the 2-meter observation well are clearly defined and sharp. With time and distance of travel, breakthrough curves are broader with more pronounced tailing due to mixing and dispersion. For a more complete discussion of the movement of the predicted hydrophilic organic solutes (benzoate, ethanol, and acetate) as tracers refer to Malcolm et al. (1980).

As predicted, the movement of ethylamine during Experiment 5 was severely retarded by sorption of the cation by cation exchange (Fig. 5). At the pH of the ground water (7.8), ethylamine was completely in the cationic form as predicted by a pK_b of 3.25. Even though the injected concentration of aniline (25 mg C/L) was 2.5 times that of benzoate and 12 times that of phenolphthalein, and the pulse injection time for aniline was over twice as long as that for benzoate or phenolphthalein, the Co/Ci maximum for ethylamine was only 19, which was only 56 percent and 37 percent of the Co/Ci for phenolphthalein and benzoate, respectively. The long tailing peak for ethylamine was evident for greater than 36 hours after the cessation of the injection pulse.

The extensive retention of the organic base, ethylamine, was in marked contrast to the breakthrough for aniline which is also an organic base (Fig. 5). The pK_b for aniline is 9.4, which results in aniline being an uncharged or neutral solute at the pH (7.8) of the injected ground water. As predicted, the breakthrough curve for aniline demonstrated only slight retention of aniline by aquifer solids as compared to benzoate, an unretained solute. In a comparative 3-hour injection with benzoate, the Co/Ci maximum was only slightly less than benzoate and the peak maximum only slightly lagged that of benzoate.

The simulated breakthrough curve for aniline (Fig. 5) is believed to be accurate, because in the actual 8-hour pulse injection of aniline (Fig. 6), a 50 percent Co/Ci for aniline was attained 1.5 hours after that of 50 percent breakthrough for benzoate in the same sampling point, at the same water

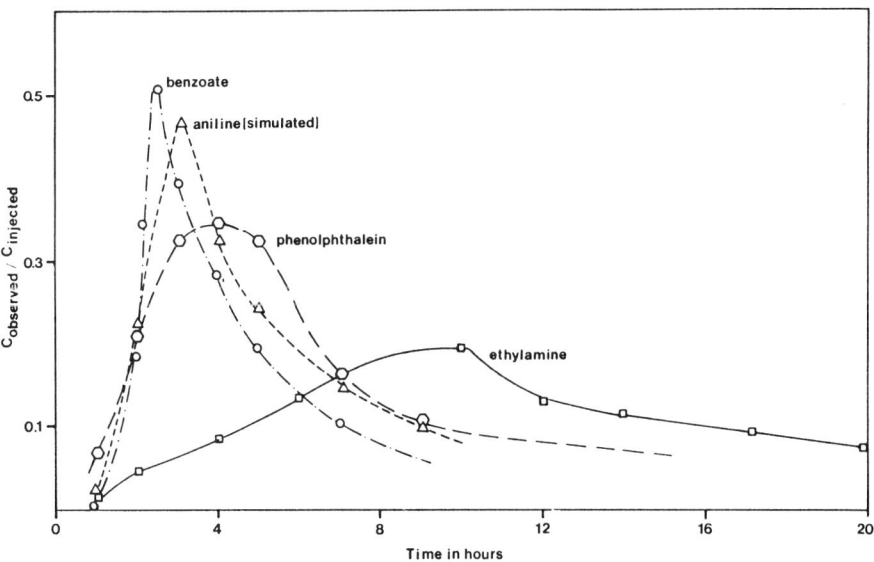

Fig. 5. Breakthrough curves for benzoate and phenolphthalein during Experiment 4, for ethylamine during Experiment 5, and simulated for 3-hour pulse injection of aniline during Experiment 5. All breakthrough curves are for the 2-meter orange sampling point

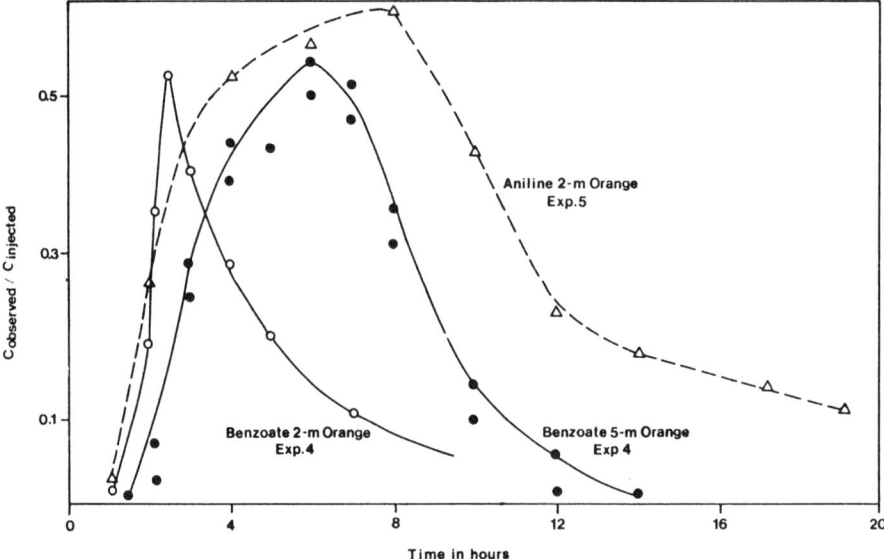

Fig. 6. Breakthrough curves for benzoate and aniline during Experiments 4 and 5.

injection rate, and at essentially the same injected concentration for both solutes (approximately 10 mg C/L).

Even after an 8-hour pulse injection for aniline (2.7 times longer injection time than for benzoate), the Co/Ci maximum was 59 percent, which is only 9 percent higher than for benzoate. Therefore, the simulated breakthrough for aniline during a 3-hour injection should peak after benzoate and at a slightly less Co/Ci ratio.

The breakthrough for phenolphthalein (Fig. 5) shows considerable sorption as compared to benzoate by a maximum Co/Ci ratio for phenolphthalein of only 68 percent that of benzoate, by a time lag of least 1 hour or more of the benzoate peak maximum, and by excessive tailing of the breakthrough curve in excess of 12 hours after cessation of solute pulse injection. Phenolphthalein was approximately 90 percent in the uncharged, neutral form at pH 7.8, because the pK_a for phenolphthalein is 8.7.

Ethanol, aniline, and phenolphthalein are all neutral uncharged solutes in these experiments. The family of breakthrough curves for these solutes represents a progression in decreasing water solubilities and an increasing degree of sorption on aquifer solids. The mechanism of sorption is proposed to be hydrophobic interactions and/or Van der Waals forces. These hydrophobic interactions are believed to be weak, even less than hydrogen bonding which is 5 to 10 kilocalories per mole. This type of bonding has been discussed by Thurman et al. (1978) relative to the sorption of uncharged organic solutes by the uncharged XAD resins. The highly silaceous silt, sand, and gravel particles of the Ogallala formation are believed to be largely uncharged surfaces or large areas of uncharged surfaces which thermodynamically favor the sorption of uncharged organic solutes. The magnitude of this sorption is determined by the relative water insolubility of the organic solute. The driving force for sorption of these uncharged organic solutes is decreasing water solubility of the solutes or the increase in entropy associated with the disruption of the H-bonding structure of water in solution.

This type of hydrophobic interaction has not been recognized until recently and the potential magnitude of sorption of neutral organic solutes by this type of mechanism is not well appreciated. The chemical partitioning of neutral organic solutes into the organic phase in soils and sediments would strongly compete with this type of sorption on mineral surfaces, but chemical partitioning is not believed to be a competitive sorptive mechanism in the Ogallala aquifer, because the organic matter content is near zero ($<$ 0.1 percent by weight).

Major variations within the aquifer due to heterogeneity, minor changes due to entrapment of air during water injection, and variable aquifer response during injection well redevelopment must be considered in the interpretation of solute measurement in these experiments. Most of the data support the overall results presented in this paper. It is also quite surprising and gratifying that the solute breakthrough curves observed under field conditions are actually as definitive as they are considering the numerous problems often associated with field experiments.

A limited amount of the experimental data which are not presented in this report, but are presented in the complete data compilation (Bassett et al. 1981), appear to be inconsistent with the solute movement as presented herein. These inconsistencies are believed to be due to variability in the Ogallala aquifer and minor limitations in the manner in which the experiments were conducted. It is very evident from the large differences with depth in the solute breakthrough curves as shown in Figs. 1 and 2 that the Ogallala aquifer is not isotropic and homogeneous. The top part of the aquifer is certainly the most permeable. Even within this permeable zone there are microfeatures which may alter the flow patterns and result in large microvariations in permeability. Keys and Brown (1978) observed small channels and vugs in the Ogallala aquifer by direct examination and in thin-sections. These types of microfeatures were not believed to be present or common in unconsolidated sandy aquifers prior to this finding. These types of small channels are probably the reason that during experiments 4 and 5 the solute appeared in the 5-m red sampling point before they were detected in any of the 2-m sampling points. The porous channel diverting water flow around the 2-m sampling points and directly to the 5-m red sampling point was probably increased in diameter and length during injection well redevelopment after experiment 3. In the latter experiments, particulate yeast was observed in the 5-m red sampling point within 0.5 hr after being introduced into the injection well (Wood and Ehrlich 1978).

The injection well was developed by pumping before initial experimentation and was redeveloped twice during the experiments (after Experiment 2 and

Experiment 3). The gradual decrease in water injection rates during experimentation was attributed primarily to entrapment of air within the aquifer. These redevelopments resulted in some minor overall changes in transmissivity and flow patterns near the injection well in addition to the speculated channel enlargement adjacent to the 5-m red sampling point. For example, during experiments 1 and 2 the Co/Ci ratios for ethanol and benzoate attained maximum values of 0.76 and 0.79, respectively, in the 2-m orange sampling point after a 54 min pulse injection, whereas the benzoate Co/Ci ratio attained a value of only 0.51 after a 3 hr pulse injection during experiment 4 in the same sampling point. For this and other reasons it was essential to inject either ethanol or benzoate as a conservative tracer in all experiments and the rate of movement of all other solutes could be compared to these internal standards.

14.5 Conclusions

Small, nonionic organic solutes, such as ethanol, or negatively charged, such as acetate or benzoate, are extremely water soluble and move conservatively in ground waters. (Such solutes would also be expected to move rapidly through soils.) Organic solutes which are positively charged, such as ethylamine, are water soluble but are strongly adsorbed onto porous media by a cation exchange process. As neutral or uncharged organic solutes increase in size and decrease in water solubility, often due to the addition of electron withdrawing groups such as Cl, Br, and NO_2, the solutes are more strongly sorbed by hydrophobic interactions onto uncharged or slightly charged mineral surfaces. The very insoluble neutral organic solutes are also sorbed by chemical partitioning into the organic phase of soil, sediments, and ground waters. It was shown in this study that the sorption of uncharged organic solutes onto uncharged to slightly charged mineral surfaces increased with decreasing water solubility (phenolphthalein was more strongly sorbed than aniline).

The charge of the organic solute and the pH of the water media are major factors affecting their adsorption or nonretention, their sorption capacity, as well as their mechanism of sorption. Most natural media are net negatively charged; therefore, negatively charged organic solutes are repulsed and positively charged organic solutes are attracted or retained. Many organic solutes have the potential to be charged or uncharged depending on their pK_a or pK_b. Therefore, the pH of the natural environment becomes very important as to the charge on such solutes.

A first approximation of the movement of organic solutes in sediments and aquifers low in organic matter can be made very quickly with a minimum of information about the organic solute and the site. The pH of the medium (soil or sediment) must be determined. The properties of the solute(s) of interest at this given pH such as charge, pK_a or pK_b, and the relative water solubility can be found in a chemical handbook. After these parameters are determined, the relative rate of movement for a given organic compound may be approximated by comparison with the most appropriate one of the compounds which were tested in this study. Because the solutes which were included in these experiments encompassed a range of positively charged, negatively charged, and uncharged (neutral) organic solutes; an appropriate model compound for comparative movement should be available. The first approximation estimate is most suitable in short time periods from days to a few weeks because it does not consider microbial transformation of organic solutes. In soils and sediments high in organic matter, retention of neutral organic solutes of low water solubility also may be greater due to chemical partitioning.

References

Basset RL, Weeks EP, Ceazan ML (1981) Preliminary data from a series of artificial recharge experiments at Stanton, Texas: US Geol Surv Open-File Report 81-149, 235 p

Brown E, Skougstad ME, Fishman MJ (1970) Method for collection and analysis of water samples for dissolved minerals and gases. In: Techniques of Water-Resources Investigations of the US Geol Surv, Bk 5, Chap 1, US Government Printing Office, Washington DC

Keys WS, Brown RF (1978) The use of temperature logs to trace the movement of injected water. Ground Water 16(1):32-48

Leenheer JA, Malcolm RL, McKinley PW, Eccles LA (1974) Occurrence of dissolved organic carbon in selected ground-water samples in the United States. J Res US Geol Surv 2(6):361-369

Malcolm RL, Aiken GR, Thurman EM, Avery PA (1980) Hydrophilic organic solutes as tracers in groundwater recharge studies, contaminants and sediments, Vol 2. Edited by Baker RA, Ann Arbor Science, Ann Arbor, MI

Signor DC (1978) Gas-driven pump for ground-water samples. Water Resour Invest 78:1-25

Thurman EM, Malcolm RL, Aiken GR (1978) Prediction of capacity factors for aqueous organic solutes adsorbed on a porous acrylic resin. Anal Chem 50:775-779

Wood WW (1973) A technique using porous cups for water sampling at any depth in the unsaturated zone. Water Resour Res 9:486-488

Wood WW, Ehrlich GG (1978) Use of bakers yeast to trace microbial movement in ground water. Ground Water 16:398-403

15 Photochemical Inactivation of Organic Pollutants from Water

A. ACHER and S. SALTZMAN[1]

15.1 Introduction

Due to the ever-increasing depletion of the natural resources of water, the reuse of treated municipal wastewater for crop irrigation, for artificial groundwater recharge, or even for potable purposes, has become a real necessity for countries of the arid and semi arid zones. The removal of organic pollutants from effluents is a prerequisite for their safe reuse. In the present work, a photochemical procedure for the inactivation of industrial and biological organic pollutants from water is described. In this procedure sunlight is used as the energy source, oxygen dissolved in water as the oxidizing agent, a dye (sensitizer) as an intermediary for the transfer of the light energy to the oxygen molecule, and the organic matter as the oxidation target.

Use of sunlight for wastewater treatment has several advantages. First, sunlight is a free energy source; therefore, energy costs with such systems are minimal. Secondly, unlike chlorination, toxic organics such as chlorinated hydrocarbons are not formed. Thirdly, the structural facilities and instrumentation associated with these systems are relatively simple to build, maintain and operate, and finally, with sensitizing agents, the contact time necessary for photochemical treatment is relatively short.

15.1.1 Dye-Sensitized Photooxidation Reactions

These reactions, also referred to as "photodynamic action" (Gollnick 1968; Kearns 1971; Foote 1968), are responsible for the oxidative processes which take place in surface waters exposed to solar radiation. They consist of the combined action of visible light and molecular oxygen (O_2) dissolved in water upon organic matter (OM) through the intermediary of an appropriate photosensitizer (S). The S is an organic molecule having a special electronic structure which enables it to absorb, and then to transfer, some of the light radiated energy. The S is added to the aerated and light exposed effluents and its absorbed energy (S*) is made available to the oxidation of OM. Either one or both of the following mechanisms can operate in aerobic photosensitized oxidations (Kautsky 1939):

---

[1] Institute of Soils and Water, ARO, The Volcani Center, Bet Dagan 50250, Israel

1. Primary interaction of the electronically excited S* is with OM to generate reactive, short-lived intermediates which subsequently react with O_2:

$$S + h\nu \rightarrow S^* \tag{1}$$
$$S^* + OM + O_2 \rightarrow \text{transient species} \rightarrow \text{oxidation products} + S \tag{2}$$
(transient species = free radicals, ion pairs, etc.)

2. The presence of O_2 will compete successfully with OM on receiving the excitation energy from S*. The addition of this energy to O_2 changes its ground electronic state (triplet state, $^3\Sigma gO_2$) to the first excited singlet state ($^1\Delta gO_2$) which has a higher energy by 22.5 kcal mole^{-1}. When more energy is imparted to O_2, another electronic state is formed ($^1\Sigma gO_2$) which corresponds to a level of 37.5 kcal mole^{-1} above the $^3\Sigma gO_2$. From the properties of singlet oxygen (exceedingly short lifetimes of $^1\Sigma gO_2$) it seems likely that only $^1\Delta gO_2$ is important in solution photooxidations:

$$S^* + {}^3\Sigma gO_2 \rightarrow S + {}^1\Delta gO_2 \tag{3}$$
$$^1\Delta gO_2 + OM \rightarrow \text{oxidation products} \tag{4}$$

In both mechanisms the sensitizer is regenerated and undergoes hundreds of cycles so that only minute amounts of it are required.

In view of the diversity of OM present in wastewaters, it is very difficult to decide which mechanism operates in the present process. The presence of singlet oxygen in natural waters was proven ten years ago (Zepp et al. 1977) and it is well known that it oxidizes unsaturated organic compounds (UC) to peroxides. The subsequent thermal and photochemical decomposition of these peroxides can further initiate free radical oxidation reactions which will also affect saturated compounds (R'H) found in wastewaters:

$$^1\Delta gO_2 + UC \rightarrow ROOR \rightarrow 2RO^\bullet \tag{5}$$
$$RO^\bullet + R'H \rightarrow ROH + R'^\bullet \tag{6}$$
$$R'^\bullet + O_2 \rightarrow ROO^\bullet, \text{etc.} \tag{7}$$

Beside $^1\Delta gO_2$, other oxidative chemical species like hydroxyl radical ($OH^\bullet$), superoxide radical anion (O_2^-) and hydrogen peroxide (H_2O_2), might be generated in the aerated and irradiated effluents (Draper and Crosby 1983). As a result of the above reactions vital biological components (proteins, lipids, polysaccharides) and industrial organic materials undergo oxidative degradations (Spikes and Livingston 1969; Spikes and MacKnight 1970).

15.2 Photochemical Inactivation of Industrial Organic Pollutants

One of the main potential sources of surface and ground water pollution is provided by pesticides. They could reach surface and ground water following direct applications to control aquatic insects and weeds, or indirectly, from soil runoff and leaching, accidental spillage of industrial waste waters, and clean-up of pesticide application equipment. The presence of such biocides in water can damage the aquatic environments and represents a health hazard (Aly and El-Dib 1971).

The persistence of industrial organic pollutants in aquatic environments is affected mainly by their chemical structure, which determines their susceptibility to degradation, by the properties of the water such as pH, temperature, and the presence of various soluble compounds and suspended materials, as well as by general environmental conditions determined by the geographical location. The main dissipation routes of pesticides from water are by biological, chemical and photochemical processes.

Photochemical reactions of pollutants in water can be either direct photolysis, in the case of direct light absorption by pollutant molecules, or indirect (sensitized) photolysis, when the process is accelerated by the presence of specific organic compounds (sensitizers). Both reactions occur in natural aquatic environments and were experimentally demonstrated. Direct photolysis has been demonstrated for a large number of pollutants (e.g. Crosby and Li 1969; Wolfe et al. 1976; Draper and Crosby 1984; Ruzo and Casida 1985; Quistad and Staiger 1984). However, except for a few pollutants, like some pyrethroid pesticides, direct photolysis is not likely to be the predominant degradation pathway in natural aquatic environments.

Sensitized photolysis processes in natural aquatic environments are due to the presence of organic materials that strongly absorb light, such as humic substances and dissolved organic compounds (Crosby 1970; Khan and Schnizter 1978; Khan and Gamble 1983). Some salts of zinc, iron, cobalt, could also enhance photochemical processes (Crosby and Li 1969).

The intentional promotion and enhancement of such photooxidation reactions in surface water was used for detoxification of water from herbicide residues (Crosby and Li 1969; Wolfe et al. 1976; Acher and Dunkelblum 1979; Acher et al. 1981; Saltzman et al. 1982; Rejto et al. 1983, 1984; Freeman and Ndip 1984; Freeman and McCarthy 1984; Draper and Crosby 1984), for treatment of organic matter in sewage effluents (Acher and Rosenthal 1977), for disinfection (Acher and Juven 1977; Acher et al. 1979; Gerba et al. 1977) and for algal destruction (Acher and Elgavish 1980). The most frequently used sensitizers for the UV range are acetone, benzophenone, acetophenone, benzonitrile, and for visible light range are mainly methylene blue (MB), riboflavin (RF), rose bengal (RB) and acridin orange (AO). A mathematical model, based on the factors influencing the efficiency of this photochemical method was developed and can be used for the design and operation of a pilot photooxidation lagoon for effluent disinfection (Watts 1983).

Studies carried out by our research group were aimed at optimizing the reaction conditions of the photochemical processes, to determine their kinetics and reaction pathways, and to define the photodegradation products, both chemically and toxicologically.

The laboratory treatment of municipal sewage effluents by oxidation of organic matter by photosensitization resulted in a decrease in the chemical oxygen demand by 67%, and of Methylene Blue Active Substances (MBAS) by 90% of their initial values. The optimal reaction conditions were obtained by the addition of about 12 mg L^{-1} MB to the aerated effluents exposed to a UV-lamp or solar radiation. The radiation time required for the completion of the reaction was dependent on the light intensity supplied (Acher and Rosenthal 1977).

The sensitized photolysis of several herbicides belonging to three chemical classes — uracils, *s*-triazines and anilides — was studied. All these compounds are photostable in environmental conditions, but fast chemical transformations occurred in sunlight, upon photosensitization. The common features of our studies were irradiation of aerated, aqueous pesticide solutions by sunlight, or artificial white light, and the use of dye-sensitizers, mainly MB and RF. The photooxidation products were isolated and identified by using a combination of column chromatography, gas chromatography, proton and carbon-13 nuclear magnetic resonance, infrared and mass spectrometry.

15.2.1 Photolysis of Uracil Derivatives

The sensitized photolysis of two uracil derivatives — bromacil (5-bromo-3-*sec*-butyl-6-methyl uracil) and terbacil (3-*tert*-butyl-5-chloro-6-methyl uracil) — was studied. The choice of compounds with different halogen and alkyl groups in the uracil moiety enabled the assessment of the influence of these constituents on the photodegradation mechanism.

Bromacil (I) solutions exposed to sunlight in the presence of dye sensitizers and dissolved oxygen, undergo fast chemical transformations. The rate of the reaction is pH-dependent, being faster at alkaline pH. In optimal reaction conditions the reaction was completed after about 1 h. The major product formed in 83% yield has been identified as a mixture of diastereoisomers of 3-*sec*-butyl-5-acetyl-5-hydroxyhydantoin (II). The mechanism proposed is based on singlet oxygen ($^1\Delta gO_2$) reactions with the double bond of the uracil ring, followed by a fast OH$^-$ catalyzed rearrangement, with concomitant loss of bromine (Scheme 1), (Acher and Dunkelblum 1979; Acher and Saltzman 1980).

The first step of the reaction is the formation of the intermediate compounds A and B. In the second step A and B are transformed by an OH$^-$ catalyzed rearrangement to the intermediate C, which undergoes ring opening (D) and further ring closure, to give the major photodegradation product (II). Further irradiation of the reaction mixture resulted in the decomposition of this product and the formation of unidentified polar compounds.

Under similar reaction conditions as for bromacil, terbacil (I) photolysis (Scheme 2) also produced a hydantoin compound as the main degradation product (3-*tert*-butyl-5-acetyl-5-hydroxyhydantoin, II), (Acher et al. 1981). The

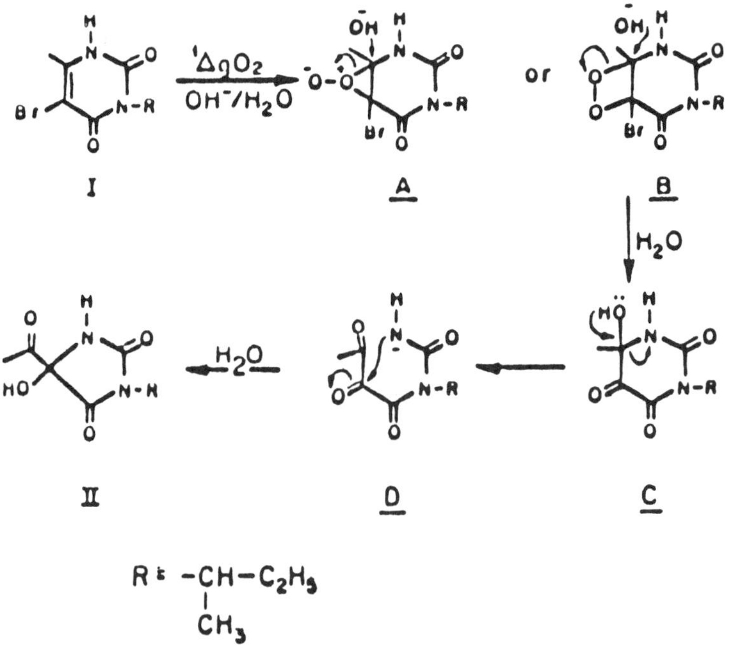

Scheme 1.

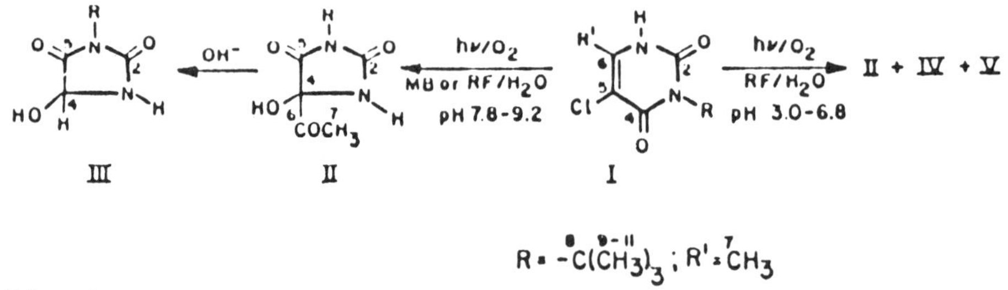

Scheme 2.

formation rate of the photooxidation products was slower in the case of terbacil, as the C-Cl bond present in this compound is stronger than the C-Br bond of bromacil. It has been demonstrated that the presence of the halogen atom in the uracil derivatives is needed to promote the photooxidation process. Unlike bromacil, terbacil photolysis proceeded also under acidic conditions, down to pH 3; the reaction was sensitized by RF, but not by MB. Under these reaction conditions, besides the hydantoin compound, a mono-N-dealkylated terbacil dimer (IV) and a nonidentified water — insoluble oligomer (V) were isolated.

The proposed mechanism for dimer formation (Scheme 3) is based on the homolytic cleavage of the tertiary C-N bond (a), reaction with terbacil (b), C-N ring closure (c), and dimer formation (IV).

Scheme 3.

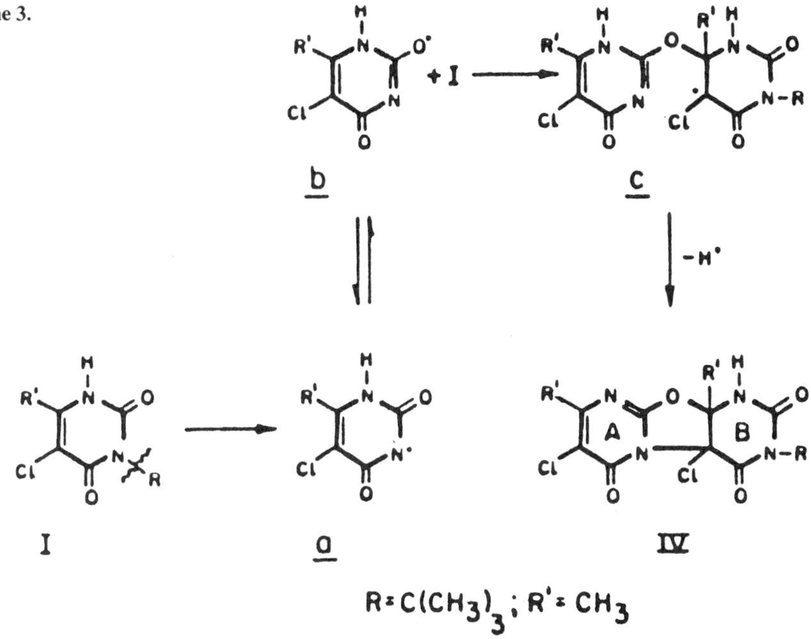

R = C(CH$_3$)$_3$; R' = CH$_3$

The fact that dimerization did not occur in bromacil photolysis is due to the more difficult homolitic cleavage of the *sec*-butyl group, as compared to *tert*-butyl present in the terbacil molecule.

Both bromacil and terbacil were rapidly decomposed by sunlight in frozen aqueous solutions, containing dye sensitizer. The degradation rate was pH-dependent; at pH 9.1 about 70–75% of the initial amounts of bromacil and terbacil were decomposed in the first 2 h of irridation. At neutral pH, bromacil decomposition was practically negligible, but the photolysis rate of terbacil was rather fast (half-life of about 7 h in the presence of MB). When the frozen solutions were covered by ice blocks up to 11 cm thick, the amount of incident light reaching the samples was attenuated, but other effects, like light scattering and/or lens effects increased the diffusiveness of light through the ice. Consequently, the photodecomposition rate of both bromacil and terbacil remained high enough to be practically significant.

By a similar photooxidation mechanism as in aqueous solutions, the sensitized irradiation of frozen bromacil and terbacil solutions yielded the respective hydantoin derivatives. At irradiation periods longer than 2 h, new dehalogenated compounds were formed. The halogen atoms release into the frozen solutions as hydroacids was observed (Acher 1982).

The efficiency of the dye-sensitized photooxidation of uracil derivatives as a procedure for water detoxification was tested by bioassay (Saltzman et al. 1982). The phytotoxicity of the photoreaction mixtures, and of the main, isolated photodegradation products of bromacil and terbacil was tested with sorghum, which is very sensitive to uracil derivatives. At equal or higher concentrations than those expected to result following regular use of bromacil and terbacil, neither the main photooxidation products, nor the photoreaction mixtures were phytotoxic. The dye sensitizer MB, at concentrations 20 times higher than those generally used in photoreactions, inhibited the development of the radicles of the sorghum seedlings. However, its leuco-form, resulting from irradiation, was non-phytotoxic.

The attempts to apply the dye-sensitized photooxidation technique for the detoxification of bromacil to effluents from a pesticide plant were successful. Yet, this procedure failed to detoxify the industrial effluents completely, as they were of a complex chemical composition (e.g. high salts concentration).

15.2.2 Photolysis of s-Triazine Derivatives

The s-triazine derivatives used as herbicides are 4,6-bis (alkyl-amino)-s-triazines having an electrophilic substituent at C-2.

$$HN-C^6\begin{matrix} & R_1 \\ & | \\ & C_2 \\ N & N \\ | & || \\ & ^4C-NH \\ N & \\ & | \\ & R_2 \end{matrix}$$
$$R_3$$

s-Triazine Derivatives

The compounds studied are listed in Table 1.

According to the pertinent literature these compounds have a very low reactivity towards singlet oxygen, so it has been assumed that they are not decomposed by sensitized photolysis. However, in the studies carried out in our laboratory significant photolysis was observed in aerated s-triazine solutions exposed to sunlight in the presence of RF (Rejto et al. 1983).

Under identical reaction conditions the degradation rate of these compounds was dependent on the electronegativity of the R_1 substituent which interferes, through the aromatic ring, with the stability of the substituents at C-4 and C-6. Compounds having a chlorine atom in the C-2 position (atrazine, simazine, propazine) were more stable to photolysis than the compounds with O-CH_3 or S-CH_3 substituents (atraton and ametryne, respectively). Following 1 h of irradiation the amount of parent compound remaining in solution, relative to the initial concentration, was about 80% for the chlorine-containing compounds, but only about 50% and 30% for atraton and ametryne, respectively. The rate of the reaction was also affected by pH; the reaction rate was faster at acid than at neutral pH.

Table 1. s-Triazine derivatives used in this study

s-triazine	desig-nation	substituents		
		R_1	R_2	R_3
ametryne	I	SCH_3	$CH(CH_3)_2$	CH_2CH_3
atraton	II	OCH_3	$CH(CH_3)_2$	CH_2CH_3
atrazine	III	Cl	$CH(CH_3)_2$	CH_2CH_3
simazine	IV	Cl	CH_2CH_3	CH_2CH_3
propazine	V	Cl	$CH(CH_3)_2$	$CH(CH_3)_2$

The proposed photolysis mechanism of s-triazines is based on the photodecomposition products, which were isolated and identified (Scheme 4).

The major product was an N-deethylated compound (a); a minor, oxidated compound (b) was simultaneously formed. The concentration of these compounds decreased upon further irradiation, and a completely N-dealkylated compound (c) appeared in the reaction mixture. Further irradiation led probably to the formation of a deaminated and unidentified products. Similar to the biochemical degradation of s-triazines, it has been observed that in the photolysis process, the N-ethyl group is photolyzed first, followed by the removal of the N-isopropyl group.

The proposed mechanism of sensitized photolysis of s-triazines was thought to be photooxidation mediated by excited-state complexes or "exciplexes". An alternative mechanism can be described by a sequence of reactions in which hydrogen is abstracted from the amine by excited RF. The fact that the presence of oxygen is necessary for the photolysis to proceed suggests the possible participation of hydroperoxide radicals in the photoreaction mechanism. Hydroperoxide radicals may also be formed in the regeneration of reduced RF in the presence of oxygen. This assumption is supported by the fact that less than a stoichiometric amount of sensitizer could be effective for the complete degradation of s-triazines. However, an alternative explanation could be the sensitizing effect of bleached RF.

It was observed that no correlation existed between the change in concentration of RF, as measured by the intensity of the yellow chromophor, and its sensitizing effect, and that this effect persisted even after complete decoloration. In addition, it has been proved that bleached RF can initiate the photolysis process of s-triazines. This effect may be explained by the changes in the UV-VIS absorption spectrum of RF, following irradiation (Fig. 1).

In the range of 300–600 nm, the non-irradiated RF has two absorption maxima, at 447 and 376 nm, The absorption at 447 nm rapidly decreases, even after 10 min of irradiation, which explains the fast RF bleaching. The absorption at 376 nm was much more stable, and a significant shift of this absorption band to lower frequencies was observed (352 nm). This suggests that this band was involved in the sensitizing process. It is known that although the UV radiation

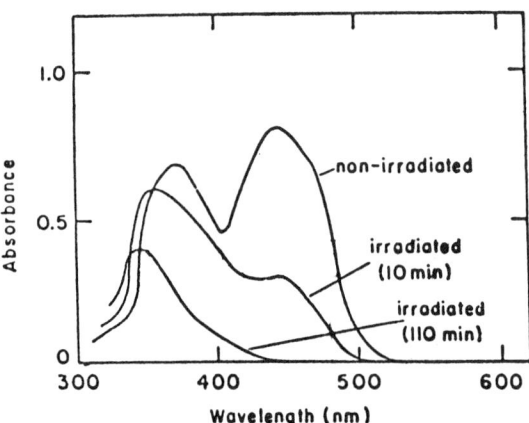

Scheme 4.

Fig. 1. The irradiation effect on RF absorbance

intensity in the near visible sunlight spectrum (350–400 nm) is very low (less than 5% of the total sunlight intensity), each quantum in this range is almost twice as energetic as a quantum at maximum intensity of sunlight (Crosby and Li 1969).

The residual phytotoxic effect of irradiated mixtures containing traces of s-triazines and various metabolites was studied by a bioassay. Oat was used as the test plant, as it is known to respond reasonably well to s-triazines (Horowitz 1976). The response of oat seedlings grown in standard solutions of ametryne, atrazine and atratone was significant at 1 mg L^{-1}. No injury was observed in the irradiated mixtures of ametryne and atrazine, but the mixture of atratone had some phytotoxic effect as it still contained 0.75 mg L^{-1} of the parent compound and 1.5 mg L^{-1} of the N-deethylated photoproduct. The results obtained indicate that no phytotoxic compounds resulted following the irradiation of s-triazines in the presence of RF, up to total decomposition of the parent compounds and their main metabolites.

15.2.3 Photolysis of Anilide Derivatives

Propachlor (2-chloro-*N*-isopropylacetanilide) was selected as a representative of the anilide herbicide group for the photodecomposition studies (Rejto et al. 1984). The sensitized photolysis was achieved by either UV or visible light. However, the photodegradation pathway was completely different, and the products formed under UV irradiation were not detected in RF — sensitized solutions exposed to visible light.

Propachlor irradiation with UV light for 5 h resulted in about 80% decomposition and gave a complex mixture of photoproducts, the composition of which was dependent on the irradiation time. The photodegradation products isolated were *N*-isopropyloxindole (II), *N*-isopropyl-3-hydroxyoxindole (III), and a spiro compound (IV) (Scheme 5).

It is suggested that propachlor decomposition under UV light involves, as a first step, a homolytic C-Cl cleavage, with subsequent reaction of the radical formed, either with benzene ring, leading to oxindole derivates, or with molecular oxygen, leading eventually, to a spiro derivative.

Aqueous solutions of propachlor exposed to visible light, were photostable. However, RF was very efficient in sensitizing the photodecomposition reaction; MB did not have a significant sensitizing effect.

The decomposition rate of propachlor in the presence of RF and dissolved oxygen was dependent on the initial pH of the reaction mixture and on the ratio sensitizer:propachlor. The fastest rate was obtained at pH 3.4 and a ratio RF:propachlor of 1:10. In such conditions, exposure to visible light for 9 h resulted in 60% propachlor decomposition. The only photodegradation product which could be isolated and identified was *m*-hydroxypropachlor (V) (Scheme 5). That shows that the primary step was the hydroxylation of the aromatic ring leading to this phenol derivative. The low yields of the degradation products might be explained by their lability under these photoreaction conditions.

The phytotoxicity of the reaction mixtures obtained following irradiation by visible light was checked with young seedlings of susceptible plants. Dose-response curves showed that the order of sensitivity of the test plants was oat >

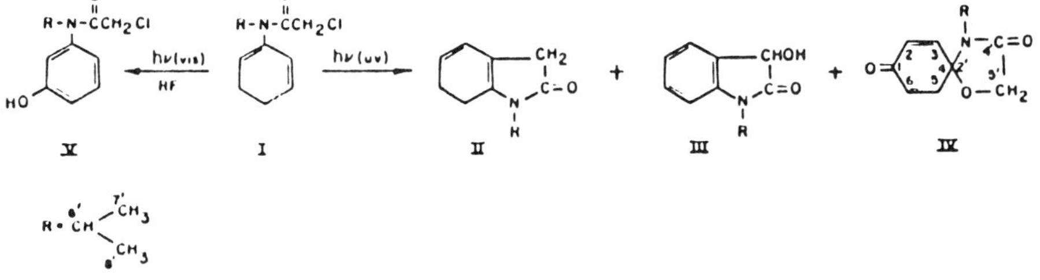

Scheme 5.

sorghum > tomatoes. Oat was sensitive at 0.5 mg L^{-1} propachlor and had an almost quantitative response. The phytotoxicity test using oat showed that the photodegradation products in the tested solutions had no growth-inhibiting effect.

15.3 Photochemical Inactivation of Biological Pollutants

When dye-sensitized photochemical reactions are used in connection with biological substrates, like microorganisms, they are also defined as "photodynamic action". The photodynamic oxidation reactions have been known since about 1890; however, detailed studies of the inactivation of proteins, nucleic acids, bacteria and viruses were not performed until the late 1920s (Spikes and Livingston 1967). A few decades later mechanistic studies were carried out on the free nucleic acids of the intact viruses. The results of these studies indicated that the site of attack was the purine base guanine, a moiety of the nucleic acid (Hiatt 1967).

Ten years ago two research groups had, almost simultaneously, the idea of using these photodynamic oxidations in water disinfection procedures. One group, in Houston, Texas (Melnick et al. 1976) performed feasibility experiments of this procedure for water and tertiary effluents disinfection from viruses. The effluents were of high quality; their pretreatment included filtration and pH change to 10. The irradiation experiments were carried out (after a dark period of MB sensitization of 4 to 24 h) using either specially built light sources (Westinghouse, 40-W, RR special SHO tubes) with an emission spectra overlapping the sensitizer (MB) absorption (670 nm) or sunlight irradiation. The disinfection results showed up to 3–4 log viruses decrease after a few minutes of irradiation, at 2–5 mg MB L^{-1} and about 4 mg L^{-1} dissolved oxygen. However, this disinfection procedure was too complicated and costly to successfully compete with the existing procedures such as chlorination, ozonization, etc. Another group, at the Volcani Center, Israel, began the research on sunlight photodynamic inactivation of microorganisms (including algae) in secondary effluents. The aim was to develop economical, safe and low technology disinfection procedures which would result in the reuse of the treated effluents for crop irrigation. Research on this procedure has continued since 1976 (with interruptions) and the most relevant results obtained are described.

15.3.1 Laboratory Experiments

Samples of tap water and of municipal oxidation pond effluents, to which an innoculum of fecal *Escherichia coli* had been added, were exposed to solar irradiation in graduated glass cylinders (100 mL) in the presence of MB, under continuous aeration (Acher and Juven 1977). Table 2 shows the effect of MB

Table 2. The effect of MB concentration on the destruction of coliforms*

MB concentration mg L^{-1}	Viable Coliforms (MPN/100 m)	
	Sewage	Tap Water
0.00	5.6 × 10^8	7.5 × 10^8
0.25	–	4.8 × 10^3
0.50	2.9 × 10^4	4.0 × 10^3
1.00	9.5 × 10^3	2.1 × 10^3
2.00	1.1 × 10^3	2.0 × 10^3
4.00	–	< 20
5.00	< 20	–

*Initial coliform density: sewage 1.3 × 10^9, inoculated tap water 9.2 × 10^8; MPN = Most Probably Number.

concentration on survival of the coliforms after 28 min of irradiation (sunlight intensity: 2,030 μEm^{-2}s^{-1}; temperature: 32° ± 2° C).

It can be seen from the above table that the most efficient MB concentration was about 3 mg L^{-1}. The tap water reached 5 log coliforms destruction even at 0.25 mg MB L^{-1}, while for the effluent the amount of sensitizer which has a similar effect is about 2 mg MB L^{-1}. Under pilot plant field conditions (effluents of about 15 Nephelometric Turbidity Units (NTU) the MB concentration is limited to less than 1 mg L^{-1} since higher concentrations will hinder the light penetration into a water depth of more than 20 cm.

The effect of sunlight intensity and exposure time on the coliforms' survival in effluent samples containing 2 mg MB L^{-1} is shown in Table 3 (the initial coliform count: 1.3 × 10^9 MPN/100 mL; temperature: 32° ± 3° C). The data presented in Table 3 show the existence of interaction among length of exposure, intensity of irradiation, and coliforms' survival. One of the most important conclusions regarding these data is that the intensity of solar irradiation may not

Table 3. The effect of sunlight intensity and exposure time on the destruction of coliform in sewage*

Exposure time (min)	Sunlight radiation		Viable coliforms (MPN/100 mL)
	Intensity (μEm^{-2}s^{-1})	Amount (Em^{-2})	
0	0	0	1.3 × 10^9
4	2,030	0.49	1.8 × 10^8
28	2,030	3.40	7.9 × 10^2
42	2,030	5.10	< 20
40	68	0.16	1.7 × 10^4
120	68	0.49	7.9 × 10^2
150	68	0.61	< 20

*Control sample after 150 min incubation in dark at 35° C had 1.1 × 10^9 coliforms/100 mL.

be a limiting factor when disinfection of waters by sensitized photooxidation is considered. Another observation shows that there is a time limiting factor below which the lethal cell damage is very low. For example, an energy of 0.49 Em^{-2} will have a small impact on coliforms viability after 4 min ($<$ 1 log), but a large effect after 120 min ($>$ 6 log). As far as the destruction of coliforms in sewage effluents is concerned, the results obtained by this method appear to be much more satisfactory than those reported for chlorination (Kott 1973).

15.3.2 Pilot Plant Experiments

15.3.2.1 Pilot Plant Batchwise Experiments

The first pilot plant built for the effluent disinfection was an epoxy coated steel reactor, shown schematically in the Fig. 2. The effluent used was from a municipal sewage oxidation pond, having about the same main characteristics as the effluent described later (page 316). The effluent was pumped into the reactor to which a solution of MB was continuously added (2 g MB m^{-3} effluent) in a mixing container connected directly to the reactor. When the effluent depth reached 25 cm (about 2.5 m^3), the suction from the oxidation pond was stopped and the pump switched to recirculate the treated effluent. The effluent was contaminated with laboratory culture of bacteriophages (coliphage X and F2) and polio virus type 1-L Sc (vaccine strain). Samples taken at different times were bacteriologically analyzed. The results showed that those bacteriophages which were resistant to chlorination (up to 30 mg Cl L^{-1} for 2-3 h) were inactivated in a few minutes by this method. Polio virus were completely inactivated in a period of 5-8 h depending on working conditions. Other bacteria and algae were also affected (Acher et al. 1979).

The conclusions that were reached from this first pilot plant were: the photodynamic inactivation of microorganisms is possible even into poor quality secondary effluents; the batchwise method should be replaced by a continuous method which could offer some technical advantages in comparison with the batchwise method.

15.3.2.2 Saline Water Carrier Experiments

In another series of experiments, the feasibility of photochemical disinfection was evaluated in the saline water carrier which diverts the saline water sources (1000 mg Cl⁻ L^{-1}) from entering into the Lake of Galilee. The settlements neighboring this channel use it as a sewer for their sewage effluents, therefore its water has a relatively high bacteriological count (e.g. *E. coli* 10^4 MPN), but a low turbidity (e.g. less than 2 NTU). This is an open concrete channel of rectangular profile 2.0 m wide, 1.0 m high and more than 10 km long, of which 3.5 km were suitable for use in disinfection experiments) carrying the saline water and effluents from north

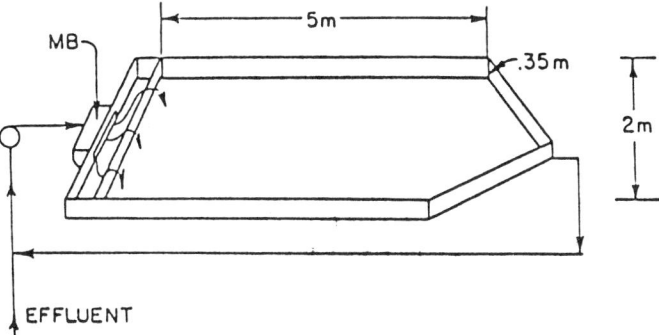

Fig. 2. Photochemical reactor

to south (Jordan River). The linear velocity of the water was about 1 m sec^{-1}, corresponding to a flow of about 0.8 m^3 sec^{-1}. The sensitizer (2% MB) was added to the water at a constant rate (0.4g MB sec^{-1}) from a barrel mounted on the channel. To ensure a rapid mixing of the dye, the water containing the dye was passed through a chicane system. The initial dye concentration was of 0.5 to 0.6 g MB m^{-3}. The bacteriological analyses showed that 3 h of sunlight irradiation (1800 μE m^{-2} s^{-1}) would be necessary for total destruction of coliforms in the channel water. However, the limited length of the channel which could be used for the disinfection experiment (3.5 km) provided about 90 minutes of irradiation time at the optimum flow rate of the saline water (3000 m^3 h^{-1}) and gave only partially disinfection results (about 2 log of coliforms destruction). Further optimization of the disinfection process in this water carrier was not possible due to its specific engineering construction characteristics.

15.3.2.3 Continuous Pilot Plant Experiments

The experience accumulated in the previous experiments was used for the construction of a pilot plant system for effluent disinfection by a continuous flow method. The pilot plant system was built from identical, galvanized steel tanks, modular units, installed in a series, in a North-South direction (to minimize the walls' shade). Each unit has a trapezoidal cross-section of 210/140 cm, a length of 500 cm and an operational depth of 20 to 30 cm. A height differential of about 10 cm between 2 adjacent units ensure the free overflowing of the treated effluent along the pilot system. The treated effluent leaves the tank in a thin layer stream which cascades down a step into the entrance compartment of the next tank (Fig. 3). The total volume of the pilot varies from 9 to 14 m^3 and the total surface exposed from 48 to 61 m^2, depending on effluent depth which can be varied from 20 to 30 cm, respectively. The operational flow rate of the treated effluent was fitted to the climatic conditions and effluent characteristics, and varied between 3 and 8 m^3 h^{-1}.

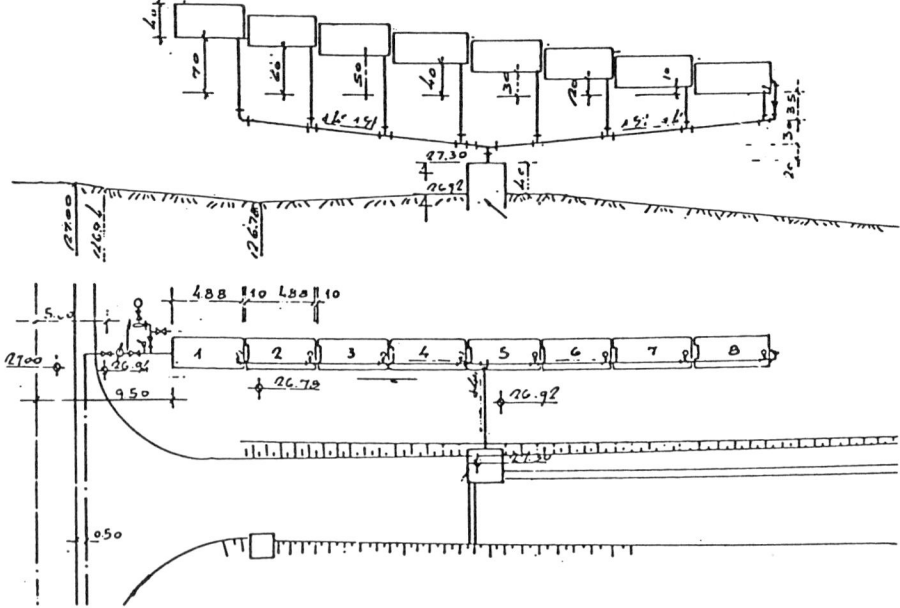

Fig. 3. Pilot plant system — side and top view

An injector mounted on a by-pass pipe of the effluent pipe entrance allows the addition and mixing of solutions (sensitizer, virus inocculum, etc.) into the effluent.

The most relevant characteristics of the effluent entering into the pilot are:

Suspended Solids	30–40 mg/L
Turbidity (NTU)	12–16
Oxygen dissolved (% saturation)	2–30%
BOD_5	30–40 mg/L
COD	80–90 mg/L
Temperature	12–30°C
pH	6.9–7.1
Coliforms (MPN)	10^6–10^7/100 mL
Fecal Coli (MPM)	10^5–10^6/100 mL
Enterococci (MPN)	10^4–10^5/100 mL
Polioviruses (PFU)*	10^3–10^4/100 mL

*(polio viruses, vaccine strains, were innoculated by us).

Table 4 shows the best disinfection results obtained during the winter season 1986/1987.

Table 4 and all other experimental data brought in this case study, show that disinfection of effluents by the proposed photodynamic action is feasible.

Table 4.

Microorganism	Entrance	Exit	Efficiency, %
Coliforms (MPN)	700,000	400	99,943
Fecal coli (MPN)	100,000	500	99,500
Enterococci (MPN)	50,000	4	99,992
Polioviruses (CFP)	5,000	60	98,750

*Winter season, 700–800 $\mu Em^{-2}s^{-1}$, residance time 130 minutes, flow rate 5 $m^3\ h^{-1}$; sensitizer 0.7 g m^{-3}.

15.4 Summary and Conclusions

The removal of organic pollutants from the treated municipal waste water is a prerequisite for their safe use for crop irrigation. This work presents results on the use of a photochemical procedure, which by oxidation of organic pollutants is aimed to achieve the detoxification and disinfection of effluents. The photochemical procedure consists of the combined action of sunlight and molecular oxygen dissolved in water upon organic materials, through the intermediary of a dye-sensitizer.

Water detoxification from nonbiological organic materials was studied by using potential agricultural pollutants (herbicides), belonging to different chemical classes: uracils, s-triazines, anilides. The photodegradation kinetics was investigated and the nature of the intermediary and final degradation products was characterized by their isolation and identification. The efficiency of the photochemical procedure for the detoxification of herbicide residues in water was tested by bioassay. For all the herbicides studied, the main photodegradation products and the photoreaction mixtures were not toxic to the test plants, indicating the suitability of this procedure for the detoxification of water from herbicides residues.

Water disinfection was studied on a laboratory scale, in a small pilot plant (2m^3), in a water carrier (an open canal of 3.5 km length, transporting up to 3000 m^3/h), and eventually in a pilot plant (12 m^3) built for continuous disinfection of effluents. The results obtained in the last system show the efficiency of this procedure: with an effluent flow rate of 5 m^3/h, a sensitizer (MB) concentration less than 1 ppm, and a residence time of 2 h, the amount of coliforms decreased from 700,000 to 400; of fecal coliforms from 100,000 to 500; of enterococci from 50,000 to 4 and of polioviruses from 5,000 to 60 (per 100 ml).

References

Acher AJ (1982) The fate of organic pollutants in frozen waters; sunlight decomposition of uracil herbicides in frozen aqueous solutions. J Water Res 16:405–410

Acher AJ, Dunkelblum E (1979) Identification of sensitized photooxidation products of bromacil in water. J Agric Food Chem 27:1164–1167

Acher AJ, Elgavish A (1980) The effect of photochemical treatment of water on algae growth. J Water Res 14:539–543

Acher AJ, Juven BI (1977) Destruction of fecal coliform in sewage water by dye-sensitized photooxidation. J Appl Environ Microbiol 33:1019–1023

Acher AJ, Rosenthal I (1977) Dye-sensitized photooxidation – a new approach to the treatment of organic matter in sewage effluents. J Water Res 11:557–562

Acher AJ, Saltzman S (1980) Dye-sensitized photooxidation of bromacil in water. J Agric Food Chem 9:190–194

Acher AJ, Marzouk J, Manor J (1979) Disinfection of effluents by a photochemical treatment. Progress Report to Israeli Council for R & D, Jerusalem (Nov 1979, in Hebrew)

Acher AJ, Saltzman S, Brates N, Dukelblum E (1981) Photosensitized decomposition of terbacil in aqueous solutions. J Agric Food Chem 29:707–711

Aly OM, El-Dib MA (1971) Studies on the persistence of some carbamate insecticides in the aquatic environment. I. Hydrolysis of Sevin, Baygon, Pyrolan and Dimetilan in waters. Water Res 5:1191–1205

Crosby DG (1970) The nonbiological degradation of pesticides in soils. In: Pesticides in the Soil: Ecology, Degradation and Movement. Int Symp on Pesticides in the Soil, 1970, Michigan State Univ, East Lansing, pp 186–94

Crosby DG, Li MY (1969) Herbicide photodecomposition. In: Kearney PC, Kaufman DD (eds) Degradation of Herbicides. Dekker, New York, pp 321–363

Draper WM, Crosby DG (1983) Photochemical generation of superoxide radical anion in water. J Agric Food Chem 31:734–737

Draper WM, Crosby DG (1984) Solar photooxidation of pesticides in dilute hydrogen peroxide. J Agric Food Chem 32:231–237

Foote CS (1968) Mechanism of photosensitized oxidation. Science 162:963–970

Freeman PK, McCarthy KD (1984) Photochemistry of oxime carbamates. 1. Phototransformations of aldicarb. J Agric Food Chem 32:873–877

Freeman PK, Ndip EMN (1984) Photochemistry of oxime carbamates. 2. Phototransformations of methomyl. J Agric Food Chem 32:877–881

Gerba CD, Wallis C, Melnick JL (1977) Disinfection of wastewater by photodynamic action. J Water Pol Control Fed 49:578–583

Gollnick K (1968) Type II photooxygenation reactions in solution. Adv Photochem 6:1–122

Hiatt CW (1967) Kinetics of virus inactivation by photodynamic action. In: Radiation Research. Wiley, New York, pp 857–868

Horowitz M (1976) Application of bioassay techniques to herbicide investigations. Weed Res 16:209–215

Kautsky H (1939) Quenching of luminescence by oxygen. Trans Farad Soc 35:216–219

Khan SU, Gamble DS (1983) Ultraviolet irradiation of an aqueous solution of prometryn in the presence of humic materials. J Agric Food Chem 31:1099–1104

Khan SU, Schnizter M (1978) UV irradiation of atrazine in aqueous fulvic acid solution. J Environ Sci Health B13:299–310

Kearns DR (1971) Physical and chemical properties of singlet molecular oxygen. Chem Rev 71:395–399

Kott Y (1973) Hazards associated with the use of chlorinated oxidation pond effluents for irrigation. J Water Res 7:853–862

Melnick JL, Gerba CP, Wallis C, Hobbs MF (1976) Photodynamic inactivation of virus in sewage. In: Baldwin LB, Davidson JM, Gerber JF (eds) Virus Aspects of Applying Municipal Wastes to Land. Univ of Florida Press, Gainesville, pp 25–36

Quistad GB, Staiger LE (1984) Photodegradation of fluvalinate. J Agric Food Chem 32:1134–1138

Rejto M, Saltzman S, Acher AJ, Muszkat L (1983) Identification of sensitized photooxidation products of s-triazine herbicides in water. J Agric Food Chem 31:138–142

Rejto M, Saltzman S, Acher AJ (1984) Photodecomposition of Propachlor. J Agric Food Chem 32:226–230

Ruzo LO, Cassida JE (1985) Photochemistry of thiocarbamate herbicides: oxidative and free radical processes of thiobencarb and diallate. J Agric Food Chem 33:272–276

Saltzman S, Acher AJ, Brates N, Horowitz M, Gevelbert A (1982) Removal of phytotoxicity of uracil herbicides in water by photodecomposition. Pestic Sci 13:211–215

Spikes JD, Livingston R (1967) The Molecular Biology of Potodynamic Action: Sensitized Photoautoxidations in Biological Systems. Wiley, New York, pp 857–868

Spikes JD, Livingston R (1969) The molecular biology of photodynamic action: sensitized photoautoxidations in biological systems. Adv Biol 3:29–38

Spikes JD, MacKnight ML (1970) Dye-sensitized photooxidation of proteins. Ann NY Acad Sci 171:149–155

Watts JR (1983) The development and design of dye-sensitized photooxidation waste stabilization ponds for treatment of industrial waste. Final Report to BARD-No 1-171-80, BARD-Israel

Wolfe NL, Zepp RG, Baughman GL, Fincher RE, Gordon JA (1976) Chemical and photochemical transformation of selected pesticides in aquatic systems. US Environ Protection Agency, Ecological Research Series PB, 258-848, p 141

Zepp RG, Wolfe NL, Baughman BL, Hollis RC (1977) Singlet oxygen in natural waters. Nature 267:421

16 Restoration of Aquifers Polluted with Hydrocarbons

H. SONTHEIMER, G. NAGEL and P. WERNER[1]

16.1 Introduction

Oil pollution of ground waters has been an important problem during the last 20 years in most countries. In some cases it has lead to the design and construction of special water treatment plants using for example ozone and/or activated carbon. The large importance of mineral oil products with respect to the number of accidents and the volume lost to the environment can be seen from Table 1 which contains data for the Federal Republic of Germany for the years 1982–1984.

It can be seen from these data that very large oil spills have been observed in these years, polluting ground waters in most cases. Very similar observations have also been made in the oldest water work of the city of Karlsruhe, which is located between a large railway area and parts of the city. About 500 railway cars containing oil or other mineral oil products pass through the area each day, coming from the three refineries in the vicinity of the city, and being switched in the marshalling-yard near the railway station to trains going in different directions. The situation of the water works "Durlacher Wald" where the studies to be discussed in this paper have been made can be seen in more detail in Fig. 1 (Nagel et al. 1986).

The four wells shown in the diagram were constructed in 1967 after the first water quality problems had been observed in the drinking water. The ground-

Table 1. Accidents with Water Polluting Substances in Germany between 1982–84

Pollutant	Number of Accidents			Volume lost to the Environment (m^3)		
	1982	1983	1984	1982	1983	1984
Crude Oil	39	29	13	975	1 736	22
Gasoline	86	78	83	954	212	256
Heating (Diesel) oil	828	913	1 022	5 449	1 169	997
Heavy oil	120	102	84	210	181	519
Other organics	249	273	242	315	8 668	548
Inorg. substances	38	47	81	84	118	143

---

[1] Universität Karlsruhe, Engler-Bunte-Institut, Abteilung für Wasserchemie, Postfach 6380, Richard-Willstätter-Allee 5, 7500 Karlsruhe, FRG

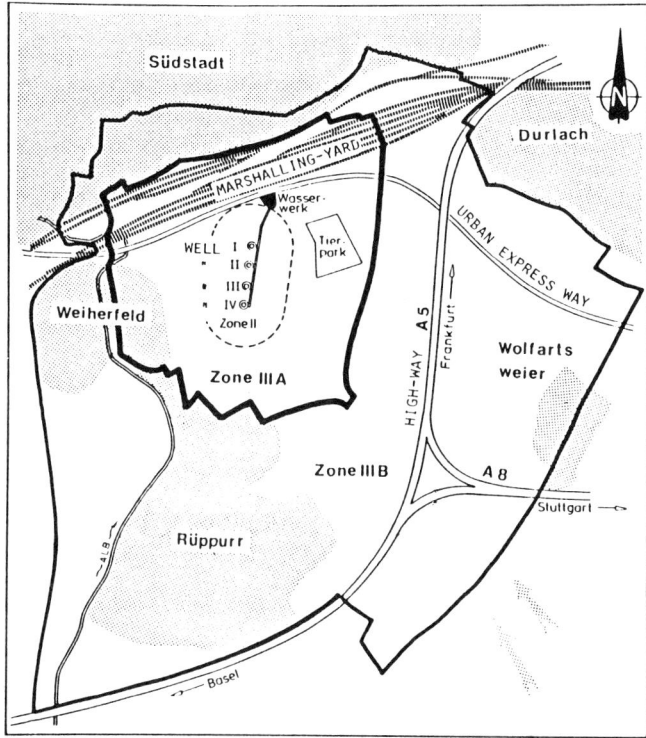

Fig. 1. Site of the water works "Durlacher Wald" in Karlsruhe

water has still not received any treatment. Moreover, disinfection with chlorine etc. has not been necessary. Each well was designed to deliver about 500 m³/h and the distance between two of the wells is about 150 m. It can also be seen, that the watershed includes most parts of the marshalling yard and some of the housing area. There are many highways and expressways near the water works, which is located within a small forest. Looking at this situation one can expect water quality problems for this water works and this has been found true many years ago.

16.2 The Water Works "Durlacher Wald"

Figure 2 contains further and more detailed information about the well structure which is very similar for all 4 wells. The information given indicates very clearly that there exists a very uniform groundwater layer with a hydraulic conductivity of about 2.0×10^{-3} m/s. Using these experimental data together with the thickness of the groundwater layer one can calculate the transmissivity to be about 6×10^{-2}

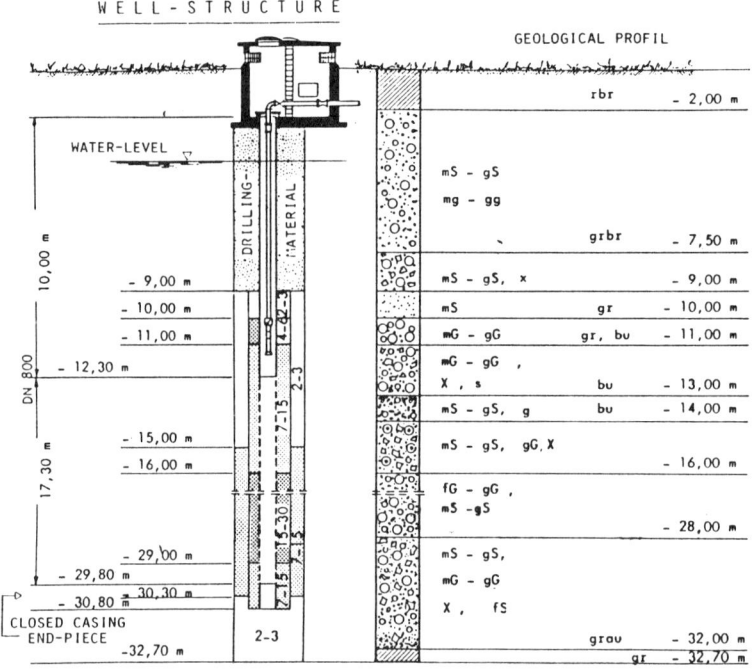

Fig. 2. Well structure of well I (rbr = red brown, grbr = grey brown, gr = grey = grau, mS = medium size sand, gS = coarse sand, G = gravel, t = time)

m^2/s. In order to confirm these important hydrogeological data and information, flowmeter measurements were made at well No. 1, which is now exclusively used as the delivery well for the water to be treated and reinfiltrated. The results obtained by the flowmeter measurements are given in Fig. 3.

The results of these measurements prove the observation that the ground water layer is very uniform and this is also true for the water flow through the aquifer. For each meter of the filterlength about 30 m^3/h can be extracted. This leads to a capacity for each delivery well of about 500 m^3/h. While the location of the described water works may lead to water quality problems it also has the great advantage, that there is only a very short distance from the works to the places with the highest water consumption.

The most important water quality measurements, shown in Fig. 4, indicate fairly high initial values for the parameters SAK (specific absorption coefficient), DOC (dissolved organic carbon), and NO_3, as well as for the oxygen consumption. The dashed lines in this figure give the mean values for the year 1978, while the straight lines indicate the effect of the treatment used here to improve the water quality.

The data for the DOC and the SAK show very similar behavior for these two water quality parameters. It can be seen that the concentration of the organic

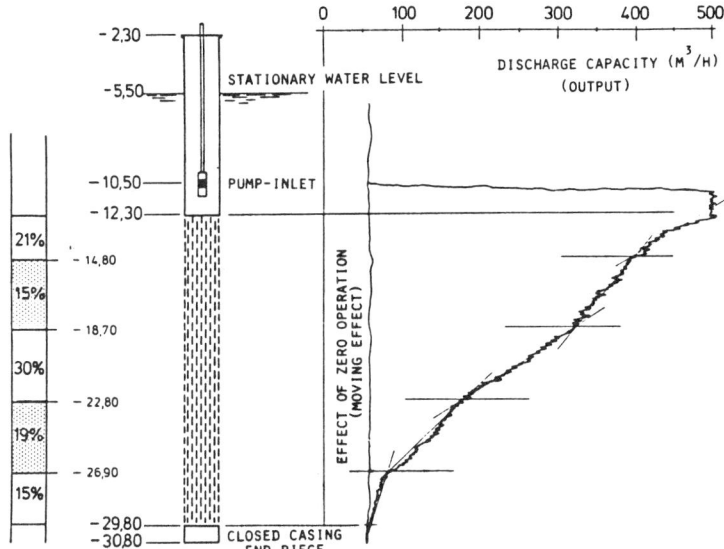

Fig. 3. Result of flow meter measurements, indicating the integrated water flow into the well from the lowest layer to the ground water level (right side) and the percentages of the flow for the different depths (left side)

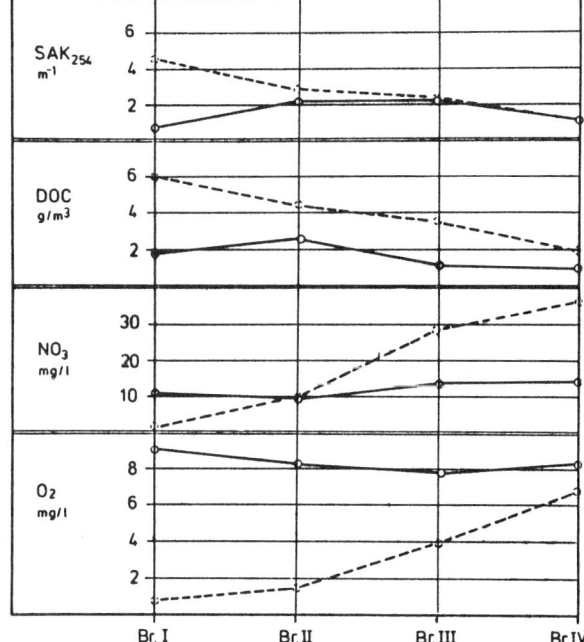

Fig. 4. Analytical data for the pollution in the 4 wells of the Durlacher Wald water works before the start of the new treatment process. (SAK = Specific Absorption Coefficient (UV-Extinction at 254 nm; DOC = Dissolved Organic Carbon; NO_3 = Nitrate concentration; O_2 = Oxygen concentration)

substances in the water of well No. I reached values up to 6 ppm DOC. Therefore it became necessary to stop the extraction from this well some years ago and to use only the water from the other three wells after the addition of some oxygen for drinking water purposes. But after only several years the water quality in well number II also became worse. At this time it became necessary to make a general decision about the future of this special water works (Nagel et al. 1982).

Besides closing, which had been proposed by many experts, one of the possibilities for a solution could be seen in the design and construction of a water treatment plant using ozone and activated carbon filters (Sontheimer et al. 1978). This solution had the disadvantage, however, that it is generally impossible to predict the future water quality changes in such a polluted aquifer. Therefore, a design for a safe water treatment plant would have been very difficult.

In order to reach a final solution which would also work in the future, it seemed necessary to combine the necessary treatment using ozone with the restoration of the polluted aquifer. It was therefore proposed to use the aquifer for the biological oxidation of the organic pollutants after an ozonation step. Such a process had never been used in a large water works before and therefore the German Ministry for Research as well as the German Railway Company subsidized the design and construction of the treatment plant, the scheme of which is given in Fig. 5.

The water to be treated is pumped from the donor well I at about 500 m^3/h and treated with 0.5–1.5 ppm of ozone. The details of the ozone reactor can be seen in Fig. 6. It was most important for the design and operation of the plant, that no additional pumping was necessary. The treated water, which is pumped from well I to the ozone treatment can then flow by gravity to the injection wells after

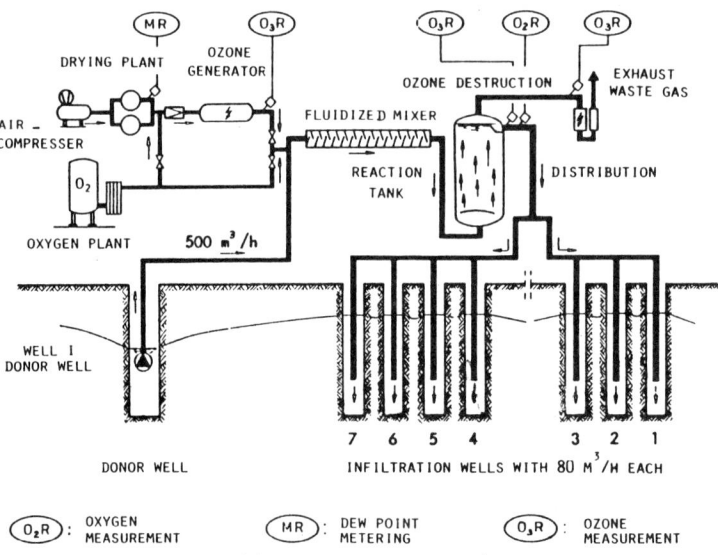

Fig. 5. Operation scheme of the new water treatment plant

Restoration of Aquifers Polluted with Hydrocarbons 325

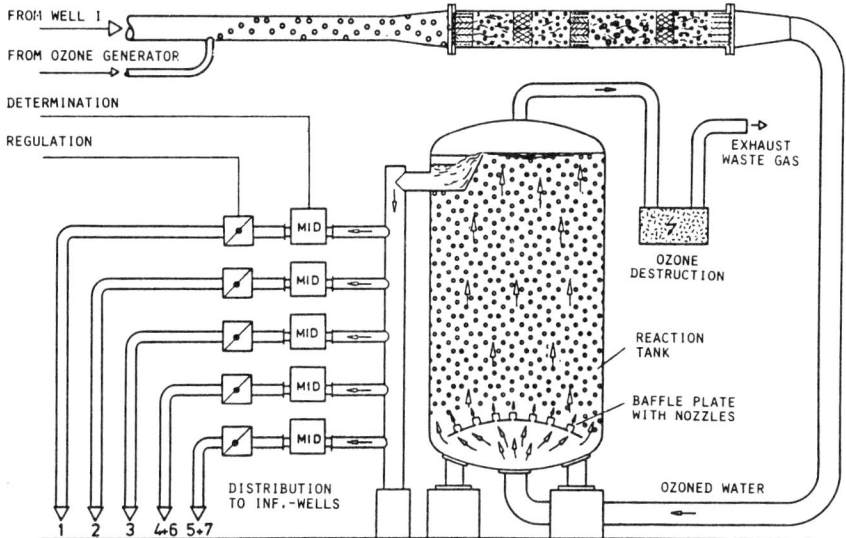

Fig. 6. Details of the ozone treatment plant

being distributed to the seven places where an injection of pretreated and oxygen containing water has been found necessary.

The ozone reactor reached 85–95% efficiency under normal operation conditions where about 1 ppm of ozone had to be added. The combination of the two mass transfer steps shown in Fig. 6 has proven worthwhile and effective. The final reactor has a volume of about 40 m^3 and has a special distribution inlet with jets for the water and gas flow. The reaction time of about 5 minutes is sufficient to reduce the ozone concentration to about 0.1–0.2 ppm. About 18 kWh/kg of ozone have been found necessary for the ozonation.

Including the investment costs, about 0.04 DM per m^3 of treated water is needed. This is much less than would have been necessary for a normal water treatment plant. Furthermore, no sludge occurs and the operation of the treatment plant is fairly easy. If some repair work has to be done, the three extraction wells can be used for drinking water purposes without any problem, as there is a large reservoir of pretreated water within the aquifer. No water quality changes have been observed in the drinking water even after 2 weeks of repair work. Another possibility for such a situation would be the introduction of oxygen for some time. This treatment might be sufficient by itself after the restoration of the aquifer.

The seven infiltration wells as well as the donor well and the extraction wells for the treated drinking water are shown in Fig. 7 together with the nearby housing area and the areas for the main contamination. In addition to the oil pollution resulting from the marshalling yards, cyanide contamination has been found in well No. 4. The cyanide results from an old chemical mill, about 3 km upstream. This mill was closed in 1910 for environmental reasons and after so many years,

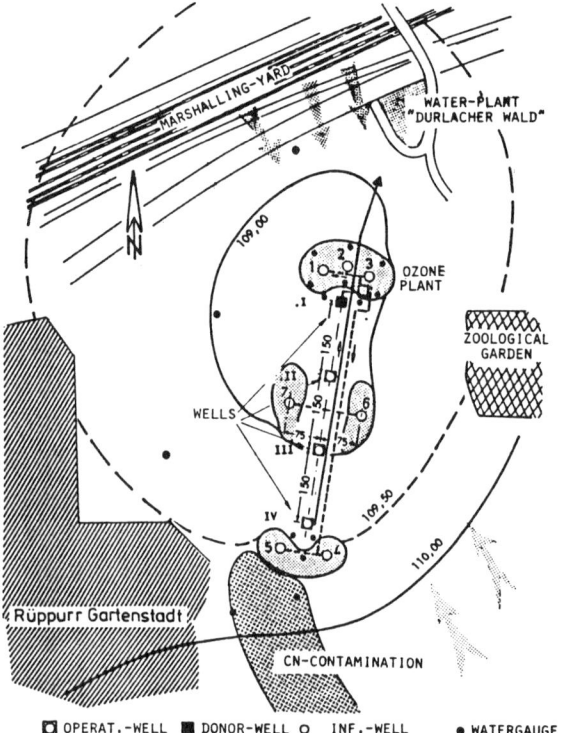

Fig. 7. Sites of the infiltration wells together with the main contaminated areas

the pollution in the ground water reached the water work area. This clearly indicates the large "memory" effect of ground water compared with surface water. The cyanide has the chemical form of the complex bound hexacyanoferrate, which is not toxic at all and was, indeed, used in the food industry in former years. However, using the German standard methods (DEW 1979) the complex is analysed as free cyanide and this may lead to some difficulties with the authorities.

Therefore, in addition to the initial wells 1, 2 and 3, two further infiltration wells, numbers 4 and 5 were used to dilute the cyanide and to prevent its introduction into the water used as drinking water. Finally the wells 6 and 7 became necessary, as highly polluted areas on the side of the well line have been found, resulting from earlier times and which are still polluting the water of well II. Altogether these seven infiltration wells can be operated at all ground water levels in such a way that the drinking water taken from the three wells 2, 3 and 4 is always of a very high quality.

In order to determine the minimal retention time of the treated water within the ground between the infiltration and the extraction wells and to be sure that the water doesn't flow in one or two layers only, tracer measurements were made in addition to the flowmeter testing. For this purpose a 10% sodium chloride solution was added to infiltration-well 2 together with the normal water quantity of about 80 m³/h, while the two other infiltration wells 1 and 3 received the same quantity as usual. It was possible to use three observation wells to analyse the electric conductivity and the chloride concentration. These wells were arranged between the injection well 2 and the donor well I. The distance between these two wells used for the tracer test was 75 m.

Figure 8 shows the groundwater levels and Table 2 gives the results of the tracer measurements. The differences found are in the usual range for such hydrogeological tests. Using these data to calculate the retention time for the maximum injection of 150 m³/h it can be predicted that the infiltrated water remains more than 15 days within the aquifer and this time should, from previous experience, be sufficient for full removal of all biodegradable organics and a good general water treatment. The results are described in the next section.

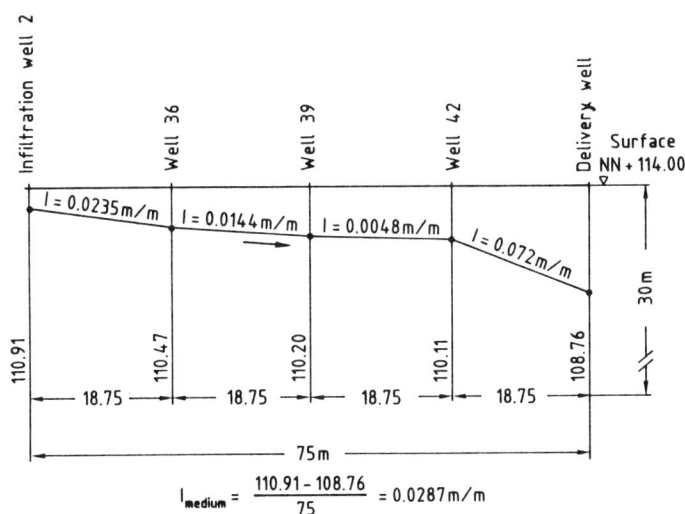

Fig. 8. Ground water levels for different test wells during the tracer experiments (I = Slope of ground water level)

Table 2. Results of the tracer experiments

Well	Initial conditions		After arrival of Tracer		Flow Time	
	Conductivity	Cl-conc.	Conductivity	Cl-conc.	Experimental	Calculated
Well	60.6	34.1	154.0	83.7	83.0	105.5
Test Well	60.6	34.1	95.2	51.0	161.0	172.2
Test Well	60.6	34.2	74.1	40.1	248.0	316.6
Delivery Well	60.7	34.0	66.8	36.2	38.0	34.4
			Cumulative Flow Time =		530.0	628.7

16.3 Treatment Results

The normally used operation scheme of the seven infiltration wells can be seen in Fig. 9.

From the data in Fig. 9, one can see that there is a nearly equal distribution to the seven wells but with somewhat higher values for the wells 1–3, those which have to prevent the further introduction of oil polluted waters from the marshalling yard. During the first months of the operation, these wells were used for the injection of about 150 m^3/h each. The water of the donor well has been always used for infiltration only. But the number of infiltration wells increased with time from 3 to 5 to finally 7. The remaining large extraction wells II, III and IV were operated for most of the time in such a way that the average amount of drinking water to be extracted and then distributed has been about the same as the water from the donor well I, which was about 12000 m^3/h. Changes in the daily amount of drinking water between 6000 and 18000 m^3/h didn't have any influence on the water quality. The result of the described treatment can be seen very clearly from the results summarized in Fig. 10.

One can see from this figure that very soon after the start of the operation of the new treatment plant, a very drastic reduction of the DOC-values could be observed. One can also see that there occurred an additional increase of the DOC data in the water of the wells II and III after some years of operation. This was due to changes in the ground water flow after the start of the injection in the wells 4 and 5, leading to additional pollution from other areas on the side of the well line. These observations were the reason for the construction of the aforementioned injection wells 6 and 7. The result of this additional injection can be observed very well from the data for the oxygen concentration as they are given in Fig. 11.

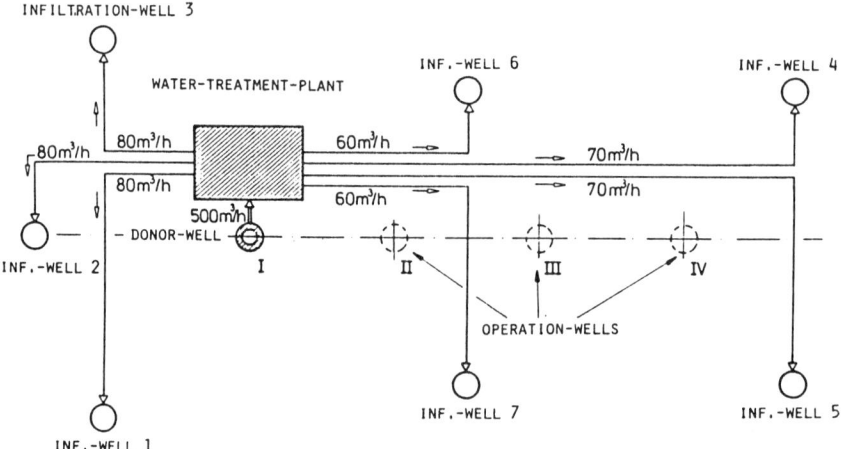

Fig. 9. Distribution of the treated water to the seven infiltration wells

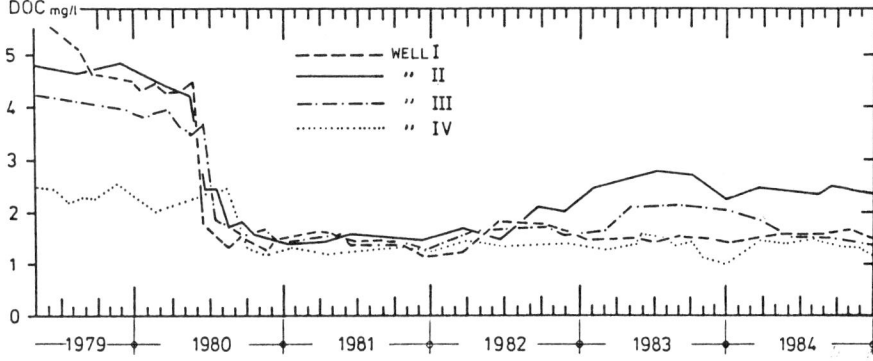

Fig. 10. DOC-concentrations in the water of the 4 large wells of the water work between 1979 and 1984

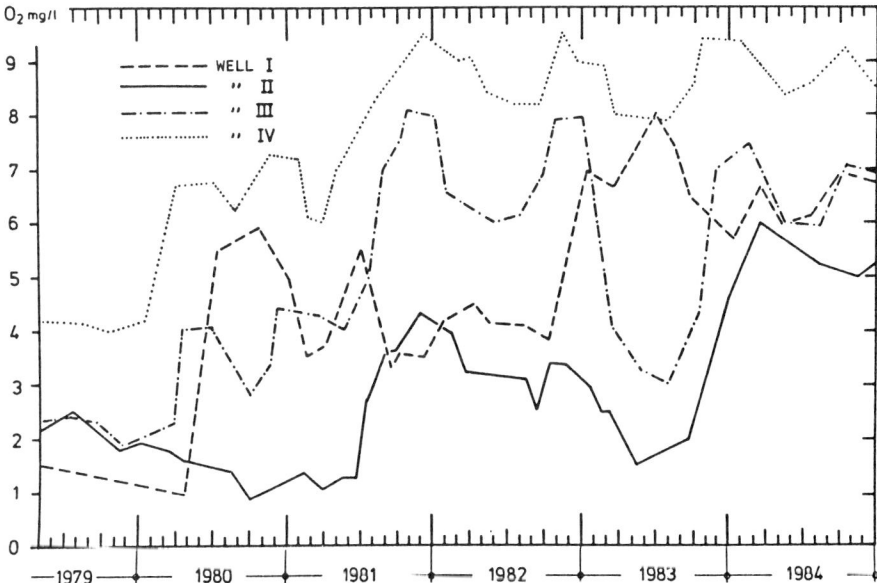

Fig. 11. Changes in the oxygen concentrations in the four wells of the water work "Durlacher Wald"

After the construction of the new infiltration wells in 1983 the oxygen concentration in well II increased again and the same was true for well III. Very similar observations were made for the UV values at 254 nm. All these and other detailed data are presented in the final report by Nagel et al. (1986).

The observations made during the now nearly seven years of operation indicate very clearly that it is very important for the restoration of the aquifer to reach all polluted areas with the treated water. However, in each case it requires a very long time for all polluted areas to be restored. Also, after 7 years of operation 2 to 3 ppm of oxygen will still be needed for the groundwater restoration. It has to be considered further that the water flowing to the water works area doesn't have a high oxygen concentration and would need some aeration in each case. In accordance with the observations made up till now it can be expected that it might be possible in time to inject the water without an additional ozone treatment. Tests for this type of treatment will be done shortly.

Besides the concentration of the organic substances measured as DOC and the oxygen concentration, other water quality parameters have also been changed during recent years. This can be seen from Fig. 12 showing the mean values for all important water quality parameters for the four wells of the water works. It should be mentioned here, that the described plant went into operation in 1980 (Nagel et al. 1986).

It is very interesting to see, how many of the usually used water quality parameters showed an increase in concentration between 1973 and 1980 together with the increase of the organic pollution. With the start of the operation of the described treatment plant this increase was stopped or even reversed. This is a

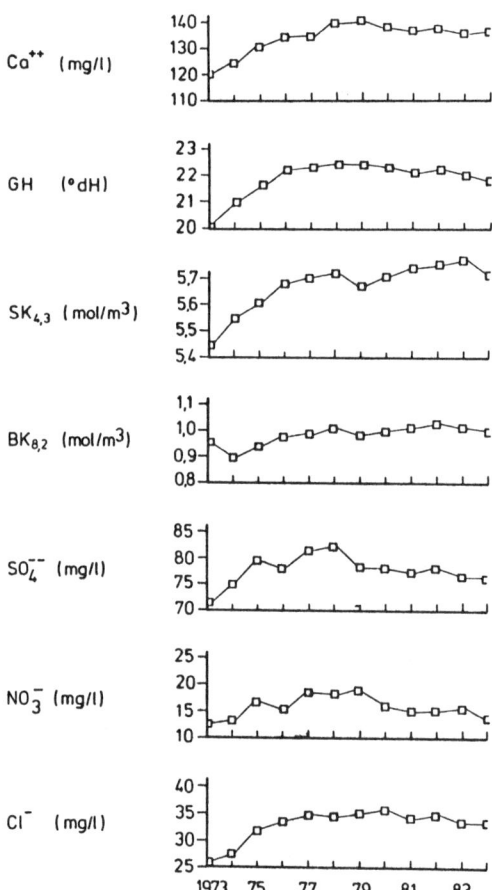

Fig. 12. Medium values for the most important water quality parameters in the water from the "Durlacher Wald" from 1973 to 1984. (GH = Total hardness; $SK_{4,3}$ = alkalinity, $BK_{8,2}$ = acidity)

very typical example of the fact, that organic pollution of groundwater does not effect the organic parameters alone but also the inorganic water composition. Most of the effects seen in this figure can be explained through the formation of inorganic carbon as CO_2 formed through the biodegradation of the organic substances. These changes in the water quality, while not very large cannot be repaired easily. Also, with a natural treatment as is now used in this special water work the recovery of the original water quality may require a very long time.

16.4 Microbiological Aspects of the Treatment Used

Within a groundwater layer polluted with mineral oil or similar products, the oxygen concentration in the ground water is normally the limiting factor for complete biodegradation of the organic pollutants. In the special case described here, the pollution had become so high, that besides the oxygen, nearly all the nitrate had been utilized as an electron acceptor for the biodegradation. If there is a long lasting continuous input of pollutants into the aquifer, the mineralization after some time can become inhibited and this may lead to a very quick increase in groundwater pollution. In this water works, the increasing pollution could be observed in well I only 10 years after the construction of the new wells. Well I is nearest to the railway area. But after closing this well it took only two years more for a decrease of the water quality in the well II to be observed.

This quick increase of groundwater pollution after a long time has been observed in other water works too. Because of the insufficient biological degradation of the polluting substances in the ground an increase of the colony counts according to the standard plate count method was observed within the distribution net. Consequently disinfection of this water before distribution became necessary.

While such a treatment has not been possible without additional installations, because the necessary reaction time before the first customer has not been available, it seemed worthwhile to activate the biological activity within the ground thus removing all biodegradable organics there. This finally should lead to water which needs no further disinfection. In order to achieve better removal of the polluting substances through biodegradation a preozonation should contribute to more effective treatment.

In order to get some more information about the amount of ozone necessary for such treatment, studies have been made using the methods developed by Werner et al. (1985a,b), which allows one to determine the growth rate of microorganisms as a function of the substrate quantity of the water. A typical result of these measurements is shown in Fig. 13. There exists a beneficial effect of the ozone dose on the biodegradability up to about 1 ppm of ozone. Higher dosages showed no further improvement. The growth rates increased from about 0.1 h^{-1} without any ozone to about 0.3 h^{-1} at 1 ppm of this oxidation mean.

Figure 14 indicates very clearly, that the preozonation leads to a better growth of the microorganisms in the water. This increase of the growth rate will be

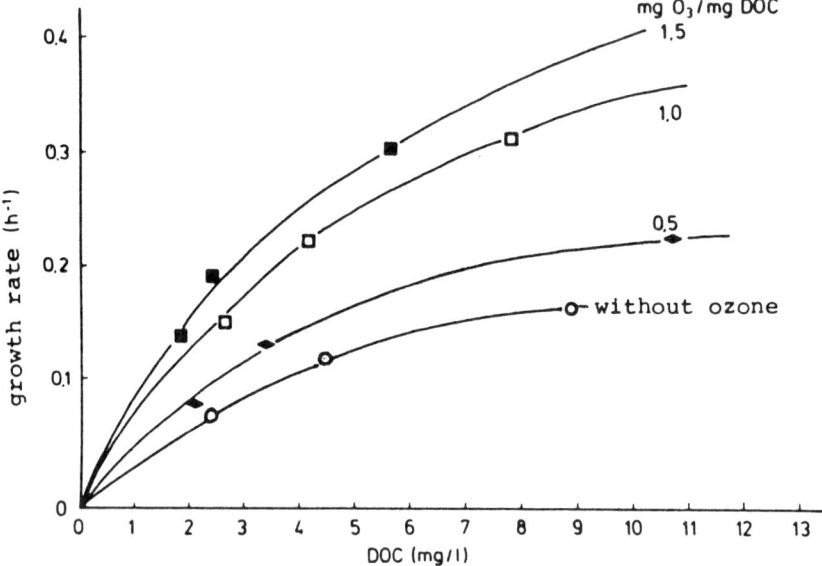

Fig. 13. Growth rate of the microorganisms for different ozone dosages and different concentrations of the organic substances

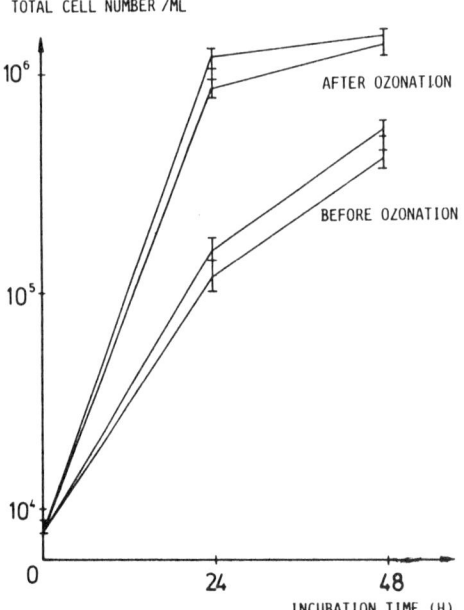

Fig. 14. Influence of ozone on the number of bacteria after different times with and without preozonation

advantageous especially in those cases, where the overall concentration of the humic substances is very low and where it is necessary to remove very small amounts of organic pollutants. This is the case for mineral oil products as very very small amounts of these substances can pollute drinking water. In these cases the activity of the microorganisms should be increased through ozonation to give a complete treatment.

If on the other hand, the concentration of the biodegradable organic pollutants within the aquifer is high enough it might be sufficient to add just water with a higher oxygen concentration. This can be accomplished by an aeration step using pure oxygen instead of air. If these concentrations are still too low, nitrate can be added as an additional oxygen source. This allows the restoration of highly polluted aquifers through biodegradation. Experience with this interesting type of treatment has been described by Werner (1985b) and Battermann and Werner (1984).

16.5 Summary and Conclusions

The experiences presented in this paper indicate an interesting, economic and safe way of solving a very difficult and important water quality problem. The water works "Durlacher Wald" should have been closed in the opinion of many experts for being located in the vicinity of various sources of pollution and exhibiting the effects of this through water quality deterioration. Today the water works delivers water of excellent quality without any treatment of the extracted well water. Furthermore, it is evident from the observations, that it will be possible in this case to clean the polluted aquifer, thus providing additional safety for the future.

The complete success of this clean-up campaign, on the other hand, cannot be used as an argument for fewer preventive measures against groundwater pollution. The experience made with the treatment process also indicates that it takes a very long time for a full restoration of the polluted aquifer. Therefore, measures to prevent pollution of the aquifer are still preferable.

On the other hand, one can conclude from the described observations that groundwater pollution with biodegradable organics can be removed through biological oxidation, that is, by natural means. This is not the case for non-biodegradable organics and in the presence of substances which are toxic to bacteria. In addition, good hydraulic conductivity and a fair solubility of the pollutants is necessary. If we also have a homogenous distribution of the pollutant within the aquifer and can achieve a full flushing of the contaminated surface, we can then expect that this natural treatment will lead to full restoration of the aquifer and after some time to very good groundwater quality.

References

Battermann G, Werner P (1984) Beseitigung einer Untergrundkontamination mit Kohlenwasserstoffen durch mikrobiellen Abbau. GWF Wasser/Abwasser 125:366–373
Deutsche Einheitsverfahren zur Wasseruntersuchung (1979) Chemie, D1
Nagel G et al. (1982) Grundwassersanierung durch Infiltration von ozontem Wasser. GWF Wasser/Abwasser 123:399–407
Nagel G, Sontheimer H, Kühn W, Werner P (1986) Das Karlsruher Verfahren zur aktivierten aeroben Grundwassersanierung. Veröffentlichungen des Bereichs und Lehrstuhls für Wasserchemie am Engler-Bunte-Institut der Universität Karlsruhe, Heft 29
Sontheimer H et al. (1978) The Mülheim process. JAWWA 70:393–396
Werner P et al. (1985a) Eine Methode zur Bestimmung der Verkeimungsneigung von Trinkwasser. Vom Wasser 65:257–270
Werner P et al. (1985b) A new way for the decontamination of polluted aquifers by biodegradation. Water Supply 3:41–43

Subject Index

Page numbers in *italics* refer to citations in figures.

Abiotic reactions 91, 93, 96, 100, 108, 125, 131, 136–148
– and heavy metals 98
– effect of temperature on reaction rates 105
– enhancement by adsorption 101
– evaporation of hydrocarbons 214–215
– in ground water 144
– in sediments 142
Accelerated degradation 193–205
– antimicrobial agents 204, *204*
– – chloramphenicol 198
– – cycloheximide 198
– control and management 202–204
– cross-enhancement 194, 201–202
– extenders 203
– mechanisms 198–200
– microbial degraders 198
– problem (history) soil 194, *197*
– self-enhancement 194
– soil sterilization
– – methylbromide 198, *204*
– – solarization 198, 203, *204*
Acceptable daily intake 20, 30
Acetanilides 60
Acetate
– as tracer in groundwater 291–300
Adsorption 55, 68, 96
– and activity of organophosphates 49
– and kinetics of degradation 116–122
– and K_{ow} 69–71, 152–156, 172
– and solubility 71, 152–156, 172
– and surface enhanced reactions 91–92, 95, 101–104
– by humic substances 30–31
– competitive adsorption 183, 184
– desorption 16
– effect of heavy metals 98
– effect on degradation processes 101, 108, 159
– effect on hydrolysis 140
– from aqueous systems 165–166
– from organic solvents 167–170
– from complex mixtures 180–183

– kinetics *221*
– mechanisms 58–68, 158, 163–173
– multilayer 128
– methods of predicting 69–70, 151
– of cationic TOC 56
– of multi-sorbate mixtures 183–187
– of PAHs 46
– unified description 122–128
– vs. partitioning 70–71
– vs. sorption 163
Adsorption isotherms
– BET 128, 170, 171
– C-type 56–58, *56*
– Freundlich 56–58, 220
– H-type 56–58, *56*
– Langmuir (L-type) 56–58, *56*
– linearity 70, 116–118, 164–165, *166*
– vapor-sorption isotherms 170–171, *171*
Alachlor 45, 49
– accelerated degradation 203
– hydrogen bonding 60
– van der Waals adsorption 60
Aldicarb (Temik) 17–19, 45
– adsorption variability 159
– aqueous solubility 18
– effect of organic matter on movement and adsorption 50
– half life 19
– hydrolysis 18, 19
– in groundwater 18
– – in Florida 19
– – in Long Island 19
– LD_{50} 18
– Redox reactions 141
Aldicarb sulfoxide 18, 19, *141*
Aldicarb sulphone 18, *141*, 189
Aldrin 20, 45
– activity 48
– degradation 48, 49
– in the biosphere 44
– leaching 49
– LD_{50} 30
– residues in water supplies 21
– vapor pressure 45

Aldrin (cont.)
- volatilization 48
Algae destruction by photolysis 304, 312
Allophane 96, 109, 129
Ametryne photolysis 308–310
Amiben 48
3-Aminotriazole 113
Amitrole 43, 59, 63, 105, 194
Aniline
- as tracer in groundwater 291–300, *297, 298*
- effect of humic substances on photolysis 52, 77
- enhanced reactivity 67
Anion exchange capacity 94, 97
Anthracene
- effect of humic substances on bioaccumulation of 53
- solubility in mixed solvents *178*
Aquifers
- fresh water 6
- Ogallala 288–300
- polluted 6, 321–327
- rehabilitation 11, 331–333
- saline 6
Aroclors 25
Aromatic compounds 212, 231–234, *232*
Arthrobacter sp.
- accelerated degradation of EPTC 200
A. paradoxus
- accelerated degradation of 2,4-D 200
Asulam 60
Atraton phytolysis 308–309
Atrazine
- adsorption 57
- from biphasic solvents 182–183
- hydrogen bonding 113
- hydrolysis 73–74, 75, 102
- hydroxylation 109
- ionic bonding 59
- mobility 22, 188–190
- photolysis 52, 77–78, 308–310
- solubility 23
Attapulgite *see* palygorskite
Azobenzine 77

Bendiocarb 194
Benefin 45–46, 49
Benomyl 194
Bentonite *see* montmorillonite
Benzanthracene 53
Benzene
- adsorption 220
- vapor phase 164, 171
- carcinogenicity and health hazard 232–233
- diffusion in soils 225–226, *226*
- evaporation *216*
- solubility 25
- transport 188–190
Benzidine 106
Benzoate 291–300, *295, 296, 297, 298*
Benzophenone 27, 79
Benzopyrene 53, 54
Benzthiazuron 45
BHC 21
Bioaccumulation 53–54
Biodegradation *see also* microbial activity
- enhanced, for cleanup of the vadose zone 283–286
- in restoring polluted groundwater 331–333
- of hydrocarbons in groundwater 246–247
- of hydrocarbons in soils and sediments 251–270
- – anaerobic conditions 252–255, 263–266
- – biochemical considerations 252–253
- – hexadecane as model compound 256–270
- – lag phase 256, 259
- – oxygen dependency 252–254
- – role of dinitrification 252, 254–255, 263–264
Biolysis 138–139
BOD 26
Boron 26
Bound residues
- and organic matter 79
- DCB 28
- defined 112
- formation 130
- persistence 131
Breakthrough curves
- in Ogallala aquifer 295–300, *295, 296, 297, 298*
- of mixed pollutants 188–190, 189f
Bromacil 23, 44, *57*, 60, 305
Bromide 293–295, *295*
Bromoform 24, 144
Bromomethane 144
Bromoxynil 44
Butralin 45
Butylate 194, 201
Butylbenzene
- adsorption *233*
- – kinetics *221, 222*
- – desorption *224*
- evaporation *217*
- transport 226–228, *227*

N-butylbenzene sulfonamide 27

^{13}C-NMR 40, 80
Capillary zone
– and hydrocarbon contamination of groundwater 234–238, *235*, *239*
Captan 30
Carbamates 59–60, 68
Carbaryl 60, 194
Carbofuran 45, 50, 194, 195, *197*, 201
Carbontetrachloride 24, 28, 144, 273
Case histories
– polluted groundwater
– – Florida 10
– – Long Island, N.Y. 19
– – Michigan 27
– – New Jersey 21, 24, 25
– – Puerto Rico 279
– – South Carolina 21
– restoration of polluted aquifers
– – Karlsruhe, FRG 210, 222
– – Twin Cities Army Ammunition Plant, MN 228, 279, *279*
Cation exchange capacity 91, 96, 97
– and adsorption 108
– as affected by metal oxides 96
– of clay minerals 93–94
– of organic matter 92, 123
CDAA 45, 49–50
CDEC 49–50
Cetyl alcohol 258–267, *260*, *265*, *266*
Charge density
– and adsorption 92, 97, 108
– of iron oxides 95
– of kaolinite 94
Charge transfer complexes 60–63
– phenol-herbicides 42
– s-triazines 59
Chemical degradation *see also* surface enhanced transformations and abiotic processes 16, 282–283
Chemodynamic properties 176
Chloracetamide 45
Chloramben 44
Chloradan(e) 21, 44, 45, 48
Chlordimeform 56, 58, 61
Chlorbenzene 25, 28, 171
bis(2-chloroethyl)ether 25
Chloroform 25, 144, 276
Chloromethane 144
Chloropicrin 164, 170
Chlorpropham 45, 49, *57*
Chlorpyrifos 137
Chlortoluron 45
Chrysene *178*

Ciordin 101, 107, 108, 113
Clay(s) minerals 16, 92–112
Clay-organic matter complexes 91, 127, 129
CMOS (completely miscible organic solvents) 176, 177–180, 181–183, 187–188
Complex formation 16
Complex mixtures 150
Conjugation 16
Contamination criteria and assessment 231–234, 274
Cosolvents *see* mixed solvents
Covalent binding 63–65
Crude oil
– composition 212–213, *215*, *233*
– distillates
– – heavy 234, *233*
– – light 232–234, *233*
– – medium 233–234, *215*, *233*
Cumene 216f, *217*, *222*, *223*, *224*, 226–228, *227*
Cyanamide 95
Cyanide 325–326
Cycloate 49, 60
Cycluron 45

1,2-D 20
2,4-D 22, 44, 57, *57*, 60, 65, 66, 74, 138, 200, 201
Dalapon 44, 194
Daphnia magna 53
DBCP 20
DDD (TDE) 21
DDE 21, 102, 109
DDT 20, 21, 44, 45, 48, 49, 51, 53, 60, 67, 71–72, 77, 103, 109, 172–173
Decane 226–228, *217*, *221*, *222*, *223*, *224*, *227*
Detergents 26, 42, 54, 75
Diazinon 26, 49, 101, 113, 172–173
Dibromochloromethane 25, 30
Dibromethane 24, 144
Dibromomethane 25, 144
Di-t-butyl-p-benzoquinone 27
Dibutylphthalate 27, 28
Dicamba 44, *57*, 60, 68
Dichlobenil 46, 49–50
Dichlofenthion 172–173
Dichloroaniline 52, 78, 79
Dichlorobenzene 24, 25, 27, 28, 54, 171, *171*
Dichlorobenzidine (DCB) 27, 64
Dichlorobromoethane 24, 25
Dichloroethane 25, 54, 144
Dichloroethylene 24, 25, 144

Dichloromethane 144
Dichlorophenol 184–185, *185*, 188–190, *189*
Dieldrin 20, 21, 30, 44, 45, 48, 170, 172
Di-2-ethylphthalate (DEHP) 47, 53
Diffuse double layer 97, 99, 103, 104, 108, 125
Diiodoethane 144
Diiodomethane 25
Dimefox 59
Dimethylfuran 52, 77
Dimethylphenol 184–185, 188–190, *185*
Dimethylphthalate 27
Dinitrophenol 184–185, 188–190
Dinoseb 44, *57*
Dioctylphthalate (DOP) 47
Dioxins 53
Diphenamid 46, 49, 194, 199, 203
Diquat 42, 48, 56, *57*, 58
Dissolved organic matter (DOM) 38
Dissolved organic carbon (DOC)
– and photolysis of TOCs 76, 78
– and transport 188
– as water quality indicator 322–330, *323*
– effect on TOC behavior 50, 52
– effect on K_{oc} 71
– in Ogallala aquifer 289, 294
– reduction of, in contaminated aquifer 328, *329*
Disulfaton 52, 77
Diuron 23, 45, 49, 167, 172, 188–190
DNOC 44
Dodecane 220, 226–228, *217*, *221*, *222*, *223*, *224*, *227*

EDB 20
Electroosmosis 280–282, *281*
Electrophoresis 280–282, *281*
Endosulfan 48
Endothal 22, 194
Endrin 21, 45, 48
Enhanced degradation *see* Accelerated degradation
Enteroccoci 316, *317*
Enzymatic reactions 129–130
– as affected by adsorption 129
– in cleanup of vadose zone 285–286
– incorporation of residues in humic acid by polymerization 66–67
– role in accelerated degradation 199, 203
Enzymes
– acylamidase 203
– adsorbed 108
– extracellular 129, 130
– glucose oxidase

– – adsorption by minerals 129
– laccase 65–67
– oxygenase 65
– peroxidase 65
– phenoloxidase 66
– phosphatase 129
– urease
– – adsorption by montmorillonite 129
Enzyme mediated bonding 65–67
EPTC 49, 194, 200, 201
Ethanol
– as tracer in groundwater 291–300, *296*
Ethoprop 159, *160*, 194
Ethylamine 297
– as tracer in groundwater 291–300, *297*
Ethylbenzene 28, *216*
Ethylene dibromide 164, 170
Ethylenthiourea 52
bis(2-ethylhexyl)phthalate 27, 28
Exchangeable cations 92, 97, 110, 111, 125, 126

Fecal coliforms 316, 317
– coliform inactivation 313, 319
– – effect of methylene blue 313
– – effect of sunlight 313
– – exposure time 313
– removal from wastewater 26
Fenarimol 102
Fenuron 45, 49, *57*, 62–63
Fertilizers 6, 111
Fluometuron 45, 49, *57*
Fluoroform 25
Fonofos 49, 57
Free radicals
– and dealkylation of s-triazines 51
– and photolysis 78
– in humic acid-triazine complexes 61–62
– of humic substances 41, 65, 66
Fugacity 156
Fulvic acid 37
– and free radical content 65
– and ion exchange 58–59
– catalyst for hydrolysis 138
– C/N ratio 40
– definition 39
– effect on pesticide transformations 49
– effect on TOC solubility 51, 72
– functional group content 40
– in DOM 38
– surface tension 41

Glyphosate 45
Groundwater
– alternatives 3

Subject Index   339

- legislation 8–10
- memory effect 326
- modelling of contamination 28–29
- pollution 4–7, 10, 12–14
- – by industrial chemicals 24–28
- – by insecticides 17–23
- quality 5
- restoration 10–11, 277–278, 321–333
- tracer movement in 295–301

Halloysite 93–94
Health criteria for priority pollutants 27, 28
Heavy metals 11
- and electrokinetics 282
- as electron acceptors/donors 109
- in New Jersey groundwater 24
- role as catalysts 98
- role in ligand exchange 60
Henry's law 169–170, 276
Heptachlor 21, 30, 44, 45, 48, 172–173
Heptachlor epoxide 21
Herbicides 17, 21–23, 49
Hexachloroethane 144
Hexodecane
- degradation
- – in marine sediments 256–266, *257, 258, 259, 260, 261, 262, 264*
- – in soils 266–267, *267, 268, 269*
Holding ponds 5
Humic acids 37
- and free radical content 65
- and ion exchange 58
- and photolysis of TOCs 76–77
- and catalyst for hydrolysis 138
- C/N ratio 40
- definition 39
- effect on pesticide transport 49
- effect on TOC solubility and uptake 51, 53, 72
- functional group content 40
- interaction with metal oxides 96
- surface tension 41
Humic substances 37–80
- and chemical reduction 52
- and photolysis 41, 52–53
- and TOC bioaccumulation 53–54
- clay mineral complexes 48
- definition 39
- effect on TOC solubility 51, 71–72
- elemental composition 39
- free radical content 41
- functional group composition 40
- in DOM 38
- surface activity 41
- TOC interactions 50

Humin 39
Hyamine 56
Hydraulic conductivity 243, 245, 321
Hydrogen bonding 59–60, 73–74, 113
Hydrolysis 16, 114
- acid catalyzed 75
- aldicarb 18
- base catalyzed 74, 75
- by humic substances 37, 51, 73–75
- effect of adsorption 108
- effect of moisture content 106, 110
- effect of surface pH 99–100
- half lives 136
- in groundwater 143
- in sediment 140
- in water column 137–138
- kinetic model 74–75
Hydrophobic adsorption 67–68, 123, 140, 165, 183–184, 298
Hydrophobic surface area 126, 181

Illite 93, 129
Immiscible compounds
- contaminants 231
- solvents
- – microemulsification 188
- – partitioning 68
- – transport 187–188, *214*
Industrial chemicals 24–26, 110
Ion exchange 16, 58–59
Iprodione 194, *197*
Isocil 44
Isofenphos 194, 198

Kaolinite
- hydrolysis of pirimiphos-ethyl 105
- properties 93–94
- surface enhanced hydrolysis of parathion 101–102, 106, 107, 110, 111, 114, 121, 125
  surface enhanced polymerization of styrene 102
Kerosene 233–234
- in cleanup of vadose zone 276
- in crude oil 215
- simulated 216–233, *227*
K_{oc}
- and K_{ow} 69, 71, 165, 172, *154, 155*
- and molecular connectivity 158–159
- and partitioning 69, 70–71
- and solubility 69, 71, 165, 172, *154, 155*
- as affected by DOM 71, 72
- as affected by organic matter source 73
- TOCs adsorbed by humic substances 56–57

K_{oc} (cont.)
- polarity of humic substances 68

K_{om} see K_{oc}

K_{ow}
- adsorption, estimated by 69–70, 152–156
- and K_{oc} 69, 165
- and solubility in mixed solvents 177, 179, *179*, 188
- of PAHs 69
- relationship to binding by humic substances 72–73

Landfills 5, 7, 42, 136

LD_{50}
- herbicides 22
- of selected compounds 18, 19, 30
- oral 30

Legislation 8–9
- Clean Water Act 9
- Comprehensive Emergency Response Compensation and Liability Act 9
- Resource Conservation Act 9
- Safe Drinking Water Act 8–9
- Surface Mining and Reclamation Act 9
- Toxic Substances Control Act 9

Leponis machrochinus 53

Leptophos 67

Ligand exchanges 60, 127

Lindane 21, 44, 45, 48, 49, 51, 71, 72, 164, 165, *166*, 167, 169–170, 172, 194

Linuron 49, 57

Malathion 45, 60, 113

Marasmius ordeadis 79

MBC 200

MCPA 194, 200, 201

Metalaxyl 194

Metal oxides 94–96
- amorphous iron oxides 94–96
- coatings on clay minerals 96
- ferrihydrite 95
- goethite 94, 95
- pH dependence of exchange capacity 97
- surface induced reactions 103

Methabenzthiazuron 45, 79

Methazole 67

Methiocarb 45

Methomyl 19–20

Methoxychlor 21, 65, 79

Methylbromide 164

Methylchloride 25

Methylene chloride 24

Methylparathion 64, 79

Methylthiobenzothiazole 27

Metolachor 60, 67

Microbial activity
- and transformation of pollutant mixtures 188–190
- as affected by DOC 331, *332*
- as affected by mineral surfaces 104

Microbial degradation *see also* biodegradation
- aerobic and anaerobic 16, 19
- and transport 190
- in the cleanup of a polluted aquifer 331–333

Migration potential 280–282, *281*

Mirex 21, 30, 52

Mixed solvents
- and transport 187–188
- adsorption from 181–182
- defined 176

Molecular connectivity 150, 152, 156
- and K_{oc} 158–159
- and K_{ow} 158–159
- and QSAR 158–159
- and solubility 158
- definition 156–157

Montmorillonite
- adsorption of enzymes 129
- adsorption of hydrocarbons 220, 224
- binding of nucleotides 125
- catalytic capacity 101–102, 109
- hydrolysis of pirimiphos ethyl 105
- hydrolysis reactions 106, 107, 111, 113
- properties 93
- surface transformation of DOT 103, 109

Monuron 30, 45, 49

Multiphase flow
- capillary pressure 244–247
- fingering 236
- flow field characteristics 244–247
- hydrocarbon spreading 236, 240
- lens (pancake) formation 236, 237, *237*, 240
- modelling 240–244
- relative permeabilities 236, *236*
- Richards equation 243

Multiphase solvents 176, 179–180, 182–183

Naphthalene 27, *178*, 188–190

Naphthenes 212, 231–234, *232*

Napropamide 22, 23, 155

Neburon 49

Nitralin 45–46, 49

Nitrobenzene 25

Nitrofen 30

Nitrophenol 200

Subject Index 341

N-nitrosamine 22
NNA 22–23
Norflurazon 23, 67

Ogallala aquifer 21, 288–301
Olefins (alkenes) 231–234, *232*
Organic carbon *see* Organic matter
Organic matter 16
 – and adsorption 152, 163–164, 172
 – and clay minerals 96, 97
 – and hydrocarbon solubility 219
 – definition 39
 – effect on pesticide activity 48–50
 – effect on TOC removal 54
 – humus 38, 48, 68
Organic pollutants 302, 303, 312
Organochlorinated insecticides 20–21, 37
Orizalyn 49
Oxadiazon 67
Oxidation reactions 16, 64, 108, 141
Ozone
 – effect on bacteria in polluted groundwater 332f
 – for treating polluted water 324–326
 – reactor 325, *325*

Palmitic acid 258–269, *260, 265, 266*
Palygorskite 94, 102
Pantoporeia hovi 53
Parachor 69
Paraffins (alkanes) 212, 231–234, *232*
Paraoxon 101
Paraquat 22, 42, 48, 56, *57*, 58, 61
Parathion
 – adsorption 113, 164, 165, 167, 169–170, *166, 169*
 – bioactivity 49, 172–173
 – hydrophobic adsorption 68
 – microbial transformations 64
 – persistence 26, 45
 – solubility 169
 – surface catalysis 101–103, *112*
 – Van der Waals adsorption 60
Partitioning 68–70, 299
 – of hydrophobic compounds between phases 213–223
 – into organic matter 166, 167–168, 171, 172
 – versus adsorption 70–71
PCBs 28, 51, 53, 60, 68, 71–72, 282–283
Pebulate 49
Pentachlorophenol I (PCP) 28, 44, 48, 66, 184–186, *185*, 188–190
Perthane 30

Pesticides *see also* individual compounds and classes
 – detoxification 304, 317
 – in groundwater 17–23, 25
 – interactions with humic substances 42–46
 – risk 14
 – usage 6
Petroleum products 126, 209
 – as pollutants 11, 14, 231, 321–324
 – biodegradation of model compounds 252–270
 – composition of crude oil 212–213
 – fractionation *215, 233*
 – hydrocarbons 211–228, 232–234
 – migration 234–240
 – modelling their movement 240–247
 – short and long term processes in the vadose zone 239–240
 – simulated kerosene
 – solubilities in water 218
 – surface degradation 223–225
 – volatilization 214–217
Phenacridane chloride 43, 48, 58
Phenol 25, 28, 184–185, *185*, 188–190, *189*
Phenophthalein 298
 – as tracer in polluted groundwater 281–300, *297*
Phorate 49
Phosmet 107
Phosphon 43, 48, 56, 58
Photochemical inactivation 302
Photochemical reactions 304
 – direct photolysis 138, 304
 – indirect photolysis 138
 – mathematical model 304
 – sensitized photolysis 304
Photolysis
 – anilide derivatives 311
 – as influenced by humic substances 52–53
 – in water columns 138
 – oxidative species 303
 – pilot plant experiments 314–316
 – s-triazine derivatives 43, 308–310
 – uracil derivatives 305–307
Photooxidation reactions
 – intentional promotion 304
 – mechanisms 302
 – photosensitizer 302
 – singlet oxygen 75, 76, 139, 303, 305, 308
 – triplet state 303
Photooxidation products 76, 305

Photosensitization *see also* sensitizers and photolysis
- by humic substances 38, 41, 52–53, 75–79
Phthalates 47
Phytotoxicity 62, 307, 310, 311
Picloram 30, 44, 48, 57, *57*, 60, 67
Pirimiphos-ethyl 105
pK_a 300
- and adsorption of HIOC 184–186
- and s-triazine adsorption 58–59
- of fulvic acids 74
- phenolphthalein 298
- stomach 22
Plasmids 200, 205, 285
PMOS (partially miscible organic solvents)
- defined 176
- solubility in multiphase solvents 179–180
- sorption from 182–183
- transport 187–188
Poliovirus 316, 317
Polynuclear aromatic hydrocarbons (PAH) 46, 50, 53, 63, 69, 77, 158, *178*
Prodiamine 22
Profluralin 45, 46
Prometon 23, *57*, 62
Prometryn 52, *57*, 59, 77
Propachlor 45, 49, 311
Propanil 45, 79
Propazine *57*, 309
Propham 45, 49
Protonation 113
Pyrazone 50
Pyrene 73, *178*

QSAR (Quantitative Structure-Activity Relationships) 149, 151, 158–159

Rate equations 115–122, 137, 138
Redox reactions
- denitrification 145
- Eh 142, *143*
- in groundwater 144–145
- in sediments 141–142
- in water columns 139–140
Reduction reactions 16, 52, 100, 102, 110, 141, 144
Restoration of contaminated media
- groundwater 328–333
-- polluted by hydrocarbons 247, 251, 320–333
- vadose zone
-- by biological methods 283–286
-- by chemical methods 282–283

-- by electrokinetics 280–282, 286
-- by enhanced volatilization 276–280
-- by radio frequency heating 283
Rhizoctonia pracicola 66
Ronnel 101

SAK (specific adsorption coefficient) 322, *323*
Salting out 219
Sensitizers 302
- acetone 79, 304
- acetophenone 304
- acridine orange 304
- and photolysis of TOCs 75, 77, 138
- benzonitrite 304
- benzophenone 79, 304
- bleached RF 309
- methylene blue 77, 79, 304, 305–308, 312–315
- riboflavin (RF) 77, 79, 304, 307–310, *310*
- rose bengal 79, 304
Sepiolite 94
Sewage effluents 305, 312, 317
Simazine 59, 308–309
Singlet oxygen 75, 76, 139
Smectites 93, 96, 101, 102
Solubility
- aqueous 16–19, 23, 25, 126
-- and adsorption 71, 172
-- and cleanup of the vadose zone 176
-- and estimation of K_{oc} 70, 152–156, 165
-- and movement of TOCs 288–301
-- enhanced 180
-- of petroleum hydrocarbons 218, 219
- mixed solvents 177–179, 180
Solubilization 37, 71–73, 177–180, 219
Solvophobic theory 150, 181
Sorption *see* adsorption
Steric hindrance 59, 110, 111, 124
Streaming potential 280–282, *281*
Styrene 102, 109
Substituted ureas 59–60, 62, 68
Surface area
- EDB adsorption 170
- of clay minerals 92, 93–94
- of metal oxides 95–97
Surface catalysis 101–103, 106–108
Surface enhanced transformations 91–92, 95, 99, 100–104, 116
- and exchangeable cations 111, *112*, 113, 126
- effect of moisture content 106–108, 110, 128

Subject Index

- effect of temperature 105
- of petroleum hydrocarbons 223–225
- kinetics 118
Suspended solids 26

2,4,5-T 44, 66, 285
TBA 44
TCA 44
Temik *see* Aldicarb
Terbacil 44, 182–183, 305
Terbufos 45
Terbutol 45
Terbutryne 58
Tetrachloroethane 25
Tetrachloroethene 24, 26–28, 54, 144, 276
Tetramethylbenzidine 102
p-(tetramethylbutyl)phenol 27
Thiram 204
Threshold limit values 30
Thyamine 43, 48
Toluene 25, 27, 28, 188–190, *216*, 220, 233, 253
Toxaphene 30, 44, 48
Tracers 290–292, 294, 326–327
Trametes versicolor 66
Transport
- as influenced by microbial pollution 188–190
- of organic pollutants 188–190
- of pesticides in soil 48–50
- of petroleum hydrocarbons 225–228

Triazines 43, 51, 58–59, 60, 65, 68, 102
Trichlorobenzene 25, 171
Trichloroethane 24–26, 28, 54, 144
Trichloroethylene (TCE) 24, 25, 28, 54, 144, 276, 278, 284
Trifluralin 30, 45–46, *57*, 67, 101

Vacuum extraction of the vadose zone 277–280
Vadose zone 275–286
Van der Waals
- attractions 60, 298
- surface area 158
Vermiculite 93
Vernolate 201
Vinclozolin *197*
Vinylchloride 25
Volatile organic compounds (VOCs) 276, 282, *279*
Volatilization 16, 214–217, 276–280, 286

Waste sites 5, 12, 13, 136, 150, 183, 187, 278
Wastewater 26, 302
Wells 277–280, *277*, *280*, 288–290, *290*, 299, 321, *322*, *323*, 324–325, 327–328, *324*, *326*, *328*

Xylenes 27, 188–190, 220, *216*, *217*, 226–228, *221*, *222*, *224*, *227*, 233

343